How does a quantum well detector, a silicon BIB or a CCD work? How are heterodyne techniques used in visible and infrared detectors and submillimeter and millimeter receivers? And how do you specify the performance of any detector system? This volume answers all these questions with an up-to-date review of all the techniques for the detection of light.

This presentation approaches light detectors from the perspective of the underlying physics; and in this way it provides a unified understanding of the detection of radiation in the ultraviolet through to the submillimeter. Clearly worked examples demonstrate the physics involved, and problems are provided to increase the reader's knowledge of how each system works.

This clearly written and authoritative review of modern detector systems will develop the understanding of final year undergraduate and graduate students. It will provide a valuable reference for professionals in astronomy, engineering, and physics.

DETECTION OF LIGHT: FROM THE ULTRAVIOLET TO THE SUBMILLIMETER

DETECTION OF LIGHT: FROM THE ULTRAVIOLET TO THE SUBMILLIMETER

Edited by Karen Visnovsky

Illustrated by Karen Swarthout

G. H. RIEKE

Departments of Astronomy and Planetary Sciences
University of Arizona

Published by the Press Syndicate of the University of Cambridge
The Pitt Building, Trumpington Street, Cambridge CB2 1RP
40 West 20th Street, New York, NY 10011-4211, USA
10 Stamford Road, Oakleigh, Melbourne 3166, Australia

First published 1994
Reprinted 1996
First paperback edition 1996

Printed in Great Britain at the University Press, Cambridge

A catalogue record for this book is available from the British Library

Library of Congress cataloguing in publication data

Rieke, George Henry.
Detection of light: from the ultraviolet to the submillimeter /
George Henry Rieke.
p. cm.
Includes bibliographical references and index.
ISBN 0-521-41028-2
1. Photon detectors. I. Title.
~~QC787.P46R54 1994~~
621.36′2–dc20 94-6358 CIP

ISBN 0 521 41028 2 hardback
ISBN 0 521 57674 1 paperback

Contents

Preface *page* xi

1 Introduction 1
1.1 Radiometry 1
1.2 Detector characteristics 8
1.3 Solid state physics 9
1.4 Example 16
1.5 Problems 18
Notes 19

2 Photography 20
2.1 Basic operation 20
2.2 Underlying processes 22
2.3 Characteristic curve 29
2.4 Performance 32
2.5 Examples 46
2.6 Problems 47
Note 49

3 Intrinsic photoconductors 50
3.1 Basic operation 50
3.2 Limitations and optimization 58
3.3 Performance specification 72
3.4 Example: design of a photoconductor 75
3.5 Problems 76
Notes 79

4 Extrinsic photoconductors 81
4.1 Basics 81
4.2 Limitations 87
4.3 Variants 92
4.4 Problems 101
Note 102

5 Photodiodes 103
5.1 Basic operation 103
5.2 Quantitative description 109
5.3 Photodiode variations 122
5.4 Quantum well detectors 129
5.5 Example 133
5.6 Problems 133

6 Amplifiers and readouts 136
6.1 Building blocks 136
6.2 Load resistor and amplifier 139
6.3 Transimpedance amplifier (TIA) 140
6.4 Integrating amplifiers 145
6.5 Performance measurement 155
6.6 Examples 159
6.7 Problems 163

7 Arrays 165
7.1 Infrared arrays 165
7.2 Charge coupled devices (CCDs) 171
7.3 Array properties 193
7.4 Example 197
7.5 Problems 198
Notes 200

8 Photoemissive detectors 202
8.1 General description 202
8.2 Quantitative results 209
8.3 Practical detectors 211
8.4 Example 231
8.5 Problems 232

9 Bolometers 234
9.1 Basic operation 234
9.2 Detailed theory 236
9.3 Bolometer construction and operation 247
9.4 Other thermal detectors 256
9.5 Example: design of a bolometer 258
9.6 Problems 260

10 Visible and infrared coherent detectors 262
10.1 Basic operation 262
10.2 Visible and infrared heterodyne 266
10.3 Performance attributes of heterodyne detectors 277
10.4 Test procedures 284
10.5 Example 285
10.6 Problems 287
Note 288

11 Submillimeter- and millimeter-wave heterodyne receivers 289
11.1 Basic operation 289
11.2 Mixers 293
11.3 Performance characteristics 304
11.4 Local oscillators 312
11.5 Problems 317
Notes 318

12 Summary 319
12.1 Quantum efficiency and noise 319
12.2 Linearity and dynamic range 320
12.3 Number of pixels 320
12.4 Time response 321
12.5 Spectral response and bandwidth 322
12.6 Overview 322
12.7 Problems 323
Note 324

Appendix A. Physical constants 325
Appendix B. Answers to selected problems 326
References 329
Index 341

Preface

Over the past decade, incredibly rapid progress has been made in the development of detectors for use at low light levels. These advances are not discussed in a unified manner in any available reference. In addition, most texts on this general subject are now a decade old and therefore omit many areas that are new and likely to grow in importance. I wrote this book to fill the need for a modern discussion of these issues. To avoid becoming quickly obsolete itself, the book approaches the subject from the perspective of the underlying physics of the various detector types. Emphasis is placed on the physical limits of detector performance and on the physical grounds for choices of detector design. Brief discussions of current performance levels are sometimes included, but it can be expected that these portions will soon be of merely historical interest. Those interested in the state-of-the-art should consult the technical journals, conference proceedings, and manufacturers' catalogs.

My goal is to provide a comprehensive overview of the important technologies for photon detection from the millimeter-wave through the ultraviolet spectral regions. The reader should gain a good understanding of the similarities and contrasts, the strengths and weaknesses of the multitude of approaches that have been developed over a century of effort to improve our ability to sense photons.

The book is aimed toward a number of audiences. It is suitable as the basis for a single semester course at either the advanced undergraduate or graduate level. It can be used for a variety of shorter courses or portions of courses. For example, Chapters 1, 2, 3, 5 up to Section 5.3.3, 6, 7, and 8 would fit into a half-semester course on optical detectors. Chapter 1, Sections 2.2.1 and 2.4.4 of Chapter 2, and Chapters 3 through 7 can provide the basis for a roughly half-semester course on modern solid state infrared and optical detectors. Chapter 1, Section 2.2.1 of Chapter 2, Section 3.3 of Chapter 3, Sections 5.1

and 5.2 of Chapter 5, and Chapters 9, 10, and 11 give an overview of submillimeter- and millimeter-wave detectors. I have tried to keep each of these sub-courses self-contained whilst maintaining the overall continuity of the book.

I hope that the book will encourage the development of courses on detectors. It illustrates a variety of physical fundamentals as applied to a specific type of goal. The devices discussed are used in a very broad range of applications, and many students can expect to use them in the course of their careers. As a result, the motivation for the physics is obvious and immediate. Low light level detectors illustrate many elegant applications of basic physical principles; as such, they are one of the highlights of our technology. This book discusses the basic physics in a fashion that unifies an extremely broad range of detection systems.

In any such course, the students usually have a range both of types of preparation and of ability to recall relevant areas. Moreover, it is tempting for instructors to concentrate on areas familiar and interesting to them – usually not including what they consider to be the prerequisites to the course. I felt it would be useful to begin the book with a review of some basic background material, and the treatment throughout attempts as much as possible to review, to refer to basic principles, and to minimize the reliance on a familiarity with advanced prerequisites.

This approach should also make the material accessible to a second audience. Low light level detectors are becoming ever more sophisticated and complex, but also more widely used. Consequently, there is an unfortunate trend for the users to become increasingly oblivious to the nature of their tools. I hope to provide them with a broad overview, not only of a particular family of devices familiar to them through direct experience, but also of how it fits in and possibly competes with other families. Because the necessary physics is developed within the discussion of detectors, the book should be self-contained for those who are outside a classroom environment.

Finally, I would like to address those who build, maintain, and repair modern detector systems. Counter to the increasing specialization in engineering training, modern detectors draw on a broad variety of physical principles. This book will provide the general understanding of these systems that is usually missing in traditional course sequences. For both users and builders, a broad knowledge may suggest better ways to approach projects, can give insights as to whether the goals of a project are realistic, and may even provide a greater sense of satisfaction.

Although the book contains a comprehensive discussion of detection techniques, readers can also skip much of the detailed discussions. Each detector type is introduced with a short paragraph that discusses its general applicability, followed by a chapter section that emphasizes important qualitative aspects of the detector performance. Only in the following sections is there a thorough and mathematical analysis, and these sections can be passed over if the reader does not have a strong interest in the detector variety in question.

The book begins with a short review chapter on foundation material, and proceeds as soon as possible with the discussion of detectors. In general, the complexity of the concepts grows as the book proceeds. At the same time, there is an attempt to discuss each concept to completion when it arises. Each chapter provides the reader with extensive references to more advanced discussions of the detectors in question. Where modern review articles or books are available, they are given as references. In many cases, however, there is no adequate review, and the reader must be referred to a large number of articles in the technical literature. By necessity in these cases, the treatment in this text is carried to a relatively advanced level.

A warning is in order about units. So far as seemed reasonable, MKS units have been used. There are some cases, however, in which some other unit is predominant in usage, frequently because it has a 'natural' size for the application. In these cases, the commonly employed system has been used in tabulations of parameters to maintain continuity with other literature. However, all formulae are in MKS. Errors can be easily avoided by rigorously carrying units through all calculations. A table of the important physical constants is included as Appendix A; it also includes selected conversions from 'conventional' units to MKS.

I am indebted to many people for making me aware of resources and for assistance in reviewing this material, including John Bieging, Mike Cobb, Rich Cromwell, John Goebbel, Art Hoag, Jim Kofron, Michael Kriss, Frank Low, Craig McCreight, Bob McMurray, Harvey Moseley, Paul Richards, Fred, Marcia, and Carol Rieke, Gary Schmidt, Bill Schoening, Michael Scutero, Ben Snavely, Chris Walker, and Erick Young. I also thank a number of classes of students at the University of Arizona for their comments on the lecture notes that evolved into this text.

It is a particular pleasure to thank Karen Swarthout for preparing the figures conscientiously, artistically, and accurately. Karen Visnovsky receives very special recognition for her careful editing of the lecture notes, and her vigorous identification of areas that were not written clearly. These two deserve much of the credit for any clarity that has crept into the presentation of the concepts in the book.

Any corrections, suggestions, and comments will be received gratefully.

They can be addressed to the author at Steward Observatory, University of Arizona, Tucson, Arizona 85721, United States of America.

George Rieke

1
Introduction

We begin by covering background material in three areas. First, we need to establish the formalism and definitions for the imaginary signals we will be shining on our imaginary detectors. Secondly, we will list general detector characteristics so we can judge the merits of the various types as they are discussed. Thirdly, because solid state physics will be so pervasive in our discussions, we include a very brief primer on that subject. This section may be a review for many readers, but it is the foundation for the subsequent chapters.

1.1 Radiometry

There are some general aspects of electromagnetic radiation that need to be defined before we discuss how it is detected. More detailed discussions of these issues can be found in Wolfe and Zissis (1978), Boyd (1983), and Wyatt (1991). Most of the time we will treat light as photons of energy; wave aspects will be important only for heterodyne detectors. A photon has an energy of

$$E_{\mathrm{ph}} = h\nu = hc/\lambda, \tag{1.1}$$

where h ($=6.626 \times 10^{-34}$ J s) is Planck's constant, ν and λ are, respectively, the frequency (in hertz) and wavelength (in meters) of the electromagnetic wave, and c ($=2.998 \times 10^{8}$ m s^{-1}) is the speed of light. In the following discussion, we define a number of expressions for the power output of sources of photons (for example, the power produced by thermal emission); conversion from power to photons per second can be achieved by dividing by the desired form of equation (1.1).

The spectral radiance per frequency interval, L_ν, is the power (in watts) leaving a unit projected area of the surface of the source (in square meters) into a unit solid angle (in steradians) and unit frequency interval (in

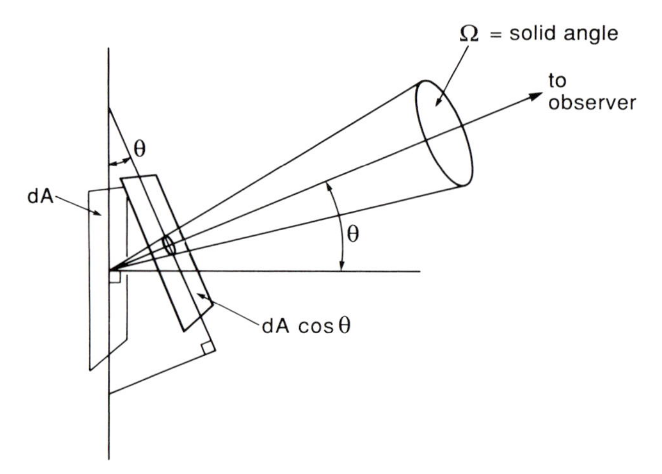

Figure 1.1. Geometry for computing radiance.

hertz = 1/seconds). The projected area of a surface element dA onto a plane perpendicular to the direction of observation is d$A \cos \theta$, where θ is the angle between the direction of observation and the outward normal to dA; see Figure 1.1. L_ν has units of $\mathrm{W\,m^{-2}\,Hz^{-1}\,ster^{-1}}$. The spectral radiance per wavelength interval, L_λ, is measured in units of $\mathrm{W\,m^{-3}\,ster^{-1}}$. The radiance, L, is the spectral radiance integrated over all frequencies or wavelengths; it has units of $\mathrm{W\,m^{-2}\,ster^{-1}}$. The radiant exitance, M, is the integral of the radiance over solid angle, and it is a measure of the total power emitted per unit surface area in units of $\mathrm{W\,m^{-2}}$.

We will deal only with Lambertian sources; the defining characteristic of such a source is that its radiance is constant regardless of the direction from which it is viewed. A blackbody is one example. The emission of a Lambertian source goes as the cosine of the angle between the direction of the radiation and the normal to the source surface. From the definition of projected area in the preceding paragraph, it can be seen that this emission pattern exactly compensates for the foreshortening of the surface as it is tilted away from being perpendicular to the line of sight. That is, for the element dA, the projected surface area and the emission decrease by the same cosine factor. Thus, if the entire source has the same temperature and emissivity, every unit area of its projected surface in the plane perpendicular to the observer's line of sight appears to be of the same brightness, independent of its actual angle to the line of sight. Keeping in mind this cosine dependence, and the definition of radiant exitance, the radiance and radiant exitance are related as

$$M = \int L \cos \theta \, \mathrm{d}\Omega = 2\pi L \int_0^{\pi/2} \sin \theta \cos \theta \, \mathrm{d}\theta = \pi L. \tag{1.2}$$

The flux emitted by the source, Φ, is the radiant exitance times the total surface area of the source, that is the power emitted by the entire source. For example, for a spherical source of radius R,

$$\Phi = 4\pi R^2 M = 4\pi^2 R^2 L. \quad (1.3)$$

Although there are other types of Lambertian sources, we will consider only sources that have spectra resembling those of blackbodies, for which

$$L_\nu = \frac{\varepsilon[2h\nu^3/(c/n)^2]}{e^{h\nu/kT} - 1}, \quad (1.4)$$

where ε is the emissivity of the source, n is the refractive index of the medium into which the source radiates, and k ($= 1.38 \times 10^{-23}$ J K^{-1}) is the Boltzmann constant. The emissivity (ranging from 0 to 1) is the efficiency with which the source radiates compared with that of a perfect blackbody, which by definition has $\varepsilon = 1$. According to Kirchhoff's law, the absorption efficiency, or absorptivity, and the emissivity are equal for any source.

In wavelength units, the spectral radiance is

$$L_\lambda = \frac{\varepsilon[2h(c/n)^2]}{\lambda^5(e^{hc/\lambda kT} - 1)}. \quad (1.5)$$

It can be easily shown from equations (1.4) and (1.5) that the spectral radiances are related as follows:

$$L_\lambda = \left(\frac{c}{\lambda^2}\right) L_\nu = \left(\frac{\nu}{\lambda}\right) L_\nu. \quad (1.6)$$

According to the Stefan–Boltzmann law, the radiant exitance for a blackbody becomes:

$$M = \pi \int_0^\infty L_\nu \, d\nu = \frac{2\pi k^4 T^4}{c^2 h^3} \int_0^\infty \frac{x^3}{e^x - 1} dx$$

$$= \frac{2\pi^5 k^4}{15 c^2 h^3} T^4 = \sigma T^4 \quad (1.7)$$

where σ ($= 5.67 \times 10^{-8}$ W m^{-2} K^{-4}) is the Stefan–Boltzmann constant.

For Lambertian sources, the optical system feeding a detector will receive a portion of the source power that is determined by a number of geometric factors as illustrated in Figure 1.2. The system will accept radiation from only a limited range of directions determined by the geometry of the optical system as a whole and known as the field of view. The area of the source which is effective in producing a signal is determined by the field of view and the

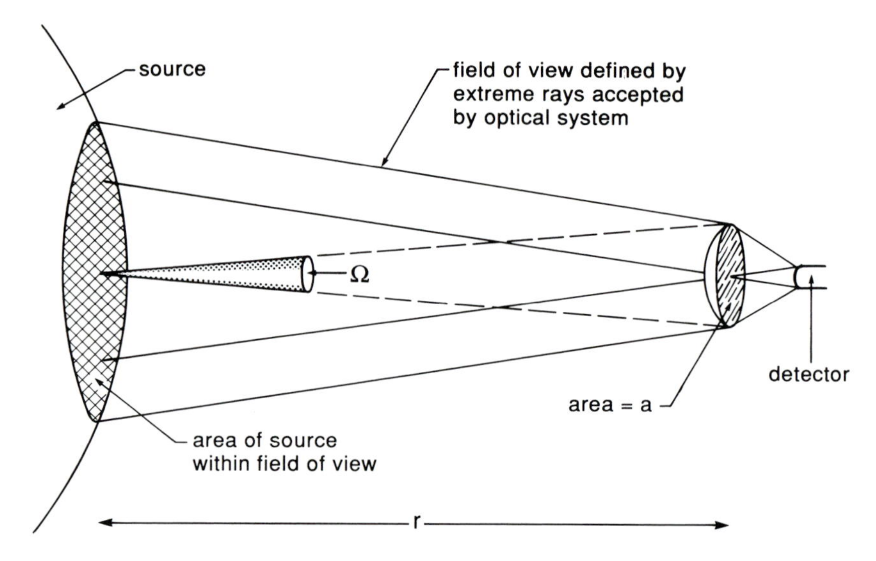

Figure 1.2. Geometry for computing power received by a detector system.

distance from the optical system to the source (or by the size of the source if it all lies within the field of view). This area will emit radiation with some angular dependence. Only the radiation that is emitted in directions where it is intercepted by the optical system can be detected. The range of directions accepted is determined by the solid angle, Ω, that the entrance aperture of the optical system subtends as viewed from the source. In addition, some of the emitted power may be absorbed or scattered by any medium through which it propagates to reach the optical system. For a Lambertian source, the power this system receives is then the radiance in its direction multiplied by the source area within the system field of view, multiplied by the solid angle subtended by the optical system as viewed from the source, and multiplied by the transmittance of the optical path from the source to the system.

Although a general treatment must allow for the field of view to include only a portion of the source, in many cases of interest the entire source lies within the field of view, so the full projected area of the source is used. For a spherical source of radius R, this area is πR^2. The solid angle subtended by the detector system is

$$\Omega = \frac{a}{r^2}, \tag{1.8}$$

where a is the area of the entrance aperture of the system (strictly speaking, a is the projected area; we have assumed the system is pointing directly at the source) and r is its distance from the source. For a circular aperture,

$$\Omega = 4\pi \sin^2(\theta/2), \tag{1.9}$$

where θ is the half-angle of the right circular cone whose base is the detector system entrance aperture, and whose vertex lies on a point on the surface of the source; r is the height of this cone.

It is particularly useful when the angular diameter of the source is small compared with the field of view of the detector system to consider the irradiance, E, which is the power in watts per square meter received at a unit surface element at some distance from the source. For the case described in the preceding paragraph, the irradiance is obtained by first multiplying the radiant exitance by the total surface area of the source to get the flux, which is $A\pi L$. The flux is then divided by the area of a sphere of radius r centered on the source to give

$$E = \frac{AL}{4r^2}, \tag{1.10}$$

where A is the total surface area of the source, and r is its distance from the irradiated surface element. The spectral irradiance, E_ν or E_λ, is the irradiance per unit frequency or wavelength interval. It is also sometimes called the flux density, and is a very commonly used description of the power received from a source. It can be obtained from equation (1.10) by substituting L_ν or L_λ for L.

The radiometric quantities discussed above are summarized in Table 1.1. Equations are provided for illustration only; in some cases, these examples apply only to specific circumstances. The terminology and symbolism vary substantially from one discipline to another; for example, the last two columns of the table translate some of the commonly used radiometric terms into astronomical nomenclature.

Only a portion of the power received by the optical system is passed on to the detector. The system will have inefficiencies due to both absorption and scattering of energy in its elements, and because of optical aberrations and diffraction. These effects can be combined into a system transmittance term. In addition, the range of frequencies or wavelengths to which the system is sensitive (that is, the spectral bandwidth of the system in frequency or wavelength units) is usually restricted by a combination of characteristics of the detector, filters, and other elements of the system as well as by any spectral dependence of the transmittance of the optical path from the source to the entrance aperture. A rigorous accounting of the spectral response requires that the spectral radiance of the source be multiplied by the spectral transmittances of all the spectrally active elements in the optical path to the

Table 1.1. *Definitions of radiometric quantities*

Symbol	Name	Definition	Units	Equation	Alternative name	
L_ν	Spectral radiance (frequency units)	Power leaving unit projected surface area into unit solid angle and unit frequency interval	$\mathrm{W\,m^{-2}\,Hz^{-1}\,ster^{-1}}$	(1.4)	Specific intensity (frequency units)	I_ν
L_λ	Spectral radiance (wavelength units)	Power leaving unit projected surface area into unit solid angle and unit wavelength interval	$\mathrm{W\,m^{-3}\,ster^{-1}}$	(1.5)	Specific intensity (wavelength units)	I_λ
L	Radiance	Spectral radiance integrated over frequency or wavelength	$\mathrm{W\,m^{-2}\,ster^{-1}}$	$L=\int L_\nu\,d\nu$	Intensity or specific intensity	I
M	Radiant exitance	Power emitted per unit surface area	$\mathrm{W\,m^{-2}}$	$M=\int L(\theta)\,d\Omega$		
Φ	Flux	Total power emitted by source of area A	W	$\Phi=\int M\,dA$	Luminosity	L
E	Irradiance	Power received at unit surface element; equation applies well removed from the source at distance r	$\mathrm{W\,m^{-2}}$	$E=\dfrac{\int M\,dA}{(4\pi r^2)}$		
E_ν, E_λ	Spectral irradiance	Power received at unit surface element per unit frequency or wavelength interval	$\mathrm{W\,m^{-2}\,Hz^{-1}}$, $\mathrm{W\,m^{-3}}$		Flux density	S_ν, S_λ

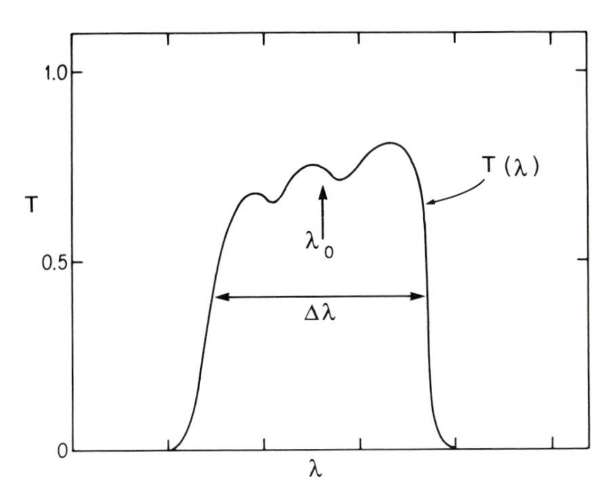

Figure 1.3. Transmittance function $T(\lambda)$ of a filter. The FWHM $\Delta\lambda$ and the effective wavelength λ_0 are indicated.

detector, and by the detector spectral response, and the resulting function subsequently integrated to determine the total power effective in generating a signal.

In many cases, the spectral response is intentionally restricted to a narrow range of wavelengths by placing a bandpass optical filter in the beam. It is then useful to define the effective wavelength of the system as

$$\lambda_0 = \frac{\int_0^\infty \lambda T(\lambda)\,d\lambda}{\int_0^\infty T(\lambda)\,d\lambda}, \tag{1.11}$$

where $T(\lambda)$ is the transmittance of the system. Often the spectral variations of the other transmittance terms can be ignored over the restricted spectral range of the filter. The bandpass of the filter, $\Delta\lambda$, can be taken to be the full width at half maximum (FWHM) of its transmittance function (see Figure 1.3). If the filter cuts on and off sharply, its transmittance can be approximated as the average value over the range $\Delta\lambda$:

$$T_F = \frac{\int_{\Delta\lambda} T(\lambda)\,d\lambda}{\Delta\lambda}. \tag{1.12}$$

If $\Delta\lambda/\lambda_0 \lesssim 0.2$ and the filter cuts on and off sharply, it is usually adequate to approximate the action of the bandpass filter by taking the spectral radiance at λ_0 and multiplying it by $\Delta\lambda$, the average filter transmittance over the range $\Delta\lambda$, and the various geometric and transmittance terms already discussed for the remainder of the system to derive the power effective in generating a signal. However, if λ_0 is substantially shorter than the peak wavelength of the blackbody curve (that is, one is operating in the Wien region of the blackbody) or there is sharp spectral structure within the passband, then this approximation can lead to significant errors, particularly if $\Delta\lambda/\lambda_0$ is relatively large.

Continuing with the approximation just discussed, we can combine equations (1.8) and (1.11), the definition of radiance, and the other corrections we have discussed, to derive an expression to be used in many situations to estimate the power falling on the detector:

$$P_D \approx \frac{A_{\text{proj}} a T_{\text{P}}(\lambda_0) T_{\text{O}}(\lambda_0) T_{\text{F}} L_\lambda(\lambda_0)\, \Delta\lambda}{r^2}. \tag{1.13}$$

Here A_{proj} is the area of the source projected onto the plane perpendicular to the line of sight from the source to the optical receiver. T_{P}, T_{O}, and T_{F} are the transmittances, respectively, of the optical path from the source to the receiver, of the receiver optics (excluding the bandpass filter), and of the bandpass filter. The area of the receiver entrance aperture is a, and the distance of the receiver from the source is r. An analogous expression holds in frequency units. The major underlying assumptions for equation (1.13) are that: (a) the field of view of the receiver includes the entire source; (b) the source is a Lambertian emitter; and (c) the spectral response of the detector is limited by a filter with a narrow or moderate bandpass that is sharply defined.

1.2 Detector characteristics

Nearly all detectors act as transducers that receive photons and produce an electrical response that can be amplified and converted into a form intelligible to suitably conditioned human beings. There are three basic ways that detectors carry out this function:

(a) *Photon detectors* respond directly to individual photons. An absorbed photon releases one or more bound charge carriers in the detector that may (1) lead to a chemical change; (2) modulate the electric current in the material; or (3) move directly to an output amplifier. Photon detectors are used throughout the X-ray, ultraviolet, visible, and infrared spectral regions. Examples that we will discuss are photographic plates (Chapter 2), photoconductors (Chapters 3 and 4), photodiodes (Chapter 5), and photoemissive detectors (Chapter 8).

(b) *Thermal detectors* absorb photons and thermalize their energy. In most cases, this energy changes the electrical properties of the detector material, resulting in a modulation of the electrical current passing through it. Thermal detectors have a very broad and nonspecific spectral response, but they are particularly important at infrared and submillimeter wavelengths; they are also coming into use as X-ray detectors. Bolometers and other thermal detectors will be discussed in Chapter 9.

(c) *Coherent detectors* respond to the electric field strength of the signal and can preserve phase information about the incoming photons. They operate by interference of the electric field of the incident photon with the electric field from a coherent local oscillator. These detectors are primarily used in the radio and submillimeter regions and are sometimes useful in the infrared. Coherent detectors for the infrared are discussed in Chapter 10, and those for the submillimeter are discussed in Chapter 11.

Good detectors preserve a large proportion of the information contained in the incoming stream of photons. A variety of parameters are relevant to this goal:

(a) *Quantum efficiency* – the fraction of the incoming photon stream that is converted into signal.

(b) *Noise* – the uncertainty in the output signal. Ideally, the noise consists only of statistical fluctuations due to the finite number of photons producing the signal.

(c) *Linearity* – the degree to which the output signal is proportional to the number of incoming photons that were received to produce the signal.

(d) *Dynamic range* – the maximum variation in signal over which the detector output represents the photon flux without losing significant amounts of information.

(e) *Number and size of pixels* – the number of picture elements the detector can record simultaneously and the physical size of each element on the detector.

(f) *Time response* – the minimum interval of time over which the detector can distinguish changes in the photon arrival rate.

(g) *Spectral response* – the total wavelength or frequency range over which photons can be detected with reasonable efficiency.

(h) *Spectral bandwidth* – the wavelength or frequency range over which photons are detected at any one time; some detectors can operate in a narrow band that can be placed within a broader range of spectral response.

1.3 Solid state physics

The electrical properties of a semiconductor are altered dramatically by the absorption of an ultraviolet, visible, or infrared photon, which makes this class of material well adapted to a variety of photon detection strategies. Metals, on the other hand, have high electrical conductivity that is only insignificantly modified by the absorption of photons, and insulators require

Table 1.2 *Periodic table of the elements*

Ia	IIa	III b	IV b	Vb	VIb	VII b	VIII			Ib	IIb	III a	IV a	V a	VI a	VII a	0
1 H													↓				2 He
3 Li	4 Be											5 **B**	6 **C**	7 N	8 O	9 F	10 Ne
11 Na	12 Mg											13 **Al**	14 **Si**	15 **P**	16 **S**	17 **Cl**	18 Ar
19 K	20 Ca	21 Sc	22 Ti	23 V	24 Cr	25 Mn	26 Fe	27 Co	28 Ni	29 Cu	30 **Zn**	31 **Ga**	32 **Ge**	33 **As**	34 **Se**	35 **Br**	36 Kr
37 Rb	38 Sr	39 Y	40 Zr	41 Nb	42 Mo	43 Tc	44 Ru	45 Rh	46 Pd	47 **Ag**	48 **Cd**	49 **In**	50 **Sn**	51 **Sb**	52 **Te**	53 **I**	54 Xe
55 Cs	56 Ba	57 La	72 Hf	73 Ta	74 W	75 Re	76 Os	77 Ir	78 Pt	79 Au	80 **Hg**	81 Tl	82 **Pb**	83 Bi	84 Po	85 At	86 Rn
87 Fr	88 Ra	89 Ac															

more energy to excite electrical changes than is available from individual visible or infrared photons.

In addition, adding small amounts of impurities to semiconductors can strongly modify their electrical properties at and below room temperature. Consequently, semiconductors are the basis for most electronic devices, including those used for amplification of photoexcited currents as well as those used to detect photons with too little energy to be detected through photoexcitation.

Because of these properties of semiconductors, virtually every detector we shall discuss depends on these materials for its operation. To facilitate our discussion, we will first review some of the properties of semiconductors. The concepts introduced below are used throughout the remaining chapters; more detailed treatments can be found in Ashcroft and Mermin (1976), Talley and Daugherty (1976), Sze (1985), Kittel (1986), Solymar and Walsh (1988), and Streetman (1990).

The elemental semiconductors are silicon and germanium; they are found in column IVa of the periodic table (Table 1.2). Their outermost electron shells, or valence states, contain four electrons, half of the total number allowed for these shells. They form crystals with a diamond lattice structure (note that carbon is also in column IVa). In this structure, each atom bonds to its four nearest neighbors; it can therefore share one valence electron with each neighbor, and vice versa. Electrons are fermions and must obey the Pauli exclusion principle, which states that no two particles with half-integral quantum mechanical spin can occupy identical quantum states.[1] Because of the exclusion principle, the electrons shared between neighboring nuclei must have opposite spin (if they had the same spin, they would be indistinguishable), which accounts for the fact that they occur in pairs. By sharing electrons, each atom comes closer to having a filled valence shell, and a quantum mechanical binding force known as a covalent bond is created.

The binding of electrons to an atomic nucleus can be described in terms of a potential energy well around the nucleus. Electrons may be in the ground state or at various higher energy levels called excited states. There is a specific energy difference between these states which can be measured by detecting an absorption or emission line when an electron shifts between energy levels. The sharply defined energy levels of an isolated atom occur because of constructive interference of electron wave functions within the potential well; there is destructive interference at all other energies. When atoms are brought together, the quantum mechanically permitted energy levels of an individual atom split because of the coupling between the potential wells. The 'valence states' and 'conduction states' in a material are analogous to the ground state

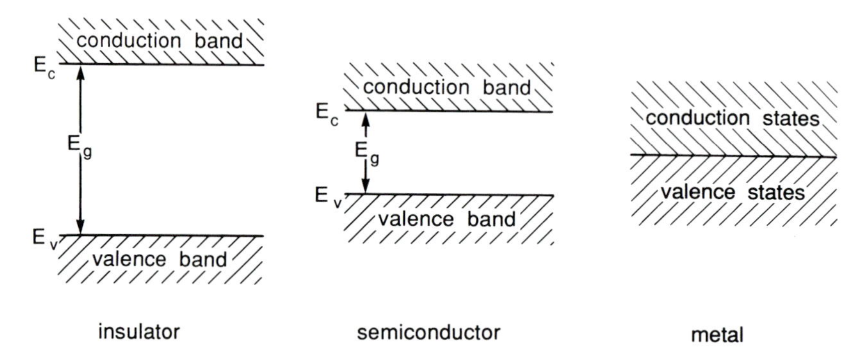

Figure 1.4. Energy band diagrams for insulators, semiconductors, and metals.

and excited states, respectively, in an isolated atom. In a compact structure such as a crystal, the energy levels split multiply into broad energy zones called bands. This situation can be represented by band diagrams such as those in Figure 1.4.

The splitting of energy levels occurs because the Pauli exclusion principle plays an increasing role in the electronic structure of the atoms. If the atoms are close enough to allow the electron wave functions to begin overlapping, then according to the exclusion principle the electrons must distribute themselves so that no two of them are in an identical quantum state.

If the material were at a temperature of absolute zero, all available states in the band would be filled up to some maximum level. The electrical conductivity would be zero because there would be no accessible states into which electrons could move. Conduction becomes possible when electrons are lifted into higher and incompletely filled energy levels, either by thermal excitation or by other means. There are two distinct possibilities. In a metal, the electrons only partially fill a band so that a very small amount of energy is required to gain access to unfilled energy levels and hence to excite conductivity. On the other hand, in a semiconductor or an insulator, the electrons would completely fill a band at absolute zero. To gain access to unfilled levels, an electron must be lifted into a level in the next higher band, resulting in a threshold excitation energy required to initiate electrical conductivity. In this latter case, the filled band is called the valence band and the unfilled one the conduction band. The bandgap, E_g, is the energy between the highest energy level in the valence band, E_v, and the lowest energy level in the conduction band, E_c. It is the minimum energy that must be supplied to excite conductivity in the material. Semiconductors have $0 < E_g < 3.5$ eV.

Metal atoms have a small number of loosely bound, outer shell electrons that are easily given up to form ions. In a bulk metal, these electrons are contributed to the crystal as a whole, creating a structure of positive ions immersed in a sea of free electrons. This situation produces metallic bonding, and the crystal contains electrons in its conduction states even with extremely small external excitation. The band diagrams for insulators and semiconductors are similar to each other, but the insulators have larger values of E_g because the conduction electrons are more tightly bound to the atoms than they are in semiconductors. It therefore takes more energy to break these bonds in insulators so the electrons can move through the material. A common kind of insulator is a compound containing atoms from opposite ends of the periodic table (one example is NaCl). In this case, the valence electron is taken from the metal atom and added to the outer valence band of the halide atom; both atoms then have filled outer electron shells. The electrostatic attraction of the positive metal and negative halide ions forms the crystal bond. This bonding is called ionic.[2]

Despite their differing electrical behavior, the band diagrams for semiconductors and insulators are qualitatively similar. Semiconductors are partially conducting under typically encountered conditions because the thermal excitation at room temperature is adequate to lift some electrons across their modest energy bandgaps. However, their conductivity is a strong function of temperature (going roughly as $e^{-E_g/2kT}$; $kT \approx 0.025\,\mathrm{eV}$ at room temperature), and near absolute zero they behave as insulators. In such a situation, the charge carriers are said to be 'frozen out'.

When electrons are elevated into the conduction band of a semiconductor or insulator, they leave empty positions in the valence state. These positions have an effective positive charge provided by the ion in the crystal lattice, and are called holes. As an electron in the valence band hops from one bond position to an adjacent, unoccupied one, the hole is said to migrate (such a positional change does not require that the electron be lifted into a conduction band or receive any appreciable additional energy). Although the holes are not real subatomic particles, they behave in many situations as if they were. It is convenient to discuss them as the positive counterparts to the electrons and to assign to them such attributes as mass, velocity, and charge. The total electric current is the combination of the contributions from the motions of the conduction electrons and the holes.

The key to the usefulness of semiconductors for visible and infrared photon detection is that their bandgaps are in the energy range of a single photon; for example, a visible photon of wavelength 0.55 μm has an energy of 2.26 eV. Absorption of energy greater than E_g photoexcites electrical conductivity in the material. Of course, other forms of energy can also excite conductivity

Table 1.3. *Semiconductors and their bandgap energies*

Columns	Semiconductor	E_g (eV)
IV	Ge	0.67
	Si	1.11
	SiC	2.86
III–V	AlAs	2.16
	AlP	2.45
	AlSb	1.6
	GaAs	1.43
	GaP	2.26
	GaSb	0.7
	InAs	0.36
	InP	1.35
	InSb	0.18
II–VI	CdS	2.42
	CdSe	1.73
	CdTe	1.58
	ZnSe	2.7
	ZnTe	2.25
I–VII	AgBr	2.81[a]
	AgCl	3.33[a]
IV–VI	PbS	0.37
	PbSe	0.27
	PbTe	0.29

[a] Values taken from James (1977). All other values are taken from Streetman (1990), Appendix III.

indistinguishable from that due to photoexcitation; a particularly troublesome example is thermal excitation.

In addition to elemental silicon and germanium, many compounds are semiconductors. A typical semiconductor compound is a diatomic molecule comprising atoms that symmetrically span column IVa in the periodic table, for example an atom from column IIIa combined with one from column Va, or one from column IIb combined with one from column VIa. Table 1.3 lists some simple semiconductor compounds and their bandgap energies. Elements that are important in the formation of semiconductors are shown in boldface in Table 1.2.

Pure semiconductors are termed intrinsic. Their electrical properties can be modified dramatically by adding impurities, or doping them, to make extrinsic semiconductors. When the impurity (or dopant) is an element whose valence shell contains surplus electrons (after the normal crystal bonds are

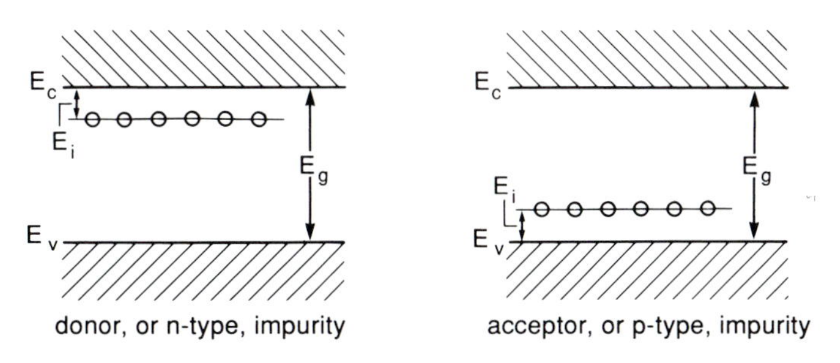

Figure 1.5. Energy band diagrams for semiconductors with n-type and p-type impurities.

accounted for), the surplus electrons are detached relatively easily to become conduction electrons. Such impurities are called donors, and material in which they dominate is called n-type. This situation is illustrated by the band diagram in Figure 1.5 in which a donor level has been added at the appropriate excitation energy, E_i, below E_c. When the valence shell of the dopant has too few electrons to form all the bonds required for the crystal structure, its atoms tend to capture, or accept, electrons from the semiconductor atoms, thus creating holes in the semiconductor valence band and so increasing conductivity. Material in which holes dominate is called p-type. In the band diagram (see Figure 1.5), we have added an acceptor level at the appropriate excitation energy above E_v. Because all semiconductors have residual impurities, weak donor or acceptor levels (or both) are always present. In an extrinsic semiconductor, the majority carrier is the type created by the dominant dopant (for example, holes for p-type). The minority carrier is the opposite type and usually has a much smaller concentration.

Impurities can also act as traps or recombination centers. Consider a p-type semiconductor. Thermal excitation, while not always supplying enough energy to raise electrons into the conduction band, may provide sufficient energy to move electrons from the valence band into an intermediate energy state, $E_v < E_t < E_c$, provided by the impurity. After the electron combines with the impurity atom, it will either be released by thermal excitation (after a delay), or the net negative charge will attract holes that will neutralize the impurity atom by recombining with the electron. If the electron is most likely to be released from the atom by thermal excitation, the impurity atom is called a trapping center, or trap; otherwise, it is termed a recombination center. Similar effects can occur at crystal defects, where some of the bonds in the regular lattice structure are broken, providing sites where the intrinsic crystal atoms can attract and combine with charge carriers.

A band diagram can oversimplify the requirements for transitions between valence and conduction bands. In many semiconductors, the energy levels corresponding to the minimum energy in the conduction band and to the maximum energy in the valence band do not have matching quantum mechanical wave vector values. In these cases, an electron must either make a transition involving greater energy than the bandgap energy or undergo a change in momentum as it moves from one band to the other. In the latter case, recombination must be by means of an intermediate state, which absorbs the excess momentum and therefore allows decay directly to the valence band. The transition can occur at any crystal atom, but the probability is frequently enhanced at recombination centers or traps. Semiconductors exhibiting this behavior (including silicon and germanium) are said to have indirect energy bands. Another class of semiconductors, of which GaAs is one example, allows direct energy band transitions; minimum energy electron transitions are permitted without a previous change of electron momentum or the presence of intermediate recombination centers. The difference between these two classes is well demonstrated by the behavior of their absorption coefficients (Chapter 3). In general, semiconductor light emitters such as light emitting diodes (LEDs) and lasers are based on materials capable of direct band transitions.

1.4 Example

A 1000 K spherical blackbody source of radius 1m is viewed in air by a detector system from a distance of 1000 m. The entrance aperture of the system has a radius of 5 cm, and the optical system has a field of view half-angle of 0.1°. The detector operates at a wavelength of 1 μm with a spectral bandpass of 1%, and its optical system is 50% efficient. Compute the spectral radiances in both frequency and wavelength units. Calculate the corresponding spectral irradiances at the detector entrance aperture, and the power received by the detector. Compare the usefulness of radiances and irradiances for this situation. Compute the number of photons hitting the detector per second. Describe how these answers would change if the blackbody source were 10 m in radius rather than 1 m.

The refractive index of air is $n \sim 1$, so the spectral radiance in frequency units is given by equation (1.4) with $\varepsilon = n = 1$. From equation (1.1), the frequency corresponding to 1 μm is $\nu = c/\lambda = 2.998 \times 10^{14}$ Hz. Substituting into equation (1.4), we find that

$$L_\nu = 2.21 \times 10^{-13}\ \mathrm{W\,m^{-2}\,Hz^{-1}\,ster^{-1}}.$$

Alternatively, we can substitute the wavelength of 1×10^{-6} m into equation

(1.5) to obtain

$$L_\lambda = 6.62 \times 10^7 \text{ W m}^{-3} \text{ ster}^{-1}.$$

The solid angle subtended by the detector system as viewed from the source is given by equation (1.8). The area of the entrance aperture is 7.854×10^{-3} m^2, so

$$\Omega = 7.854 \times 10^{-9} \text{ ster}.$$

The 1% bandwidth corresponds to $0.01 \times 2.998 \times 10^{14}$ Hz $= 2.998 \times 10^{12}$ Hz, or to $0.01 \times 1 \times 10^{-6}$ m $= 1 \times 10^{-8}$ m. The radius of the area accepted into the beam of the detector system at the distance of the source is 1.745 m, and, since it is larger than the radius of the source, the entire visible area of the source will contribute to the signal. The projected area of the source is 3.14 m^2 (since it is a Lambertian emitter, no further geometric corrections are required for its effective emitting area). Then, computing the power at the entrance aperture of the detector system by multiplying the spectral radiances by the source area (projected), spectral bandwidth, and solid angle received by the system, we obtain $P = 1.63 \times 10^{-8}$ W.

Because the angular diameter of the source is less than the field of view, it is equally convenient to use the irradiance. The surface area of the source is 12.57 m^2. Using equation (1.10) and frequency units, we obtain

$$E_\nu = 6.945 \times 10^{-19} \text{ W m}^{-2} \text{ Hz}^{-1}.$$

Similarly for wavelength units,

$$E_\lambda = 2.08 \times 10^2 \text{ W m}^{-3}.$$

Multiplying by the bandpass and entrance aperture area yields a power of 1.63×10^{-8} W, as before.

The power received by the detector is reduced by optical inefficiencies to 50% of the power incident on the entrance aperture, so it is 8.2×10^{-9} W. The energy per photon can be computed from equation (1.1) to be 1.99×10^{-19} J. The detector therefore receives 4.12×10^{10} photons s^{-1}.

If the blackbody source were 10 m in radius, the spectral radiances, L_ν and L_λ, would be unchanged. The irradiances, E_ν and E_λ, would increase in proportion to the surface area of the source, so they would be 100 times larger than computed above. The field of view of the optical system, however, no longer includes the entire source; therefore, the power at the system entrance aperture is most easily computed from the spectral radiances, where the relevant surface area is that within the field of view and hence has a radius of 1.745 m. The power at the entrance aperture therefore increases by a factor of only 3.05, giving $P = 4.97 \times 10^{-8}$ W, as do the power falling on the detector (2.48×10^{-8} W) and the photon rate (1.25×10^{11} photons s^{-1}).

1.5 Problems

1.1 A spherical blackbody source at 300 K and of radius 0.1m is viewed from a distance of 1000 m by a detector system with an entrance aperture of radius 1 cm and field of view half-angle of 0.1 degree.

(a) Compute the spectral radiances in frequency units at 1 and 10 μm.

(b) Compute the spectral irradiances at the entrance aperture.

(c) For spectral bandwidths 1% of the wavelengths of operation and assuming that 50% of the incident photons are absorbed in the optics before they reach the detector, compute the powers received by the detector.

(d) Compute the numbers of photons hitting the detector per second.

1.2 Consider a detector with an optical receiver of entrance aperture 2 mm diameter, optical transmittance (excluding bandpass filter) of 0.8, and field of view 1° in diameter. This system views a blackbody source of 1000 K with an exit aperture of 1 mm and at a distance of 2 m. The signal out of the blackbody is interrupted by a shutter at a temperature of 300 K. The receiver system is equipped with two bandpass filters, one with $\lambda_0 = 20\ \mu$m and $\Delta\lambda = 1\ \mu$m and the other with $\lambda_0 = 2\ \mu$m and $\Delta\lambda = 0.1\ \mu$m; both have transmittances of 0.8. The transmittance of the air between the source and receiver is 1 at both wavelengths. Compute the net signal at the detector, that is compute the change in power incident on the detector as the shutter is opened and closed.

1.3 Show that for $h\nu/kT \ll 1$ (setting $\varepsilon = n = 1$),

$$L_\nu = 2kT\nu^2/c^2.$$

This expression is the Rayleigh–Jeans law and is a useful approximation at long wavelengths. For a source temperature of 100 K, compute the shortest wavelength for which the Rayleigh–Jeans law is within 20% of the result given by equation (1.5). Compare with λ_{max} from Problem 1.4.

1.4 For blackbodies, the wavelength of the maximum spectral irradiance times the temperature is a constant, or

$$\lambda_{max} T = \mathbb{C}.$$

This expression is known as the Wien displacement law; derive it. For wavelength units, show that $\mathbb{C} \sim 0.3$ cm K.

1.5 Derive equation (1.9). Note the particularly simple form for small θ.

1.6 From the Wien displacement law (Problem 1.4), suggest suitable semiconductors for detectors matched to the peak irradiance from
(a) stars like the sun ($T = 5800$ K),
(b) Mercury ($T = 600$ K),
(c) Jupiter ($T = 140$ K).

1.7 Consider a bandpass filter that has a transmittance of zero outside the passband $\Delta\lambda$ and a transmittance that is the same for all wavelengths within the passband. Compare the estimate of the signal passing through this filter when the signal is determined by integrating the source spectrum over the filter passband with that where only the effective wavelength and passband are used to characterize the filter. Assume a source radiating in the Rayleigh–Jeans regime. Show that the error introduced by the simple effective wavelength approximation is a factor of

$$1 + \frac{5}{6}\left(\frac{\Delta\lambda}{\lambda_0}\right)^2$$

plus terms of order $(\Delta\lambda/\lambda_0)^4$ and higher. Evaluate the statement in the text that the approximate method gives acceptable accuracy for $\Delta\lambda/\lambda_0 \leqslant 0.2$.

Notes

1 This exclusion rule does not apply to particles with integral spin, which are called bosons.
2 A fourth kind of bonding is exhibited by frozen rare gases. Atoms are attracted to each other by van der Waals forces only, which are caused by the distortions of the cloud of electrons around adjacent atoms.

2
Photography

Photography is based on chemical changes that are initiated by the creation of a conduction electron when a photon is absorbed in certain kinds of semiconductor. These changes are amplified by chemical processing until a visible image of the illumination pattern has been produced. Compared with other modern detectors, the quantum efficiency achievable with photography is low, about 1–5%. Photography also suffers from comparatively poor linearity and, in some applications, limited dynamic range. It remains, however, the unquestioned leader in pixel quantity; an 8 by 10 inch plate can have 10^{11}–10^{12} grains, providing some 10^9 potential picture elements. In addition, photographic materials are inexpensive, provide efficient information storage, and, if treated appropriately, are stable for long periods of time. For these reasons, photography remains the best detection method for many observations in the X-ray, ultraviolet, visible, and very near infrared spectral regions, in spite of the performance improvements that have been attained for individual pixels in electronic detectors.

2.1 Basic operation

Photography was invented and developed through a long series of experiments before solid state physics was understood (as described by Newhall, 1967). A variety of photographic materials have been discovered, but those based on silver halide grains (tiny crystals of AgBr, AgCl, or AgBrI) have significant advantages in sensitivity and are universally accepted and used. Although some details remain controversial, the silver halide photographic process is now explained in terms common to other types of solid state detector.

A schematic cross-section of a photographic plate is shown in Figure 2.1. The silver halide grains are the active detectors; they are suspended in a gelatin binder and coated onto glass for mechanical stability. The bonding

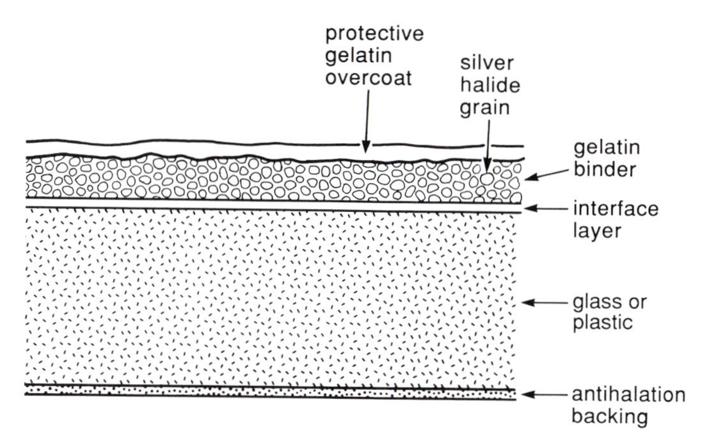

Figure 2.1. Cross-section of a typical photographic plate.

of the gelatin to the glass is improved by use of an interface layer, and the grains are often protected in red-sensitive plates by adding an overcoat of gelatin. To avoid reflections from the back of the glass, it is coated with an 'antihalation' backing that absorbs light which passes through the emulsion. Photographic film uses a flexible plastic support like Estar or Mylar instead of glass to improve handling convenience at the expense of some dimensional stability. The gelatin suspension has many important attributes: it is transparent to light (except ultraviolet); it is easily compatible with the chemical processing that is required to produce images; it swells when soaked in water, allowing the chemicals to contact the silver halide grains for development; it is readily available; and it can be made to have uniform properties.

Photographic materials are prepared in a series of steps. First, the silver halide grains are formed by chemical reactions in aqueous solution containing gelatin; because silver halides are relatively insoluble in water, they precipitate out as fine particles. These particles are prevented from coalescing by the presence of gelatin in the solution. The final size of the grains is regulated by the time allowed for precipitation and by further growth ('ripening') of the grains in the solution. Grain growth is halted by adding additional gelatin which sets up as a solid upon cooling and standing. Next, this material is shredded and washed to remove excess chemicals. After being subjected to a variety of baking and chemical processes, the emulsion is finally coated onto the glass or plastic backing. Further details are provided by James and Higgins (1960).

The photographic process is centered on the silver halide grains. When photons strike a grain, they excite its atoms by raising electrons to the conduction band. A chain of events is triggered which results in the growth of a small silver speck, known as a development center, in the grain. During chemical development, these development centers act as catalysts that can blacken the entire grain through reduction of silver ions to silver atoms. Given time, the undeveloped grains would also blacken; this process is stopped by fixing the developed image by dissolving any remaining silver halide and washing it away. It should be noted that photographic detection is binary in character; a grain either receives sufficient photons to develop or it does not. The process therefore has an intrinsically limited dynamic range, and it provides controllable 'shades of grey' in images largely because of the range of properties of the grains distributed over the plate. However, by intentionally introducing multiple types and sizes of silver halide grain, photographic materials are provided with dynamic ranges up to 400:1, and materials optimized for this performance aspect could have dynamic ranges some five times greater.

2.2 Underlying processes

The following discussion gives an abbreviated account of the absorption of a photon in a silver grain, how this event is eventually transformed into a blackening of the grain, and how the information content of the initial photon stream is affected by the steps in between. More details can be found in James and Higgins (1960), Katz and Fogel (1971), and James (1977). Many practical aspects of the use of photographic plates are discussed in Eastman Kodak (1987).

2.2.1 Photon absorption, quantum efficiency, and information content of photon stream

The inherent spectral sensitivity of photographic emulsions is usually limited to blue wavelengths. For example, the bandgap for AgBr ($E_g = 2.81$ eV) requires that a photon has a wavelength shorter than 0.44 μm to raise an electron into the conduction band. The absorption coefficient for this process, $a(\lambda)$, is shown in Figure 2.2; it is given in units of cm^{-1}, as is conventional. The absorption of a flux of photons, S, passing through a differential thickness element dl is expressed by

$$\frac{dS}{dl} = -a(\lambda)S, \tag{2.1}$$

with the solution for the remaining flux at depth l being

$$S = S_0 e^{-a(\lambda)l}. \tag{2.2}$$

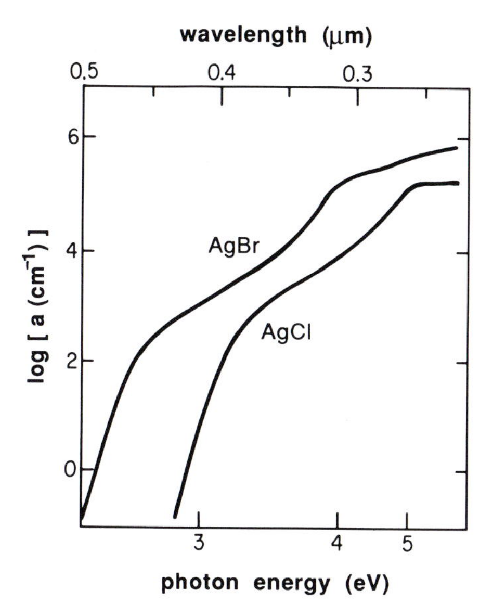

Figure 2.2. Optical absorption coefficients of AgCl and AgBr, after James and Higgins (1960). The extension of the absorption to energies below the bandgap is an artifact of the measurement process.

The quantum efficiency, η, is the flux absorbed in the detector divided by the total flux incident on its surface. Ignoring the reflection at the detector surface,

$$\eta_{ab} = \frac{S_0 - S_0 e^{-a(\lambda)d_1}}{S_0} = 1 - e^{-a(\lambda)d_1}, \tag{2.3}$$

where d_1 is the thickness of the detector, in this case of the silver halide grains. The quantity η_{ab} is known as the absorption factor.

Photons are lost by reflection from the surface before they enter the grains, leading to a reduction in quantum efficiency below η_{ab}. Minimal reflection occurs for photons striking a nonabsorptive material at normal incidence:

$$r = \frac{(n-1)^2}{(n+1)^2}, \tag{2.4}$$

where r is the fraction of the incident flux of photons that is reflected, n is the refractive index of the material, and we have assumed that the photon is incident from air, which has a refractive index of $n = 1$. In addition, reflection from the back of the grain can result in absorption of photons that would otherwise escape. If we ignore this potential gain, the net quantum efficiency is

$$\eta = (1 - r)\eta_{ab}. \tag{2.5}$$

In Figure 2.2, two absorption regimes can be distinguished. For wavelengths shorter than about 0.32 μm, absorption occurs through direct electron transitions, and $a(\lambda) > 10^5\,\text{cm}^{-1}$. For longer wavelengths, the photon energy can only be absorbed by indirect transitions, and $a(\lambda)$ is significantly reduced. The refractive index of the halide grains can be taken to be $n \approx 2$. From equation (2.5), we see that for grain diameters near 1 μm ($=10^{-4}$ cm), $\eta \approx 0.9$ in the direct absorption region (where the losses are largely from reflection), but η will depend on wavelength in the indirect absorption region.

The quantum efficiency defined in equations (2.3) and (2.5) refers only to the fraction of incoming photons converted into a signal in the first stage of detector action. Ideally, the signal-to-noise ratio attained in a measurement is controlled entirely by the number of photons absorbed in the first stage. The following discussion derives the inherent ratio of signal to noise in the incoming photon stream and then compares it with that which can be achieved in the detector as a function of the quantum efficiency.

Ignoring minor corrections having to do with the quantum nature of photons, it can be assumed that the input photon flux follows Poisson statistics,

$$P(m) = \frac{e^{-n} n^m}{m!}, \tag{2.6}$$

where $P(m)$ is the probability of detecting m photons in a given time interval, and n is the average number of photons detected in this time interval if a large number of detection experiments are conducted. The root-mean-square noise in the number of independent events is the square root of the mean, n,

$$N_{\text{rms}} = \langle N^2 \rangle^{1/2} = n^{1/2}. \tag{2.7}$$

In general, Poisson-distributed errors can be taken to be independent, and hence they add quadratically, that is the noise in two measurements, n_1 and n_2, is

$$\begin{aligned} N_{\text{rms}} &= \langle N^2 \rangle^{1/2} \\ &= [(n_1^{1/2})^2 + (n_2^{1/2})^2]^{1/2} = (n_1 + n_2)^{1/2}. \end{aligned} \tag{2.8}$$

From the above discussion, the signal-to-noise ratio for Poisson-distributed events is $n/n^{1/2}$, or

$$S/N = n^{1/2}. \tag{2.9}$$

This result can be taken to be a measure of the information content of the incoming photon stream as well as a measure of the confidence that a real signal has been detected.[1]

From the standpoint of the detector, photons that are not absorbed cannot contribute to either signal or noise; they might as well not exist. Consequently, for n photons incident on the detector, equation (2.9) shows that the signal-to-noise ratio goes as $\eta n/(\eta n)^{1/2}$, or

$$\left(\frac{S}{N}\right)_{\mathrm{d}} = (\eta n)^{1/2} \tag{2.10}$$

in the ideal case where both signal and noise are determined only by the photon statistics.

2.2.2 *Image creation and detective quantum efficiency*

The Gurney–Mott hypothesis describes the processes subsequent to absorption of a photon by a silver halide grain. Photon absorption in the silver halide results in the elevation of a valence electron into the conduction band, producing an electron/hole conduction pair:

$$h\nu + Br^- \rightarrow Br + e^-. \qquad [2.1]$$

(We have distinguished chemical equations such as [2.1] from mathematical ones by a distinct numbering sequence placed in square brackets.) In [2.1], the normal state of the bromine atom fully bonded into the crystal is shown as Br^-, indicating its attachment of an electron from a neighboring silver atom. The photo-produced hole at the neutral bromine atom is of little utility in the detection process. It can meet up with a photoelectron and recombine or it can ionize a neutral silver atom, in either case negating the detection. Fortunately, there are other possible fates: it can migrate to the grain surface and react with the gelatin, or it can unite with another hole, thereby forming a bromine molecule that can escape, eroding the crystal structure.

Usable detection occurs when the freeing of photoelectrons leads to the production of a stable silver molecule. After a photoelectron is created, it wanders through the grain by Brownian motion and may recombine with a halogen atom (the reverse of reaction [2.1]); in this case, it is lost to the detection process. However, the electron may fall into a trap such as those that occur at flaws in the crystal structure. Once the electron is fixed at a trap, it attracts mobile silver ions toward this location; eventually it may combine with one of them to form a neutral silver atom,

$$e^- + Ag^+ \rightleftharpoons Ag. \qquad [2.2]$$

The single silver atom can act as a trap for a second electron,

$$Ag + e^- \rightarrow Ag^-. \qquad [2.3]$$

The resulting negative silver ion can attract a positive ion from the crystal structure, combine with it, and produce a reasonably stable, though not yet developable, two-atom silver molecule,

$$Ag^- + Ag^+ \rightarrow Ag_2. \qquad [2.4]$$

The single silver atom in reaction [2.2], however, is very unstable; it may instead decompose back into an electron–ion pair. Alternatively, a wandering hole can oxidize the atom directly back to Ag^+. Moreover, there is a chance that the thermal excitation of the crystal will allow the electron to escape from the trap before it recombines with a silver atom. In this case, the electron must again wander through the crystal, running the gauntlet of possible recombination and loss before it falls into another trap.

After its creation, the silver molecule can continue to act as a trap for photoelectrons or silver ions or both, and in this way it can grow as the exposure to light continues. The original molecule becomes the nucleus of a development center in the crystal. If this center reaches a critical size (three to four silver atoms), it renders the grain developable by catalyzing the reduction of silver ions to silver in the subsequent chemical processing.

The pattern of exposed grains on a plate is frequently called the latent image. If it is to be useful, it must be amplified. In this case, amplification is accomplished by bathing the grains in a chemical reducing agent, or developer, that provides electrons to the silver ions in the grains, thus reducing them to metallic silver; other reaction products and any remaining silver ions are washed away. Developers are selected that utilize a strong catalytic action by the silver specks in the latent image so that the conversion proceeds much more quickly in the heavily exposed grains than in unexposed ones. The nature of this catalytic action is not entirely understood; one possibility is that the silver specks act as electrodes and conduct electrons from the developer into the grain, allowing them to combine with silver ions over the surface of the speck. After development, the amount of metallic silver in the latent image has been multiplied by as much as 10^8–10^9, an enormous gain.

These steps in the detection process can degrade the information present in the photon stream absorbed by the grain and expressed in equation (2.10), either by losing signal or by adding noise. The detective quantum efficiency (DQE) describes this degradation succinctly in terms of a decrease in the number of photons that are effective in producing the output signal. We define the detective quantum efficiency as

$$\mathrm{DQE} = \frac{n_{\mathrm{out}}}{n_{\mathrm{in}}} = \frac{(S/N)^2_{\mathrm{out}}}{(S/N)^2_{\mathrm{in}}}, \qquad (2.11)$$

where n_{in} is the actual input photon signal, and n_{out} is an imaginary input signal that would produce, with a perfect detector system, the same information content in the output signal as is received from the actual system. Converting to signal to noise, $(S/N)_{out}$ is the observed signal-to-noise ratio, while $(S/N)_{in}$ is the potential signal-to-noise ratio of the incoming photon stream, as given by equation (2.9). By substituting equations (2.9) and (2.10) in equation (2.11), it is easily shown that the DQE is just the quantum efficiency defined in equation (2.5) if there is no subsequent degradation of the signal to noise.

Unfortunately, the DQEs of photographic materials are far lower than the absorptive quantum efficiency with reflection loss in equation (2.5). The inefficiencies in going from photoelectrons to developable grains are such that about 10 to 20 photons must be absorbed by a grain in a reasonably short period of time for it to have a 50% probability of developing. From this cause alone, the DQE is reduced by a factor of 10 to 20 from the absorptive QE. Two other effects can further reduce the DQE. First, the individual grains are distributed randomly over the plate. Two areas of equal size have randomly differing numbers of grains and hence have differing response even with identical exposure and development. Secondly, there are significant variations in grain properties. As a result, two plate regions with the same number of grains can be expected to exhibit nonuniformity even with identical exposure and development. These latter two effects are expressed quantitatively as the *granularity* of the plate, which refers to the variations in response measured through a fixed aperture for a plate with uniform exposure. As a result of the combination of the phenomena discussed here, if equation (2.5) shows a quantum efficiency of 50–90%, the DQE that can be realized on a plate will be no more than 2–5%.

Dainty and Shaw (1974) discuss in detail the information content of photographic plates and its extraction.

2.2.3 Spectral response

Despite the high grain quantum efficiency at short wavelengths, the overall efficiency of films and plates is reduced there because the gelatin absorbs photons in the ultraviolet, cutting off response for wavelengths short of 0.3 μm. Response can be extended farther into the ultraviolet by using plates with extremely thin coatings of gelatin so that bare silver halide grains are exposed directly to the incoming photons. Such plates require special handling to prevent damage to the unprotected grains.

As normally practiced with silver bromide grains, the photographic process described above is only effective for blue and ultraviolet wavelengths. A slight extension can be made toward the red by the addition of iodine to the grains; silver iodobromide has a slightly reduced E_g compared with that of AgBr.

Greater extension to the red is achieved by dye sensitization. In this process, a red-absorbing dye is adsorbed onto the silver halide grains. Photons absorbed in the dye are capable of creating conduction electrons in the grain. One possibility is that the photon creates a conduction electron in the dye; the electron is then transferred directly into the conduction band of the halide crystal. Another possibility is that the excited dye molecule transfers energy to the halide, which subsequently excites an electron into the conduction band. In the spectral region of dye sensitization, only the grain surface, as opposed to the entire grain volume, takes part in the initial photon detection.

2.2.4 Color photography

A variety of methods have been used for color photography, but most modern approaches are based on a depthwise superposition of emulsions such as in Figure 2.3. We will use this figure to illustrate the action of a positive color film. The topmost layer of the emulsion in Figure 2.3(a) contains grains that are not dye-sensitized and respond only to blue light. A yellow filter removes the blue light to protect the underlying layers of grains (otherwise, the blue light would react with the silver halide in these grains to make them developable), but this filter transmits both green and red. One of the underlying layers is sensitized with a dye that responds to green but not red, and the other with a dye that responds to red but not green. Thus, when exposed to a colored scene, the three primary colors are recorded separately in the three layers of emulsion.

In the course of development, the yellow filter dye is removed and image dyes are produced in the emulsion layers, as shown in Figure 2.3(b). After initial development of the exposed grains, a second development forms the complementary dye in the unexposed grains in each layer. That is, yellow dye is produced in the unexposed grains in the blue-sensitive layer, magenta dye in the unexposed grains in the green-sensitive layer, and cyan dye in the unexposed ones in the red-sensitive layer. Magenta is a dye which absorbs only green but transmits red and blue; cyan removes red and transmits green and blue. At the end of the process, all silver has been removed and only the image dyes remain. As an example, where the film has been exposed to blue light, the top layer will be undyed and the lower ones will be dyed, respectively, magenta and cyan, which together transmit blue. Similarly, where the film has been exposed to green light, the top layer will be dyed yellow, the second layer will be undyed, and the bottom layer will be cyan; yellow and cyan together transmit green. Where the film has been exposed to red light, the top two layers will be dyed, respectively, yellow and magenta, which combine to transmit red. Thus, the colors of the illumination are reproduced directly.

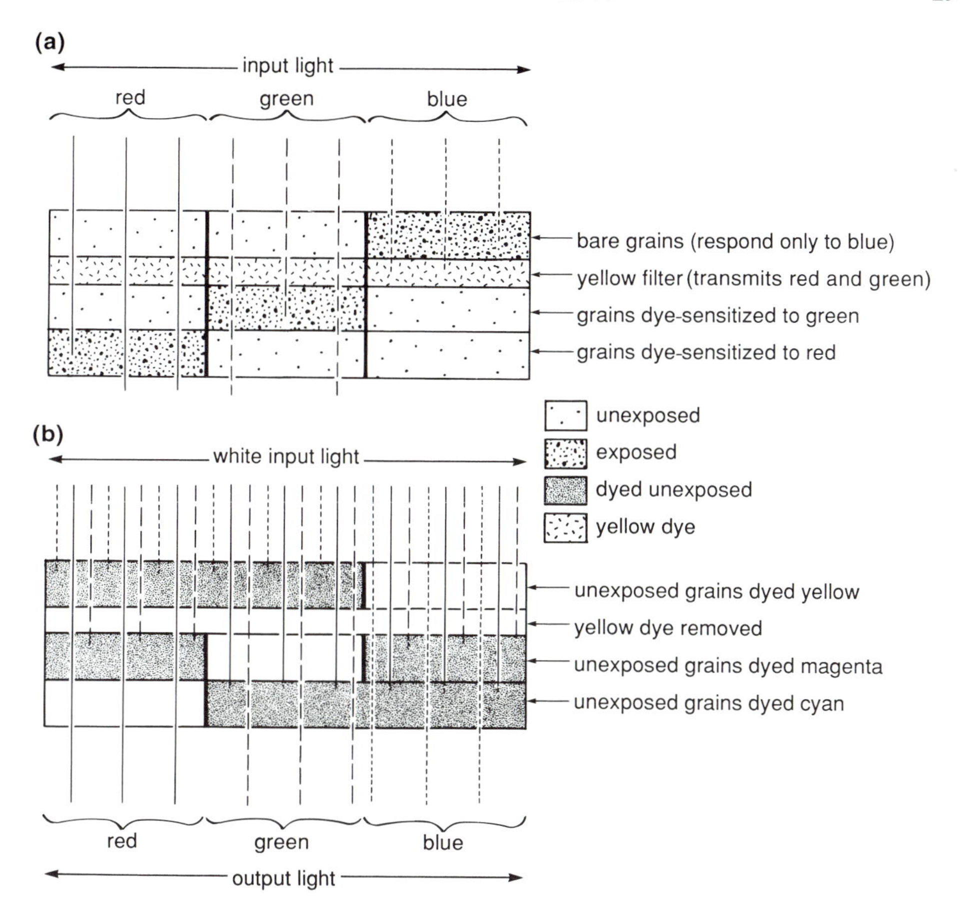

Figure 2.3. Illustration of operation of color photography: (a) shows the exposure of the film to the three primary colors; (b) shows the result after development.

There are many variations on the basic theme illustrated in this example. Negative films, for example, yield a picture which is converted to red, blue, and green only by transfer to positive photographic materials such as print paper. Further discussion can be found in James and Higgins (1960) and James (1977).

2.3 Characteristic curve

Information can be extracted from a photographic plate in a number of different ways. Density estimates are frequently done by eye. More quantitative information can be obtained by shining a light source through the plate and using a detector to convert the transmitted portion to an electrical signal, as shown in Figure 2.4. The transmittance is the ratio of the amount of light

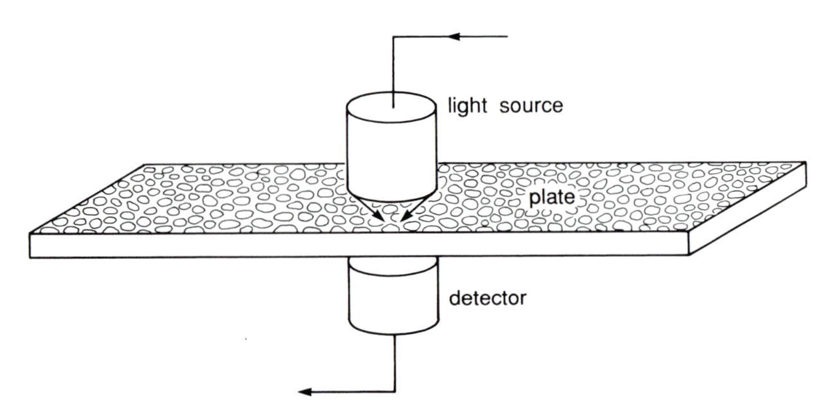

Figure 2.4. Electronic measurement of a photographic image.

that passes through the developed plate to the amount that is incident on it, that is,

$$T=\frac{I_T}{I_0}, \tag{2.12}$$

where I_0 is the intensity of light incident on the plate, and I_T is the intensity received after passage through the plate. The opacity is the reciprocal of the transmittance, $O=1/T$, and the density, D, is the logarithm of the opacity:

$$D=\log(O)=-\log(T). \tag{2.13}$$

Automated machines are used to scan the light source and detector over a plate. The output of the detector can be converted into a digital signal and used to record the entire plate as a two-dimensional digital array whose entries are proportional to the density. Because internal scattering is significant in plates, the transmittance of the plate will depend on the geometry of the detector used to measure the emergent light. The diffuse density refers to density determined by measuring all the light transmitted over a hemisphere from an incident collimated beam; hence, light that traverses the plate regardless of scattering angle is counted as transmitted. Specular density refers to density measured over only a portion of the emergent angles. Plate scanning machines that measure specular density can give accurate relative results, but only diffuse density measures can be compared absolutely from one type of plate to another.

Unlike many of the electronic detectors to be described later, the photographic plate can be quite nonlinear in response to faint light levels. This behavior is described by the characteristic curve (sometimes called the H & D curve after its originators, Hurter and Driffield), shown in Figure 2.5.

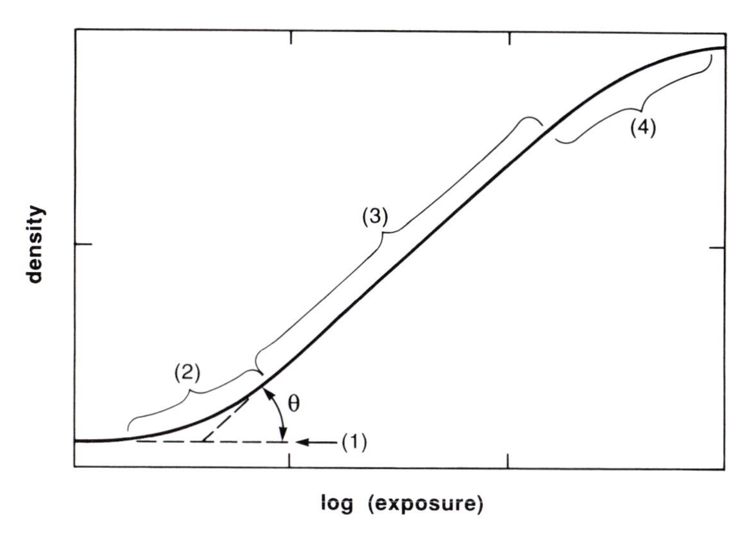

Figure 2.5. Characteristic curve. Zones of the curve include: (1) level of gross fog; (2) toe; (3) straight-line portion; (4) shoulder. $\gamma = \tan\theta$.

This curve shows the relationship between the density of the developed plate and the logarithm of the exposure, H, where $H = Et$, and E and t are the irradiance (watts per square meter) and the exposure time, respectively (a variety of other units are also used for exposure). This curve is therefore the log of the opacity versus the log of the total energy that has fallen on a unit area of the plate. The characteristic curve is influenced not only by the manufactured properties of the emulsion, but also by the handling of the emulsion after manufacture, the development, and the means used to measure the plate after development. Consequently, the characteristic curve is not uniquely determined for a given emulsion. Different batches of the same emulsion will have similar characteristic curves *only* if they are all subjected to similar handling, development, and measurement procedures.

The reciprocity law describes the condition in which the net response of a detector depends only on the total number of photons received rather than on either the photon arrival rate or the total exposure time. In the nonlinear region of reciprocity failure, the photographic emulsion produces more response for a short burst of photons than for the same number of photons distributed over a long exposure time. This behavior occurs because, at very low photon arrival rates, electrons may escape from the trapping centers before a viable development center forms. They may then recombine elsewhere in the grain, reducing the efficiency of the plate. Because of low-intensity reciprocity failure, the characteristic curve can vary strongly with exposure conditions, particularly at low exposures.

There are four distinct regimes of response; they depend on the level of exposure and are indicated by numbers on Figure 2.5:

(1) *Level of gross fog*. A minimum density will be produced even with no exposure of the plate. There are two components: base fog arises from the imperfect transmission of light by the glass or plastic support that carries the emulsion (the zero of the characteristic curve is frequently set at the level of base fog), and chemical fog results from grains that develop without receiving any light.

(2) *Underexposure*. Low levels of exposure produce a nonlinear response for reasons closely related to those producing reciprocity failure: conduction electrons escape from trapping centers and recombine before viable development centers form. This section of the characteristic curve is frequently called the toe.

(3) *Region of linearity*. The log of the number of developed grains increases more or less linearly with the log of the number of photons incident on the emulsion. This section is often described as the straight-line portion.

(4) *Overexposure*. High photon rates produce saturation and nonlinearity. Saturation occurs in part simply because the majority of grains have already become developable so additional exposure produces relatively little effect. In addition, where the illumination level is high, conduction electrons are produced in the grains more quickly than the silver ions can migrate to the trapping centers. Consequently, many electrons recombine with halide atoms before they can combine with the silver ions, leading to high-intensity reciprocity failure. The portion of the nonlinear section of the curve prior to extreme saturation is frequently called the shoulder.

2.4 Performance

The performance of a plate is described in terms of its *speed*, *contrast and signal to noise*, *calibration* properties, and *resolution*. In terms of the characteristic curve, the speed is defined as the reciprocal of the exposure required to reach some given density. High speed emulsions therefore have characteristic curves lying to the left of those of low speed emulsions. Contrast refers to the density range that distinguishes different exposure levels. It is described by the slope, gamma (or γ) $= \tan\theta$, of the relation between density and exposure (see Figure 2.5); large values of gamma correspond to high contrast. Gamma can also be used to measure the response of the plate in a quantitative way, that is, it gives the relationship between input signal (change in exposure) and output signal (change in density). Calibration refers to the steps required to obtain quantitative information about the relationship between density and exposure. For the purpose of calibration, the response of the plate should be uniform over its surface, and it is desirable that the response be reproducible from one plate to another. At low light levels, the effects of reciprocity failure make careful calibration difficult but essential. The resolution of the plate refers to its ability to distinguish different levels of illumination incident on adjacent regions. Each of these concepts is discussed in more detail below.

2.4.1 Speed

The speed of photographic plates can be understood by assuming that a grain must absorb a certain number of photons to become developable. The number of photons that must be absorbed is largely independent of the grain size, as illustrated in Figure 2.6. Now, since (a) speed goes inversely as the exposure required to reach a given density, and (b) exposure has units of number per unit area, it follows that speed increases with increasing grain area. Strictly speaking, in the blue where absorption occurs in the silver halide, the speed is proportional to grain area times the absorption factor (equation (2.3)). If $a(\lambda)d_1 \ll 1$, the absorption factor reduces to $\eta_{ab} = a(\lambda)d_1$, and the speed goes as the volume times the absorption coefficient. For very large grains which absorb most of the photons striking them, the absorption efficiency per unit

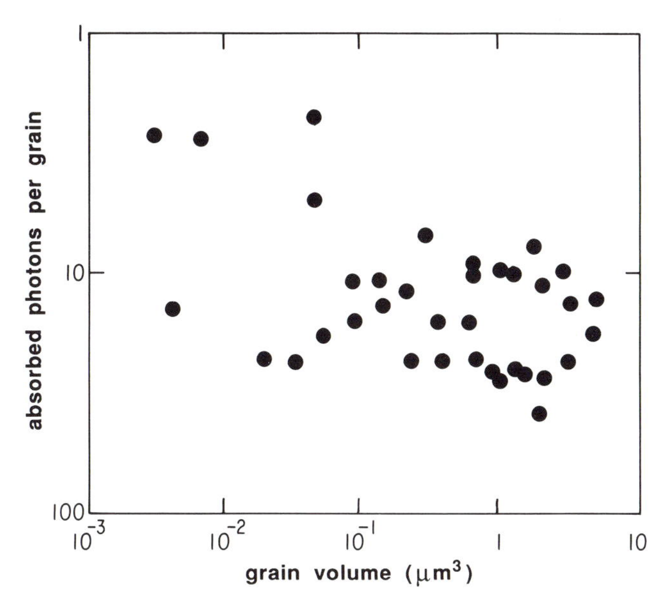

Figure 2.6. Average number of photons that must be absorbed per grain to make half the grains developable (from Tani, 1989).

volume decreases because of the reduced photon flux toward the back of the grain. In the case of dye-sensitized grains operating in the red, only the surface is effective at absorbing photons, so the speed is proportional to the grain surface area rather than to its volume. The absorption effects that reduce the speed of large grains can be overcome by producing grains that are large but thin and depositing them with the large cross-section perpendicular to the incoming light. Kodak T-grains (for example, T-max films) are an implementation of this concept; they produce extremely high speeds and good DQE. Dye-sensitized T-grains have an advantage over normal dye-sensitized grains because they have a larger surface-to-volume ratio than conventional grains.

For very small grains having diameters of the order of $1/\pi$ times the wavelength of the incoming photons, diffraction affects the absorption efficiencies, reducing the speed as compared to the simple scalings described above.

The relationship of speed with grain size is illustrated in Table 2.1 (data from James and Higgins 1960); it is easy to demonstrate that the speed increases in proportion to increasing grain area (except for the smallest sizes where diffraction becomes important).

At low light levels, the speed of an emulsion can be severely reduced because of reciprocity failure. This problem can be attacked in a variety of ways (see, for example, Smith and Hoag, 1979). Cooling the emulsion can reduce the reciprocity failure because it reduces the thermal excitation in the grains, thus inhibiting the tendency of the photoelectrons to escape from the trapping centers and slowing the decomposition of single silver atoms. On the other hand, overcooling reduces the mobility of the free silver ions that must migrate to the trapping centers, thus reducing sensitivity. Plates can also be pre-flashed with a uniform low level of light. A properly pre-flashed plate has a fog level just above the level of gross fog, so the sensitivity to faint objects is improved by lifting their images out of the region of extreme reciprocity failure. Bathing plates in ammonia or silver nitrate increases the concentration of Ag^+ ions, improving the likelihood that these ions will reach the trapped electrons to form free silver atoms before the electrons escape from the traps. Bathing plates in hydrogen gas also hypersensitizes them, evidently by reducing the hole concentration and thus inhibiting the oxidization of Ag atoms back to Ag^+. Because hydrogen gas is explosive when mixed with air, a more stable mixture of hydrogen and nitrogen called forming gas is frequently used instead of pure hydrogen.

A variety of other treatments improve the overall speed of plates. For example, it has been found that absorbed water and oxygen act to desensitize

Table 2.1. *Speed of some photographic emulsions*

Grain area (μm^2)	Grain thickness (μm)	Emulsion speed
0.14	0.062	12
0.28	0.096	52
0.50	0.18	150
0.82	0.25	250
1.5	0.33	450
2.6	0.59	910

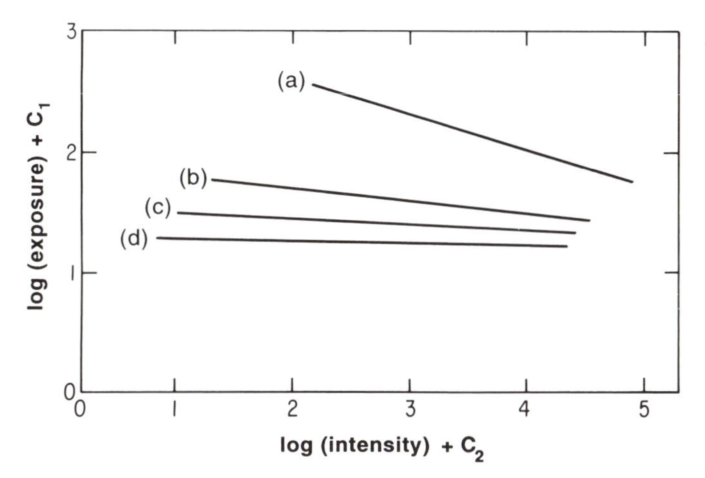

Figure 2.7. Exposure (in arbitrary units) required to reach a density of 0.6 as a function of the intensity of illumination, for type IIIa-J emulsion (from Eastman Kodak, 1987). Curve (a) illustrates the results with untreated material, while (b), (c), and (d) show, respectively, the results of hypersensitization by soaking in vacuum, baking in nitrogen, and treatment with hydrogen.

emulsions. Treatments that are effective in driving these contaminants out of the emulsion include baking the plate in a vacuum or a dry nitrogen atmosphere, or simply flushing it with dry nitrogen at ambient temperature for an extended period before exposure. To achieve adequate storage life, red-sensitive emulsions are usually treated with stabilizers, or restrainers, which substantially reduce their sensitivity. These substances can be washed out of plates just before exposure by bathing the plates in water.

The action of various types of hypersensitization is illustrated in Figure 2.7. The vertical axis shows the log of the exposure required to reach a given density level (0.6 above base-plus-fog); note that the faster the film, the lower the required exposure will be (that is, on this figure, down is good). The horizontal axis shows the intensity of the illumination. No reciprocity failure would mean that the required exposure is independent of the level of illumination, that is, the relation would be a horizontal line in Figure 2.7. Note as the intensity is decreased, the untreated emulsion shows a dramatic increase in required exposure. This behavior results from reciprocity failure; the same exposure distributed over a long time period does not produce as high a density as it would if it had occurred quickly. The reciprocity failure is reduced substantially by all the methods of hypersensitization. Treatment with hydrogen gas almost eliminates reciprocity failure for this emulsion; the required exposure becomes virtually independent of the intensity.

These procedures can reduce exposure times at low light levels significantly, but many of them also decrease the uniformity from plate to plate both because they may not be applied uniformly and because their effects (for example, driving excess water from the emulsion) decay rapidly with time. These problems are particularly severe with extremely red-sensitive emulsions. Thus, the calibration difficulties described below can be exacerbated for hypersensitized plates. In addition, hypersensitization nearly always produces an undesirable increase in chemical fog.

Further information on precautions and procedures for low light level photography can be found in Smith and Hoag (1979), and Eccles, Sim, and Tritton (1983).

The speed of a photographic plate at low light levels also depends in a complex way on the illumination conditions established by the optics used to form an image on the plate. Parameters of the optical system can vary the physical scale of the image. On a given plate, the rate at which photons fall on a grain goes inversely as the image area. For example, if the photon rate is constant but the image diameter doubles then the photon rate per grain falls by a factor of four because the photons are being spread over a four times larger area. The exposure needed to reach a given density must be increased to compensate. Thus, the speed of the plate in a particular optical system goes inversely with image area unless the plate is operating in the region of reciprocity failure. The effects of reduced illumination can be significantly amplified by reciprocity failure, further reducing the speed as the image area is increased.

To illustrate, consider the specification of exposures in photography in terms of the f/number, the ratio of the focal length of the optics to its diameter. If the focal length of a camera lens is doubled with no other changes, then the f/number is doubled and the photon rate per grain falls by a factor of four. If the focal length is doubled and the f/number kept constant (that is, the diameter of the optics is also doubled), then the photon rate per grain is unchanged. Hence, the system used to specify exposures with cameras is independent of the size of the camera and its optical system.

As another example, consider astronomical photography (most plates used in this application suffer some degree of reciprocity failure, so the following discussion is only approximately correct). The physical area of a given image formed by a telescope is proportional to the square of the focal length of the telescope, and the total light collected is proportional to the square of the diameter of the telescope. The density of illumination in the image – and hence the speed of the plate – therefore goes as the (diameter/focal length) squared, or as the inverse f/number squared. In this example, however, the number of grains across an image diameter increases proportionately with focal length. It would be more logical to select plates having grain sizes that scale with the focal length, keeping the resolution of the plate constant in terms of the number of grains within the image. In this case, and if the plate speed scales with grain area (for example, the first-order relation for dye-sensitized plates), the speed of the telescope/plate combination will go only as the telescope diameter squared (that is, as the telescope collecting area) without a dependence on f/number.

2.4.2 Contrast and signal to noise

As γ ($=\tan\theta$; see Figure 2.5) is increased, the contrast of the plate increases; that is, the signal tends to stand out more clearly from the noise. The contrast for a given plate type is controlled by the processing. As the development proceeds, more and more grains in the latent image are reduced to silver, increasing the density difference between the exposed and unexposed parts of the plate. Chemical fog also increases but much more slowly because the grains that produce fog do not have good development centers to catalyze the development action. In general, a properly developed image is one where the contrast is near its maximum but development has been stopped before there is an unacceptable increase in fog. The achievable contrast is generally higher for emulsions with smaller grains, in part because small-grained emulsions are relatively homogeneous in grain size, and variations in grain size (and hence speed) tend to reduce the slope of the composite characteristic curve.

For a perfectly uniform exposure, the density will vary over the plate because of the probabilistic nature of the chain of processes starting with absorption of photons in grains, continuing through formation of development centers, and ending with development. We have already discussed how random

Table 2.2. *Density noise versus density*

Emulsion type	103a-O	IIa-O	IIIa-J
Grain size	moderately coarse	fine	fine
γ	~1.75	~1.0	~1.9
Speed	200	200	50
Granularity	0.030	0.019	0.016
Total density		σ_D	
0.2	—	0.017	0.007
0.4	0.037	0.023	0.011
0.6	0.039	0.028	0.014
0.8	0.041	0.033	0.017
1.0	0.046	0.038	0.019
1.2	0.051	0.042	0.022
1.5	0.060	0.048	0.025
2.0	0.079	0.058	0.031
2.5	0.101	0.068	0.036
3.0	0.129	0.075	0.038

variations in the number of grains per resolution element and in the grain properties can also lead to density nonuniformities. All of these variations taken together constitute the noise of the detection process, above which the detected signals must lie. It is found empirically that this noise depends on the density alone (measured above base fog) for a given emulsion, that is it does not depend on such factors as hypersensitizing treatment or exposure time. Once the relation between density and noise has been determined for a given emulsion, it can be used to estimate the noise anywhere on plates taken with this emulsion. Table 2.2 shows the relation between density and density noise, σ_D, for three Kodak emulsions, all with blue-peaked response but with differing grain size (data from Furenlid, 1978), along with other characteristics of the emulsions from Eastman Kodak (1987).

In the table, the speed is inversely proportional to the exposure required to reach density 0.6. The effect of grain size is measured by the rms diffuse granularity, which is defined as the standard deviation in a measure of diffuse density of average value 1.00 obtained through a 48 μm diameter aperture. The granularity is usually inversely proportional to the square root of the area of the measuring aperture and is often roughly proportional to the cube root of the density. Further information can be found in Eastman Kodak (1987).

Given a characteristic curve for an emulsion and set of processing conditions, the data in Table 2.2 allow one to relate the density at any point to the uncertainty in the measurement of the exposure at that point, σ_H. If we note that $\sigma_H/\sigma_D \approx dH/dD$ for small variations in H and D and that $\gamma = dD/d(\log H)$, we can use the characteristic curve and Table 2.2 to compute the signal to noise and DQE of the emulsion. Specifically,

$$\frac{dH}{H} = d(\ln H) = 2.3026\, d(\log H), \tag{2.14}$$

and

$$\begin{aligned}\left(\frac{S}{N}\right)_{out} &= \frac{H}{\sigma_H} = \frac{H}{\sigma_D}\frac{dD}{dH} \\ &= \frac{0.4343}{\sigma_D}\frac{dD}{d(\log H)} = \frac{0.4343\gamma}{\sigma_D}.\end{aligned} \tag{2.15}$$

Equation (2.9) shows that $(S/N)_{in} = \mathbb{C}H^{1/2}$, where the constant $\mathbb{C}$ depends on the units of H. From equation (2.11), we then have

$$\mathrm{DQE} = \frac{0.1886}{\mathbb{C}^2 H}\frac{\gamma^2}{\sigma_D^2}. \tag{2.16}$$

Most photographic materials have DQEs near 1%, although hypersensitized emulsions or T-grains can achieve DQEs of 4–5%.

In general, from the discussion in Section 2.4.1, fine-grained emulsions tend to have relatively large values of γ; from Table 2.2, fine-grained emulsions also tend to have relatively small values of σ_D for a given density. On the other hand, we have also seen that the exposure required to reach a given density can have a strong inverse dependence on grain size. Since these dependencies on grain size vary from one emulsion type to another, the selection of an optimum emulsion must be made carefully.

The choice of an optimum emulsion can be discussed in terms of two extreme cases. In one extreme, there is a weak signal with a negligible flux of background photons, that is the contrast between signal and background is high. In this case, the time required for the measurement decreases with increasing grain area (or volume) and large-grained emulsions are preferred. In the other extreme, there is a weak signal that is superimposed on a stronger background, that is the contrast between signal and background is low. In this case, the limiting noise is set by the noise of the image of the background. Therefore, as shown in equation (2.16), a high DQE and a better detection limit can be achieved with a fine-grained emulsion with large γ and small σ_D, even though the required exposure time to reach an acceptable density will be increased relative to that for a large-grained emulsion.

2.4.3 Calibration

Reciprocity failure is a significant problem for quantitative reduction of data on plates and film obtained at low light levels. To extract brightness measurements, it is necessary to calibrate the plate over a range of actual illumination levels and times (and other conditions such as temperature and development) that are closely comparable to the circumstances under which the original data were obtained. To do so, a portion of the plate should be illuminated with light of a brightness level and having spectral characteristics similar to the original exposure. This type of calibration is done with spot sensitometers that control the necessary long, low light level exposures, and it is frequently conducted simultaneously or nearly simultaneously with the actual exposure of the plate by placing the sensitometer field on an isolated region near one edge of the plate. In color photography, the reciprocity failure is usually different for the different emulsion layers responsible for different colors; hence, color photography at low light levels has poor color fidelity.

A much simpler and less time consuming calibration procedure can be used for linear electronic detectors. For them, short exposures are made of bright calibration objects, and, because the linearity is good and the dynamic range is large, the signals can be related directly to those obtained in long exposures on faint objects. Moreover, because the detectors can be reset and reused, calibration can be carried out on the identical detectors used to obtain the data on the unknown objects. As a result of these advantages, electronic detectors are nearly always to be preferred over plates when precise information about illumination levels is desired.

Calibration of plates can be further complicated by adjacency effects, in which the density is a function not only of the exposure of the region in question but of the density of adjacent regions. These effects occur because of the relatively rapid exhaustion of developer where there is a high density of developable grains. As a result, a heavily exposed region next to a lightly exposed one will have more access to fresh developer than the same region would if surrounded by other heavily exposed ones, and therefore it will tend to develop more completely. Similarly, a lightly exposed region next to a heavily exposed one will have less access to fresh developer than it would if surrounded by other lightly exposed regions; it will tend to underdevelop. Additional inhibition of development may occur due to action of development byproducts from the heavily exposed region. As a result of adjacency effects, the calibration may depend on local exposure, and isolated areas of different sizes may develop to different densities even with identical exposures.

In addition to the problems introduced by nonlinear response, the calibration of photographic data is complicated because the detectors are used up in a single observation; it is not possible to reset the plate and repeat the measurement with the identical detector array! Consequently, the ability to compare results from different areas on a plate depends on the ability of the manufacturer to control the uniformity of the response over relatively large areas. The necessary process control can be held at the 1–2% level, which becomes the smallest reliably detectable variation on a single plate against a uniform background.

The region of linearity on the characteristic curve of a plate encompasses a limited dynamic range. In general, information at densities higher than the region of linearity is difficult to recover. Nonetheless, in the case of compact images of the same intrinsic shape (for example, images of stars), the diameter of the image can be used to estimate the number of photons collected even if the core of the image is overexposed. The basic dependence of diameter on illumination level arises because of scattering within the photographic emulsion that causes grains away from the spot of illumination to intercept a portion of the photons, leading to spreading of the image. This spreading is sometimes termed the 'turbidity' of the emulsion.

It is found empirically that an approximation to the effects of scattering on the diameter of an image is

$$w = a + \Gamma \log\left(\frac{H_0}{H_w}\right), \qquad H_0 > H_w, \tag{2.17}$$

where H_0 is the exposure at the core of the image, H_w is the exposure at which the diameter is measured, the logarithm is base 10, and a and Γ are constants that can be determined from measurements of plates exposed to sources of known brightness (a is closely related to the diameter of the image in the absence of spreading). For an emulsion that responds as in equation (2.17), the difference in diameters for two point-like images measured to the same H_w is proportional to the difference in the logarithms of the exposures, or (assuming exposure times are equal) to the difference of the logarithms of the intensities of the imaged sources.

Although equation (2.17) is mathematically convenient to use, image diameters frequently show a more complicated dependence on exposure, both due to effects of scattering and to other influences on the image size (for example, the optics used to produce the image). As a result, the calibration can often be improved with more complex calibration procedures than those described here.

2.4.4 Resolution

One might expect the resolution of a plate to correspond to the average grain size. Because the grains scatter light, however, the smallest images are significantly larger than this limit, and the practical limiting resolution area is 10–100 times the area of a typical grain. In fact, as just discussed, the growth of image diameter with increasing exposure is used to estimate the brightness of objects when the image cores are saturated.

The resolution of a plate can be most simply measured by exposing it to a pattern of alternating white and black lines and determining the minimum spacing of line pairs that can be distinguished. The eye can identify such a pattern if the light–dark variation is 4% or greater. The resolution of the plate is expressed in line pairs per millimeter corresponding to the highest density of lines that produces a pattern at this threshold on the plate.

Although it is relatively easy to measure resolution in this way for the plate alone, a resolution in line pairs per millimeter is difficult to combine with resolution estimates for other components in an optical system used with the photographic plate. For example, how would one derive the net resolution for a camera with a lens and plate whose resolutions are both given in line pairs per millimeter? A second shortcoming is that the performance in different situations can be poorly represented by the line pairs per millimeter specification. For example, one might have two lenses, one of which puts 20% of the light into a sharply defined image core and spreads the remaining 80% widely, whereas the second puts all the light into a slightly less well defined core. These systems might achieve identical resolutions in line pairs per millimeter (which requires only 4% modulation), yet they would perform quite differently in other situations.

A more general concept is the modulation transfer function, or MTF. Imagine that a sinusoidal input signal of period P and amplitude $F(x)$ is impressed on the plate,

$$F(x)=a_0+a_1\sin(2\pi fx), \tag{2.18}$$

where $f=1/P$ is the spatial frequency, x is the distance along one axis of the plate, a_0 is the mean height (above zero) of the pattern, and a_1 is the amplitude. These terms are indicated in Figure 2.8(a). The modulation of this signal is defined as

$$M_{\text{in}}=\frac{F_{\text{max}}-F_{\text{min}}}{F_{\text{max}}+F_{\text{min}}}=\frac{a_1}{a_0}, \tag{2.19}$$

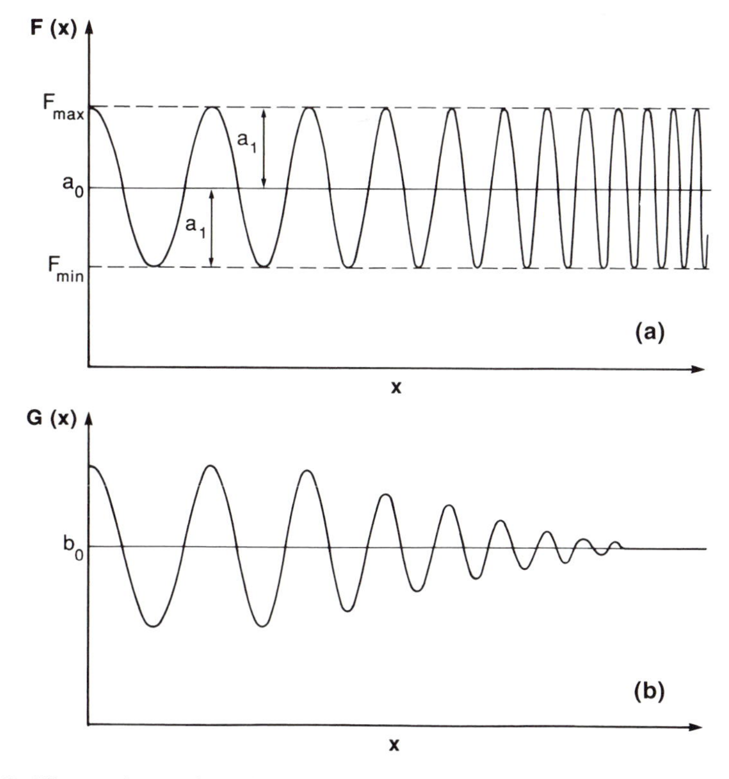

Figure 2.8. Illustration of variation of modulation with spatial frequency. (a) Sinusoidal input signal of constant amplitude but varying spatial frequency. (b) How an imaging detector system might respond to this signal.

where F_{max} and F_{min} are the maximum and minimum values of $F(x)$. Assuming that the resulting image on the plate is also sinusoidal (which may be only approximately true due to nonlinearities of the detector), it can be represented by

$$G(x) = b_0 + b_1 \sin(2\pi f x), \tag{2.20}$$

where x and f are the same as in equation (2.18), and b_0 and b_1 are analogous to a_0 and a_1. The modulation in the image will be

$$M_{out} = \frac{b_1}{b_0} \leqslant M_{in}. \tag{2.21}$$

The modulation transfer factor is

$$MT = \frac{M_{out}}{M_{in}}. \tag{2.22}$$

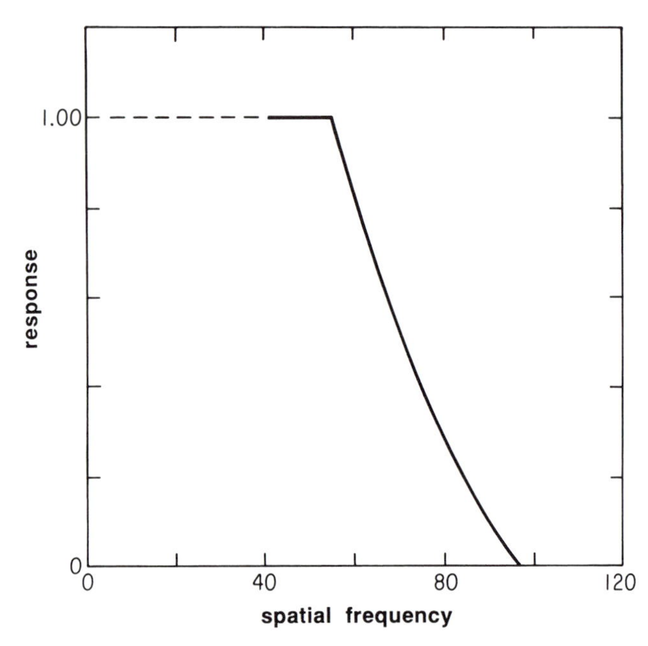

Figure 2.9. The modulation transfer function (MTF) for the response illustrated in Figure 2.8(b).

A separate value of the MT will apply at each spatial frequency; Figure 2.8(a) illustrates an input signal which contains a range of spatial frequencies, and Figure 2.8(b) shows a corresponding output in which the modulation decreases with increasing spatial frequency. This frequency dependence of the MT is expressed in the modulation transfer function (MTF). Figure 2.9 shows the MTF corresponding to the response of Figure 2.8(b).

The MTFs for a fine-grained (IIIa-J) and a moderately coarse-grained (103a-O) emulsion are shown in Figure 2.10. In a few cases, the detector may have enhanced sensitivity at some spatial frequency, resulting in an MTF that exceeds 1 there. For example, adjacency effects can produce increased response just above zero spatial frequency.

Computationally, the MTF can be determined by taking the absolute value of the Fourier transform, $F(u)$, of the image of a perfect point source. This image is called the point spread function. Fourier transformation is the general mathematical technique used to determine the frequency components of a function $f(x)$ (see, for example, Bracewell, 1986, and Press *et al.*, 1986). $F(u)$ is defined as

$$F(u)=\int_{-\infty}^{\infty} f(x)\, e^{j2\pi ux}\, dx, \qquad (2.23)$$

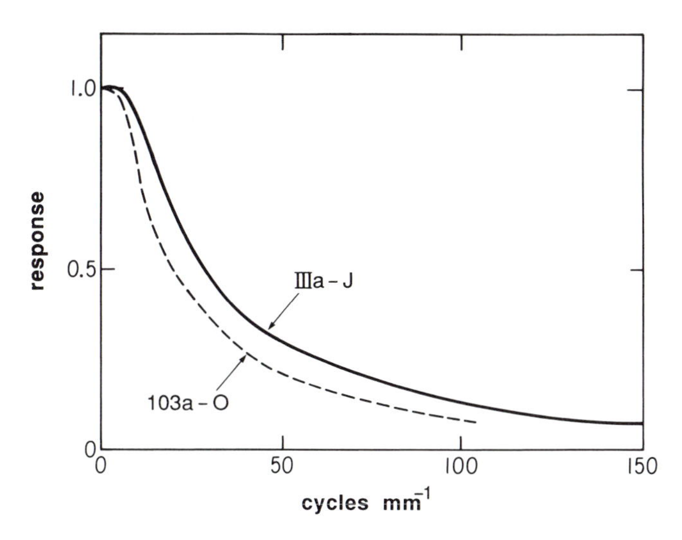

Figure 2.10. The MTF for two emulsions.

with inverse

$$f(x) = \int_{-\infty}^{\infty} F(u)\, e^{-j2\pi xu}\, du, \tag{2.24}$$

where j is the (imaginary) square root of -1. In the current discussion, $f(x)$ is a functional representation of the point spread function, and u is the spatial frequency. The Fourier transform can be generalized in a straightforward way to two dimensions, but for the sake of simplicity we will not do so here. The absolute value of the transform is

$$|F(u)| = (F(u)F^*(u))^{1/2}, \tag{2.25}$$

where $F^*(u)$ is the complex conjugate of $F(u)$; it is obtained by reversing the sign of all imaginary terms in $F(u)$. If $f(x)$ is the point spread function, $|F(u)|/|F(0)|$ is the MTF. This formulation holds because a sharp impulse contains all frequencies equally and hence the point spread function gives the spatial frequency response of the detector. Note that the MTF is normalized to unity at spatial frequency 0 by this definition. As emphasized in Figure 2.9, the response at zero frequency cannot be measured directly but must be extrapolated from higher frequencies.

Only a relatively small number of functions have Fourier transforms that are easy to manipulate. A short compilation of some of these cases is listed in Table 2.3. With the use of computers, however, Fourier transformation is a powerful and very general technique.

Table 2.3. *Fourier transforms*

$f(x)$	$F(u)$
$F(x)$	$f(-u)$
$aF(x)$	$aF(u)$
$f(ax)$	$(1/\|a\|)F(u/a)$
$f(x)+g(x)$	$F(u)+G(u)$
1	$\delta(u)^c$
$e^{-\pi x^2}$	$e^{-\pi u^2}$
$e^{-\|x\|}$	$2/(1+(2\pi u)^2)$
$e^{-x}, x>0$	$(1-j2\pi u)/(1+(2\pi u)^2)$
$\text{sech}(\pi x)$	$\text{sech}(\pi u)$
$\|x\|^{-1/2}$	$\|u\|^{-1/2}$
$\text{sgn}(x)^a$	$-j/(\pi u)$
$e^{-\|x\|}\,\text{sgn}(x)$	$-j4\pi u/(1+(2\pi u)^2)$
$\Pi(x)^b$	$\sin(\pi u)/\pi u$

[a] $\text{sgn}(x)=-1$ for $x<0$ and $=1$ for $x\geqslant 0$.
[b] $\Pi(x)=1$ for $|x|<1/2$ and $=0$ otherwise.
[c] $\delta(u)=0$ for $u\neq 0$, $\int\delta(u)\,du=1$; that is, $\delta(u)$ is a spike at $u=0$.

The MTF of an entire linear optical system can be determined by multiplying together the MTFs of its constituent elements. The multiplication occurs on a frequency by frequency basis, that is, if the first system has $\text{MTF}_1(f)$ and the second $\text{MTF}_2(f)$, the combined system has $\text{MTF}(f)=\text{MTF}_1(f)\,\text{MTF}_2(f)$. The overall resolution capability of complex optical systems can be easily determined in this way. In addition, the MTF gives a 'complete' description of a linear optical system as opposed to single parameter descriptions that may be equivalent for systems having significantly different resolution characteristics.

2.5 Examples

2.5.1 Detective quantum efficiency

Suppose the mean density, D, on a plate (in arbitrary units) is equal to the log of the exposure, the noise in the density per pixel element is equal to $7/D^2$, and $\gamma=1$. Compute the DQE for exposures of (a) 1000, and (b) 10^4 photons per pixel element.

$\gamma=1$ means that the relative noise in the density is equivalent to the relative uncertainty in the exposure. Thus, for an exposure of 1000 photons, $D=3$ and the noise in $D=7/3^2=0.78$. We then have $(S/N)_{\text{out}}=3/0.78=3.9$. From equation (2.9), $(S/N)_{\text{in}}=(1000)^{1/2}=31.6$. From equation (2.10), we find that the DQE is $(3.9)^2/(31.6)^2=0.015$. A similar calculation for an exposure of 10^4 photons yields DQE $=0.0084$.

2.5.2 *Modulation transfer function*

We will compute the MTF of a plate with image widths given by equation (2.17). We first need to describe the point spread function. To do so, let $a=0$ and solve equation (2.17) generally for x by indicating w as x and H_w as $H(x)$: $H(x)=H_0 e^{-x/\Gamma'}$, where $\Gamma'=\Gamma/2.30$ (the extra factor arises when changing from $\log_{10}$ to $\log_e$). To allow for the symmetry of the image around $x=0$, this solution should be written as

$$E(x)=E_0\, e^{-|x|/\Gamma'}. \tag{2.26}$$

From Table 2.3, the Fourier transform of equation (2.26) is

$$E(f)=E_0\left[\frac{2\Gamma'}{1+(2\pi\Gamma' f)^2}\right], \tag{2.27}$$

and since $H^*(f)=H(f)$, the absolute value of the Fourier transform of equation (2.26) normalized to 1 at $f=0$ is

$$\mathrm{MTF}=\frac{1}{1+(2\pi\Gamma' f)^2}. \tag{2.28}$$

A virtually identical calculation is illustrated in the following chapter (equations (3.21) through (3.25)) in deriving the frequency response of an electronic circuit from its behavior to an electrical impulse.

2.6 Problems

2.1 From equation (2.7a) (see Note 1), show that the correction to the rms noise $\langle N^2\rangle^{1/2}$ is less than 10% if $(5\varepsilon\tau\eta kT/h\nu)<1$. Consider a blackbody source at $T=1000$ K viewed by a detector system with optical efficiency 50% and quantum efficiency 50%. Calculate the wavelength beyond which the correction to the noise would exceed 10%. Compare this wavelength with that at the peak of the source output.

2.2 A photographic plate is exposed to a 100 W light bulb with a filament temperature of 5000 K and at a distance of 10 m. The output of the light bulb is filtered to the band of 0.5–0.6 μm, and the filter and bulb envelope together have 70% transmittance. An exposure of 1/50 s is taken. Calculate the number of silver atoms formed in a silver grain that is 0.1 μm in diameter and assuming that every photon that enters the grain creates a silver atom. If the threshold for development is seven silver atoms, what percentage of such grains will develop?

2.3 Compute the Fourier transform of $f(x)=H(x)+\text{sech}(10x)$, where $H(x)=0$ for $x<0$ and $=1$ for $x\geqslant 0$.

2.4 Consider the image formed by a series of optical elements, each of which forms an image with a Gaussian profile. Show that the final image is Gaussian with width equal to the quadratic combination of the individual widths. Do you expect this result to hold generally for different image profiles?

2.5 A photographic plate has grains that become developable upon exposure to four or more photons. The plate is illuminated at a level that corresponds to an average of one photon per grain. After development, the signal is read out with a light system/detector that illuminates a plate area containing 100 grains on average. The grains are distributed nonuniformly such that the number in the area being read out has a 1-standard-deviation variation of 25 (that is, 25% of its average value). Compute the DQE of the plate in this application.

2.6 Assume the characteristic curve of a plate is given by

$$(D+1.95)^2=4+2(\log H-2)^2,$$

where D is the density and H is the exposure calibrated in photons per area of the scanning aperture. For $H<100$, $D=0.05$. Let the detective quantum efficiency be

$$\text{DQE}=\frac{0.01}{1+H/10^4}$$

for $H\geqslant 1000$. Complete the second, third and fourth columns of the following table, using equation (2.16) for σ_D:

H	D	γ	σ_D	σ_D (ideal)
10				
10^2				
10^3				
10^4				
10^5				
10^6				

Use equation (2.11) to compute the fifth column for an ideal detector (but with the DQE given above).

2.7 Using the data in Table 2.2, compare the detective quantum efficiencies of IIa-O and IIIa-J plates with that of 103a-O ones (at a density of 0.6). Suppose you wish to detect a signal of 100 photons $\mathrm{sec}^{-1}\ \mu\mathrm{m}^{-2}$ against a uniform background of 50 photons $\mathrm{s}^{-1}\ \mu\mathrm{m}^{-2}$. Which emulsion would be the quickest? Suppose you wish to detect a signal of 10 photons $\mathrm{s}^{-1}\ \mu\mathrm{m}^{-2}$ against a background of 200 photons $\mathrm{s}^{-1}\ \mu\mathrm{m}^{-2}$, with a density $\geqslant 0.6$ to place the measurement on the linear portion of the H & D curve. Which emulsion should you select?

Note

1 A more rigorous description of photon noise takes account of the Bose–Einstein nature of photons, which causes the arrival times of individual particles to be correlated. Equation (2.7) is derived using the assumption that the particles arrive completely independently; the bunching of Bose–Einstein particles increases the noise above this estimate. The full description of photon noise shows it to be

$$\langle N^2 \rangle = n\left[1 + \frac{\varepsilon\tau\eta}{e^{h\nu/kT} - 1}\right], \tag{2.7a}$$

where $\langle N^2 \rangle$ is the mean square noise, n is the average number of photons detected, h is Planck's constant, ε is the emissivity of the source of photons, τ is the transmittance of the system optics, η is the detector quantum efficiency, ν is the photon frequency, k is Boltzmann's constant, and T is the absolute temperature of the photon source (see van Vliet, 1967). Comparing with equation (2.7), it can be seen that the term in square brackets in equation (2.7a) is a correction factor for the increase in noise from the Bose–Einstein behavior. It becomes important only at frequencies much lower than that of the peak emission of the blackbody, and then only for highly efficient detector systems. In most cases of interest, particularly with realistic instrument efficiencies, this correction factor is sufficiently close to unity that it can be ignored.

3

Intrinsic photoconductors

Intrinsic photoconductors are the most basic kind of electronic detector. They function by absorbing a photon whose energy is greater than that of the bandgap energy of the semiconductor material. The energy of the photon breaks a bond and lifts an electron into the conduction band, creating an electron/hole pair that can migrate through the material and conduct a measurable electric current. Detectors operating on this principle can be made in large arrays, and they have good uniformity and quantum efficiency. They are the basic component of CCDs (charge coupled devices), which are the most widely used two-dimensional, low light level, electronic detectors in the visible and very near infrared, and in some cases they are also used to detect X-ray and ultraviolet photons. Photoconductors made of semiconductor compounds with small bandgaps are also useful as high speed detectors over the 1–25 μm range.

These devices also illustrate in a general way the operation of all electronic photodetectors. All such detectors have a region with few free charge carriers and hence high resistance; an electric field is maintained across this region. Photons are absorbed in semiconductor material and produce free charge carriers, which are driven across the high resistance region by the field. Detection occurs by sensing the resulting current. The high resistance is essential to allow usefully large signals while minimizing various sources of noise. For the photoconductors discussed in this chapter, the photon absorption occurs within the high resistance region; in following chapters, we will encounter a variety of detectors where the absorption may occur in a separate, relatively low resistivity layer of semiconductor.

3.1 Basic operation

Assume that we arrange a sample of semiconductor as shown in Figure 3.1 or 3.2. In Figure 3.1, the device has transverse contacts on opposite faces;

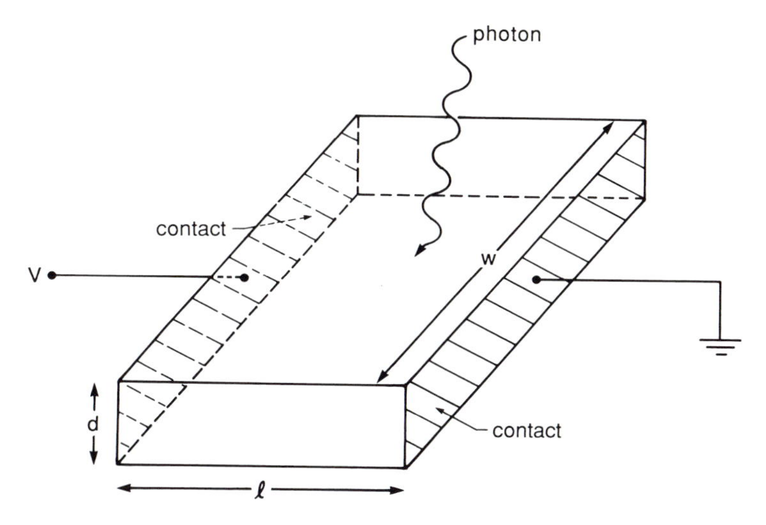

Figure 3.1. Photoconductor with transverse contacts.

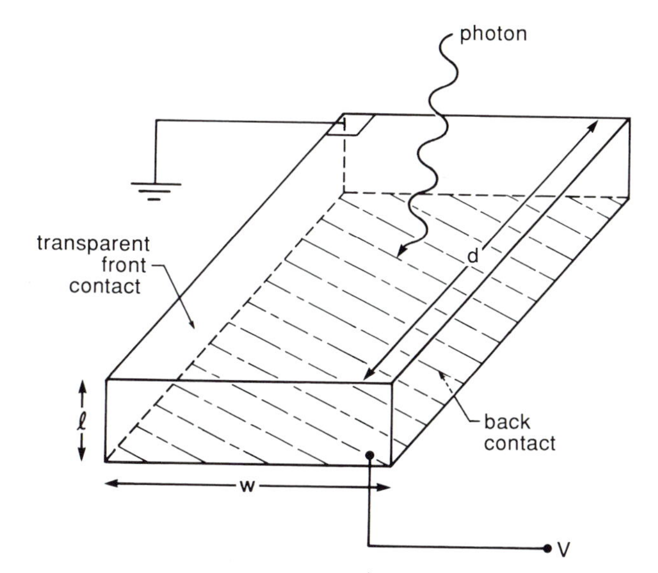

Figure 3.2. Photoconductor with transparent front contact.

these contacts are usually made of evaporated metal films. The detector is illuminated from a direction parallel to the contacts. In Figure 3.2, the device is built with a transparent contact on its front face, and the photons are received through this contact. The back contact can be a metal film as before.

Now let us apply an electric field to the detector, and at the same time illuminate it with photons having energy greater than its bandgap energy. If we know the bias voltage, V_b, across the detector and measure the current through it, we can calculate its resistance from Ohm's law, $R_d = V_b/I_d$. By definition, the conductivity, σ (conventionally given in units of (ohm centimeters)$^{-1}$; σ (ohm meters)$^{-1} = 100\ \sigma$ (ohm centimeters)$^{-1}$), is related to the resistance as

$$R_d = \frac{\ell}{\sigma w d}, \tag{3.1}$$

where w, ℓ, and d are the width, length, and depth of the detector, respectively (see Figures 3.1 and 3.2, and note that these quantities are defined with respect to the electrical contacts and not to the incoming photon direction). Charge carriers are generated both by thermal excitation and by any absorbed photons that have $E \geqslant E_g$; as a result, the conductivity comprises two components:

$$\sigma = \sigma_{th} + \sigma_{ph}. \tag{3.2}$$

The thermally induced component is an example of dark current, that is, signal produced by the detector in the absence of photon illumination. For now, we will assume that the detector is operated at sufficiently low temperature that we can ignore thermally induced conductivity. To understand the detector response, we must derive the dependence of the photoconductivity, σ_{ph}, on the arrival rate of photons at the face of the detector.

Let the coordinate running between the electrodes be x. When an electric field $\mathscr{E}_x = V_b/\ell$ is applied, the resulting current density along the x axis is the amount of charge passing through a unit area per unit time; it is given by

$$J_x = q_c n_0 \langle v_x \rangle, \tag{3.3}$$

where q_c is the electrical charge of the charge carriers, n_0 is their density, and $\langle v_x \rangle$ is their effective drift velocity. From Ohm's law and the definition of conductivity in equation (3.1), we can derive an alternative expression for J_x:

$$J_x = \frac{I_d}{wd} = \frac{V_b}{R_d w d} = \sigma \mathscr{E}_x. \tag{3.4}$$

Rearranging equations (3.3) and (3.4), the electron conductivity is

$$\sigma_n = \frac{-q n_0 \langle v_x \rangle}{\mathscr{E}_x} = q n_0 \mu_n, \tag{3.5}$$

where the electronic charge is $q_c = -q$, and

$$\mu_n = -\frac{\langle v_x \rangle}{\mathscr{E}_x} \tag{3.6}$$

is the electron mobility, which has units of centimeters squared per volt second. Note that μ is positive, which means that the electron velocity is taken to be negative. Similar expressions can be derived for the holes. The total conductivity is then

$$\sigma = \sigma_n + \sigma_p. \tag{3.7}$$

The electron component of the conductivity is usually much larger than the hole component, particularly for n-type and intrinsic material, because the electron mobility is higher than that for the holes.

From equation (3.5), it is apparent that the mobility governs the conductivity. Without an applied field, the charge carriers are in rapid motion due to their thermal energy, but this motion is thoroughly randomized in both direction and amount by frequent collisions with crystal atoms or other charge carriers. Application of an electric field adds a slight bias to these motions that results in a current, but the energy transferred to the crystal in collisions imposes a limit on this extra velocity component; $\langle v_x \rangle$ is this terminal velocity. The mobility can then be interpreted as a measurement of the viscosity of the crystal against the motion of the charge carriers.

In general, the mobility is proportional to the mean time between collisions. Impurities in the crystal scatter the conduction electrons and reduce the time between collisions, thus reducing μ_n. As the temperature, T, is decreased, the carriers move more slowly, and the effect of impurity scattering becomes larger, reducing the mobility. If, for example, the impurities are ionized, μ goes as $T^{3/2}$. If the impurities are neutral or frozen out, a much shallower temperature dependence is observed, $\mu \sim$ constant. At higher temperatures, the dominant scattering centers are distortions in the crystal lattice caused by thermal motions of its constituent atoms. The mobility resulting from lattice scattering goes as $T^{-3/2}$. For detector material operating in an optimal temperature regime, the mobility is dominated by neutral impurity or lattice scattering. The mobility of silicon as a function of impurity concentration and temperature is illustrated in Figure 3.3.

Typical values for semiconductor material parameters are given in Table 3.1. The mobilities are given in conventional units; conversion to MKS ($m^2\,V^{-1}\,s^{-1}$) requires that the values be multiplied by 10^{-4}. The values in the table apply at 300 K for high purities (impurity levels at about $10^{12}\,cm^{-3}$ for silicon and germanium). Under these conditions, the mobilities are due to lattice scattering.

We have almost lost sight of our goal to derive the dependence of detector conductivity on photon arrival rate! From equations (3.5) and (3.7),

$$\sigma_{ph} = q(\mu_n n + \mu_p p), \tag{3.8}$$

Table 3.1. *Properties of some semiconductor materials*

Material	Dielectric constant[a], κ_0	Recombination time[b], τ (s)	Electron mobility[a], μ_n (cm^2 V^{-1} s^{-1})	Hole mobility[a], μ_p (cm^2 V^{-1} s^{-1})	E_g (eV)
Si	11.8	1×10^{-4}	1.35×10^3	480	1.11
Ge	16	1×10^{-2}	3.9×10^3	1900	0.67
PbS	161	2×10^{-5}	5.75×10^2	200	0.37
InSb	17.7	1×10^{-7}	1.0×10^5	1700	0.18
GaAs	13.2	$\geqslant1\times10^{-6}$	8.5×10^3	400	1.43

[a] Values taken from Streetman (1990), Appendix III. Note that the conventional units of mobility are not MKS; multiply the entries by 10^{-4} (m^2 cm^{-2}) to convert to MKS.
[b] Values taken from the *American Institute of Physics Handbook*, 3rd edn. These values are representative; the recombination time depends on impurity concentrations as well as temperature.

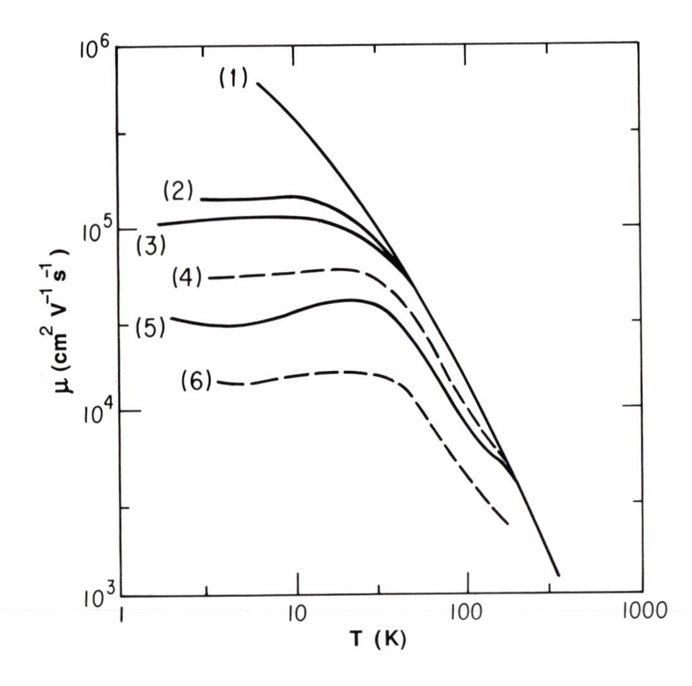

Figure 3.3. Variation of electron mobility in silicon with temperature and impurity concentration, after Norton, Braggins, and Levinstein (1973) and Canali *et al.* (1975). The solid lines are for phosphorus-doped material at concentrations of $\leqslant 1\times10^{12}$, 4×10^{13}, 2.5×10^{14}, and 1×10^{16} cm^{-3}, respectively, for curves (1)–(3) and (5). The dashed lines are for arsenic-doped material at concentrations of 8×10^{15} and 8×10^{16} cm^{-3}, respectively, for curves (4) and (6).

where n and p are the photoconductive charge carrier concentrations. For intrinsic photoconductivity, we can assume $n=p$ for the following discussion. Let φ photons s^{-1} fall on the detector. Some fraction η will create charge carriers; η is the quantum efficiency. The equilibrium number of charge carriers is then $\varphi\eta\tau$, where τ is the mean lifetime of a charge carrier before recombination. Sample values are listed in Table 3.1. The number of charge carriers per unit volume is then

$$n=p=\frac{\varphi\eta\tau}{wd\ell}. \tag{3.9}$$

Particularly for indirect semiconductors, τ is strongly regulated by the recombination centers and hence by the level of impurities; it goes roughly as N_I^{-1}, where N_I is the impurity concentration. The recombination time decreases with decreasing temperature. Theoretically, it is expected to go as $T^{1/2}$ at very low temperatures, with a $T^{5/2}$ dependence at higher temperatures (see, for example, Sclar, 1984). At least for some materials, a flatter temperature dependence is observed experimentally (see, for example, Darken, Sangsingkeow, and Jellison, 1990).

Using equations (3.8) and (3.9), equation (3.1) becomes

$$R_{ph}=\frac{\ell^2}{q\varphi\eta\tau(\mu_n+\mu_p)}. \tag{3.10}$$

We have used the subscript ph on R to emphasize that we are interested only in the photo-induced effects on the detector resistance. Thus, the detector resistance is equal to the inverse of the quantum efficiency times the photon flux combined with other properties of the detector material and its geometry. With a constant voltage across the detector, the current is proportional to the photon flux, so the detector is linear if its output is measured by sensing the current.

The output of a detector is usually quoted in terms of its responsivity, S, which is defined to be the electrical output signal divided by the input photon power. Units of S can be amperes per watt, or volts per watt depending upon the units in which the output signal is measured. Although the responsivity of a detector is sometimes given only at the wavelength at which it reaches a maximum, it is often convenient to specify the spectral response of the detector in terms of the wavelength dependence of S.

From equation (1.1), the power falling on the detector is

$$P_{ph}=\varphi h\nu=\frac{\varphi hc}{\lambda}. \tag{3.11}$$

If we define a new parameter, the photoconductive gain, as

$$G = \frac{\tau\mu\mathscr{E}_x}{\ell}, \tag{3.12}$$

we can then write (using equations (3.10), (3.11), and (3.12))

$$S = \frac{I_{ph}}{P_{ph}} = \frac{V_b}{R_{ph}P_{ph}} = \frac{\eta\lambda q G}{hc}. \tag{3.13}$$

We have also used the definition of the electric field strength in the detector.

The photoconductive gain has an interesting physical significance. Let τ_t be the time for a charge carrier to drift from one detector electrode to the other, where

$$\tau_t = -\frac{\ell}{\langle v_x \rangle}. \tag{3.14}$$

(Recall that $\langle v_x \rangle$ is less than zero.) Combining this expression with equation (3.6), we have $\tau_t = \ell/\mu\mathscr{E}_x$. or, substituting into equation (3.12),

$$G = \frac{\tau}{\tau_t}. \tag{3.15}$$

G is thus the ratio of carrier lifetime to carrier transit time, and ηG is the probability that a photon incident on the detector will produce a charge carrier that will penetrate to an electrode. The photocurrent generated by the detector is (see equation (3.13))

$$I_{ph} = \varphi q \eta G. \tag{3.16}$$

A detector can have $G > 1$ if some form of controlled charge multiplication occurs in the detector material. $G > 1$ also can result because, on average, absorption of a charge carrier at an electrode triggers the release of a carrier either of opposite type from that electrode or of the same type from the opposite electrode to maintain the overall electrical neutrality of the detector. It is a useful approximation in the theory to consider that an absorbed charge carrier reappears immediately from the opposite electrode and continues its migration through the detector. Photoconductive gains can be much larger than unity but often in an operating mode unsuited for very low light level detection – for example, where the detector is operated warm enough to excite significant thermal currents. An example of $G > 1$ in a detector well suited to very low light levels will be provided in Chapter 4.

In an ideal photoconductor, the quantum efficiency is independent of wavelength up to λ_c, the wavelength corresponding to the bandgap energy, which is given by

$$\lambda_c = \frac{hc}{E_g} = \frac{1.24\ \mu\text{m}}{E_g(\text{eV})}; \tag{3.17}$$

beyond λ_c, $\eta = 0$. Because each detected photon produces the same effect in the semiconductor (a single electron/hole pair), and because the photon energy varies inversely with its wavelength, S is proportional to λ up to λ_c (see equation (3.13)). Figure 3.4 illustrates the behavior of the responsivity for an ideal photoconductor. Near λ_c, thermal excitation can affect the production of a conduction electron when a photon is absorbed, so there is a chance that a photon with energy just above the bandgap energy will not produce an electron/hole pair. This effect and others tend to round the quantum efficiency and responsivity curves near λ_c, as shown by the dashed lines. It is also typical for the quantum efficiency to drop at wavelengths well short of λ_c.

The quantum efficiency of a detector can be measured directly in a series of laboratory experiments, as described in Chapter 6. It can also be calculated from the absorption coefficient of the material as in equation (2.3) or (2.5). Figure 3.5 shows absorption coefficients for a few materials of interest; contrast the behavior of the materials that allow direct transitions at the bandgap (InSb and GaAs) with the indirect transition semiconductors (Si, Ge, and GaP). If reflection from the detector faces is ignored and the photons traverse a distance z (centimeters) in the material,

$$\eta = 1 - e^{-az}. \tag{3.18}$$

In our transverse contact detector (Figure 3.1), $z = d$; if, however, the back

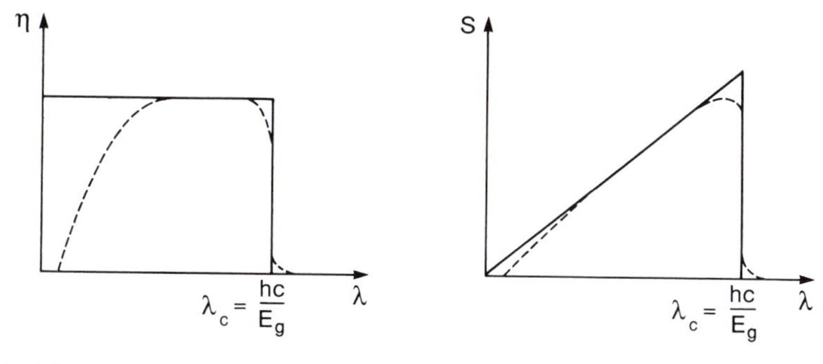

Figure 3.4. Photoconductor response. The solid lines represent the response in the ideal case; the dashed lines show typical departures from the ideal curve.

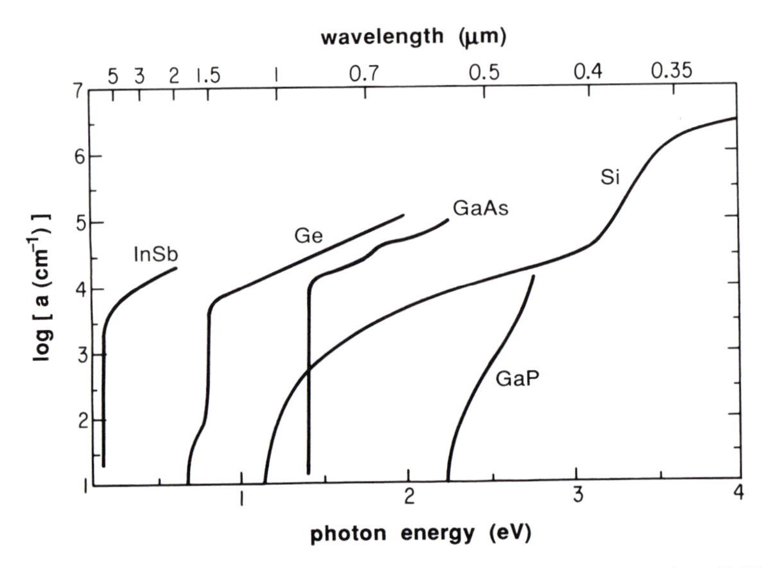

Figure 3.5. Absorption coefficients for various semiconductors, after Stillman and Wolfe (1977).

face could be made reflective, we would have $z = 2d$. Unfortunately, placing a reflective coating on the back face of a detector with transverse contacts tends to short out the detector. In addition, a detector with transverse contacts can have large nonuniformities in quantum efficiency over its face due to field nonuniformities near the contacts. For these reasons, photoconductive detectors are frequently constructed with transparent electrodes as in Figure 3.2 and with a metallized back electrode to reflect photons that are not absorbed in the first pass through the material.

3.2 Limitations and optimization

The simple form of photoconductor described above is applicable both to intrinsic materials and to extrinsic ones to be described in the next chapter. We therefore postpone discussion of the construction of these detectors and measurements of their performance to following chapters on extrinsic photoconductors and detector/amplifier systems. We will devote the remainder of this chapter to a discussion of fundamental performance parameters. More detailed discussions of many aspects of photoconductivity can be found in Ashcroft and Mermin (1976), Bratt (1977), Sze (1985), Kittel (1986) or Streetman (1990). Further information on intrinsic photoconductors can be found in Rogalski and Piotrowski (1988).

3.2.1 Spectral response

The possible spectral responses for intrinsic photoconductors are as described by Figure 3.4 and equations (3.13) and (3.17). Typical detector materials and their cutoff wavelengths at room temperature are Si ($\lambda_c = 1.1\ \mu$m), Ge ($\lambda_c = 1.8\ \mu$m), PbS ($\lambda_c = 3.3\ \mu$m), and HgCdTe. For this last material, it is possible to vary the bandgap by altering the proportions of the components. CdTe has $\lambda_c = 0.8\ \mu$m (corresponding to $E_g = 1.55$ eV), and HgTe is metallic, so it has $E_g \sim 0$. The bandgap of $Hg_{1-x}Cd_xTe$ varies monotonically with x between these values. Intrinsic photoconductors can be made with response to 25 μm with HgCdTe, although the high conductivity of the small bandgap material results in poor performance at low signal levels.

The bandgaps and hence cutoff wavelengths of these materials change slightly with operating temperature; for example, at 77 K, the cutoff wavelengths of Si and Ge are 5–10% shorter than at room temperature, whereas the cutoff wavelength for PbS becomes significantly longer with reduced temperature and the response extends beyond 4 μm at 77 K. HgCdTe has a relatively complex dependence of bandgap on temperature (Schmit and Selzer, 1969). For example, material with $\lambda_c > 2.25\ \mu$m has a bandgap that decreases with decreasing temperature, resulting in an increase in the cutoff wavelength, whereas material with a $\lambda_c < 2.25\ \mu$m shows a decrease in cutoff wavelength with decreasing temperature.

3.2.2 *Number of pixels*

Our discussion has dealt only with single detector elements which, of course, can be combined into multiple pixel detector arrays with suitable readouts (see Chapter 7). In certain applications at high illumination levels, a SPRITE device (SPRITE is an acronym for 'signal processing in the element') can provide a limited number of pixels – typically ten – with a single detector element (Elliott, Day, and Wilson, 1982).

SPRITE devices are made of n-type HgCdTe in strips that are typically $\sim 700\ \mu$m long by $\sim 50\ \mu$m wide by $\sim 7\ \mu$m thick. A constant current bias is placed across the long dimension of the element. When the detector is illuminated at the end away from the negative electrode (the cathode), holes are generated that drift toward this electrode at a rate determined by the electric field and the hole mobility. Typical values for these parameters are $\mu_p = 390\ \text{cm}^2\ \text{V}^{-1}\ \text{s}^{-1}$ and $\mathscr{E}_x = 30\ \text{V cm}^{-1}$, yielding $\langle v_x \rangle = 1.17 \times 10^8\ \mu\text{m s}^{-1}$ according to equation (3.6). Therefore, the holes require $\sim 6\ \mu$s to reach the cathode. If the detector is used with an optical system that can sweep a scene along the detector at a rate that matches the hole drift velocity, a given feature of the scene will track the holes it generates. Given a recombination time for the holes which is of the same order or longer than their transit time along the detector, the signal from the feature will grow as it approaches the cathode. This mode of operation is termed time-delay integration (TDI). Recombination times of $\tau \sim 2\,\mu s$ are typical of SPRITEs operating at $10\,\mu m$, whereas those operating at $\sim 4\,\mu m$ may have $\tau \sim 15\,\mu s$; hence, some signal integration occurs at the former wavelength and nearly complete integration occurs at the latter one. As the signals approach the cathode, a third electrode senses the local change in conductivity they produce, thus reading out the signal.

3.2.3 Responsivity

The usual first step in optimizing a detector is to obtain a large signal, that is, to maximize the responsivity. It is apparent from equation (3.13) that this goal can be achieved by increasing η and by maximizing G. From equation (3.12), we see that the photoconductive gain can be increased by reducing the distance between electrical contacts, ℓ. For transverse contacts, however, a very narrow sensitive area results if ℓ is made too small (see Figure 3.1). For a detector illuminated through a transparent contact (Figure 3.2), reducing ℓ does not affect the sensitive area but it does reduce the absorbing thickness and hence the quantum efficiency. Another possible way to maximize G is to increase the electric field in the detector by increasing the voltage. This technique works up to the point where charge carriers are accelerated sufficiently for them to excite additional carriers, leading to an avalanche that produces large fluctuations in the conductivity. This condition is called breakdown, and it seriously impairs the ability of the detector to achieve a high ratio of signal to noise.

The photoconductive gain can also be increased by adjusting the parameters of the semiconductor material itself. The carrier lifetime, τ, should be maximized; for indirect semiconductors like Si and Ge, τ is very sensitive to the presence of recombination centers, and can be increased by eliminating impurities from the material and by using processing that reduces crystal defects. In addition, G is proportional to the mobility, which can be increased by removing impurities that cause scattering. For material useful for detectors, the mobility also increases with decreasing temperature ($\propto T^{-3/2}$, that is, as for lattice scattering) until neutral impurity scattering sets in, after which μ becomes independent of temperature. The small value of the mobility at higher temperatures is an additional reason to operate the detector at a low temperature.

3.2.4 Frequency response

The response speed of a detector can be described very generally by specifying the dependence of the responsivity on the frequency of an imaginary photon signal that varies sinusoidally in time. This concept is analogous to the modulation transfer function described in Chapter 2; in this case it is called the electrical frequency response of the detector.

A variety of factors limit the frequency response of photoconductors. We begin with the RC time constant of the circuit (including the detector). The detector shown in Figure 3.1 (or Figure 3.2) is a simple parallel plate capacitor filled with a dielectric, so its capacitance is

$$C_{\mathrm{d}} = \frac{w d \kappa_0 \varepsilon_0}{\ell}, \tag{3.19}$$

where κ_0 is the dielectric constant (sample values are included in Table 3.1), and $\varepsilon_0 = 8.854 \times 10^{-12}\,\mathrm{F\,m^{-1}}$ is the permittivity of free space. This capacitance is in parallel with the detector resistance. Charge deposited on the detector capacitance bleeds off through the resistance with an exponential time constant

$$\tau_{RC} = RC, \tag{3.20}$$

where R and C are the equivalent values for the total circuit, including the detector and any other components.

Let a voltage impulse be deposited on the capacitor,

$$v_{\mathrm{in}}(t) = v_0 \delta(t), \tag{3.21}$$

where v_0 is a constant and $\delta(t)$ is the delta function (see Chapter 2, Table 2.3). We can observe this event in two ways. First, we might observe the voltage across the resistance and capacitance directly, for example with an oscilloscope. It will have the form

$$v_{\mathrm{out}}(t) = \begin{cases} 0, & t < 0 \\ \dfrac{v_0}{\tau_{RC}} e^{-t/\tau_{RC}}, & t \geqslant 0. \end{cases} \tag{3.22}$$

The same event can be analyzed in terms of the effect of the circuit on the input frequencies rather than on the time dependence of the voltage. To do so, we convert the input and output voltages to frequency spectra by taking their Fourier transforms. The delta function contains all frequencies at equal strength, that is, from Table 2.3,

$$V_{\mathrm{in}}(f) = v_0 \int_{-\infty}^{\infty} \delta(t)\, e^{-\mathrm{j}2\pi f t}\, \mathrm{d}t = v_0. \tag{3.23}$$

Since the frequency spectrum of the input is flat ($V_{in}(f)$=constant), any deviations from a flat spectrum in the output must arise from the action of the circuit. That is, the output spectrum gives the frequency response of the circuit directly. Again from Table 2.3, it is

$$V_{out}(f)=\int_{-\infty}^{\infty} v_{out}(t)\,e^{-j2\pi ft}\,dt=v_0\left[\frac{1-j2\pi f\tau_{RC}}{1+(2\pi f\tau_{RC})^2}\right]. \tag{3.24}$$

The imaginary part of the frequency spectrum represents phase shifts that can occur in the circuit. For a simple discussion, we can ignore the phase and describe the strength of the signal only in terms of the frequency dependence of its amplitude. The amplitude can be determined by taking the absolute value of $V_{out}(f)$:

$$|V_{out}(f)|=(V_{out}V_{out}^*)^{1/2}=\frac{v_0}{[1+(2\pi f\tau_{RC})^2]^{1/2}}, \tag{3.25}$$

where V_{out}^* is the complex conjugate of V_{out}. This function is plotted in Figure 3.6. It is mathematically closely analogous to the optical modulation transfer function discussed in Chapter 2. As might be anticipated from the discussion there, the effects of different circuit elements on the overall frequency response can be determined by multiplying their individual response functions together. The frequency response is often characterized by a cutoff frequency

$$f_c=\frac{1}{2\pi\tau_{RC}}, \tag{3.26}$$

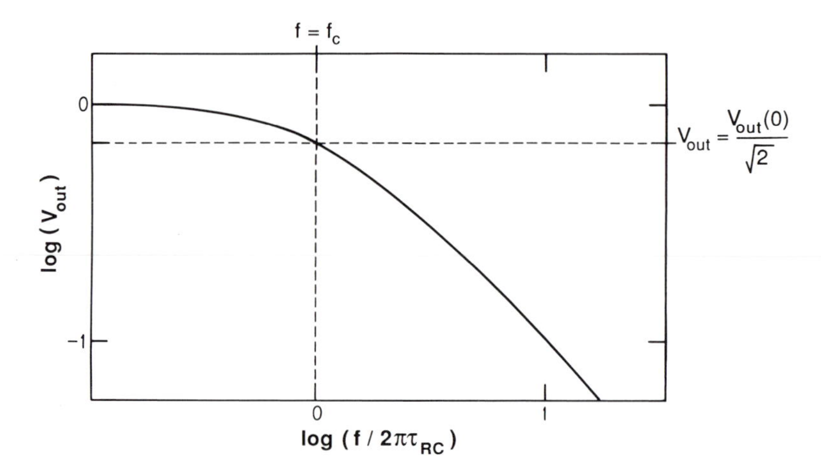

Figure 3.6. Frequency response of an RC circuit. The cutoff frequency is illustrated.

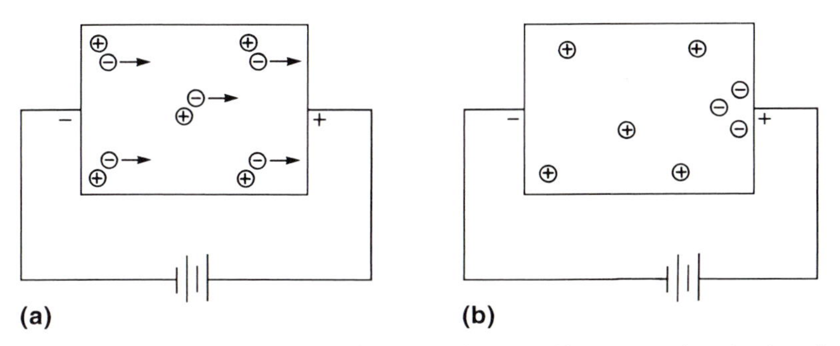

Figure 3.7. Dielectric relaxation in a photoconductor. Charge carriers begin migration from their creation sites in (a); the charge separation that results in (b) reduces the effective field in the detector and decreases its photoconductive gain.

at which the amplitude drops to $1/\sqrt{2}$ of its value at $f=0$, or

$$|V_{\text{out}}(f_c)| = \frac{1}{\sqrt{2}} |V_{\text{out}}(0)|. \tag{3.27}$$

Equations (3.25)–(3.27) define the frequency behavior for any process with an exponential response function, so long as τ_{RC} is replaced with the appropriate time constant. We will encounter such behavior frequently throughout this book.

Another factor influencing the speed of response is dielectric relaxation, which arises from space charge effects within the detector. Consider the detector in Figure 3.7(a); assume it is operating at low temperature with $\mu_n \gg \mu_p$ and high photoconductive gain ($G \approx 1$). Let the detector start in the dark at equilibrium and be illuminated by a step function. Under these assumptions, the high electron mobility results in the negative charge carriers being swept out of the detector, leaving a net positive excess charge behind. Such a distribution of excess charge is termed a space charge. The resulting charge separation creates a field that, until neutralized, opposes the field set up by the detector electrodes (see Figure 3.7(b)) and reduces the photoconductive gain.

The timescale for neutralization of the field is computed as follows. We start with Poisson's equation:

$$\frac{d^2V}{dx^2} = -\frac{d\mathscr{E}_x}{dx} = -\frac{\rho}{\varepsilon}, \tag{3.28}$$

where ρ is the charge density and $\varepsilon = \kappa_0 \varepsilon_0$ is the dielectric permittivity of the material. In the present situation, we have

$$\frac{d\mathscr{E}_x}{dx} = \frac{qN_p}{\varepsilon}, \tag{3.29}$$

where $d\mathscr{E}_x$ is the change in the field due to charge carriers being swept out, and N_p is the density of holes left in the detector. The current that neutralizes the charge has a density J_x. Consider a volume with thickness dx and unit area perpendicular to x. The rate of change in the density of holes in this volume is $-(1/q)(J_{x_2}-J_{x_1})/dx$, where $dx=x_2-x_1$, so by continuity

$$\frac{dN_p}{dt}=-\frac{1}{q}\frac{dJ_x}{dx}. \tag{3.30}$$

From equations (3.4) and (3.5), $J_x=qn_0\mu_n\mathscr{E}_x$, so, using equation (3.29), we obtain

$$\frac{dN_p}{dt}=-\mu_n n_0\frac{d\mathscr{E}_x}{dx}=-\frac{\mu_n n_0 q N_p}{\varepsilon}. \tag{3.31}$$

The solution to equation (3.31) is

$$N_p(t)=N_p(0)\,e^{-t/\tau_d}, \tag{3.32}$$

where the dielectric time constant is

$$\tau_d=\frac{\varepsilon}{\mu_n n_0 q}=\frac{\varepsilon}{\sigma}\propto\frac{1}{\varphi}; \tag{3.33}$$

the final proportionality arises from the relationship between σ and the photon flux (see equations (3.8) and (3.9)). Thus, the reduction in frequency response due to dielectric relaxation is inversely proportional to the incident photon rate, and the effect becomes increasingly important at low light levels. Again, the time response due to dielectric relaxation is exponential and the frequency response has the form of equation (3.25) and Figure 3.6.

Note that the dielectric time constant and the *RC* time constant of the detector alone are identical; equation (3.33) can be obtained by substituting equations (3.1), (3.5), and (3.19) into (3.20). The *RC* time constant, however, refers to the detector as a circuit element, and external components can increase or decrease the time response (an example is the increase in the frequency response provided by the TIA circuit discussed in Chapter 6). On the other hand, the dielectric response time is determined by the detector parameters alone.

The charge carrier lifetime sets another limit on the speed of the detector, since the conductivity can change only over the timescale for recombination. The corresponding cutoff frequency is

$$f=\frac{1}{2\pi\tau}, \tag{3.34}$$

and the behavior can be taken to be exponential. The frequency response is then similar in form to equation (3.25) and Figure 3.6.

It can be shown (Blouke *et al.*, 1972) that the photoconductive gain above the dielectric frequency cutoff ($f_d = 1/2\pi\tau_d$) is limited to

$$G_{pc} = G[1 - G(1 - e^{-1/G})], \quad (3.35)$$

where G is the low-frequency value defined in equation (3.12). For all values of G, $G_{pc} \leqslant 0.5$, and, for $G \ll 1$, this expression reduces to $G_{pc} \sim G$. Thus, a detector with a large photoconductive gain that receives a low signal can have a substantial drop in response above f_d.

3.2.5 Noise

Along with the generation of an electrical signal, detectors produce various forms of electronic noise which can hide the signal. Thus, in evaluating the use of a detector, we must understand and estimate the various noise sources. As with the photon noise discussed in the previous chapter, we will assume that these noise components are Poisson distributed (equation (2.6)) and independent, allowing them to be combined quadratically.

The fundamental noise limitation for any detector is the noise that arises because of the Poisson statistics of the incoming photon stream (see equation (2.9)). For a photoconductor, the photons are transformed into free electrons and holes, so this noise appears as a variation in the number of charge carriers. The statistical fluctuations in the number of charge carriers result in generation–recombination, or G–R, noise.

Suppose the detector absorbs N photons in a time t, $N = \eta\varphi t$. For an ideal photoconductor, these photons create N conduction electrons and N holes; we assume $\mu_n \gg \mu_p$ so we need consider the electrons only. Each conduction electron contributes an increment of conductivity until it recombines. If the detector receives a stream of photons, we can imagine that its net conductivity arises from the superposition of many randomly generated and randomly terminated contributions, one pair for each photoelectron. Since there are then two random events associated with the photoconductivity from each absorbed photon, the root-mean-square (rms) noise associated with them is, in number, $(2N)^{1/2}$. The noise current due to this sequence of generations and recombinations of photoelectrons is then this rms noise in number times the charge of an electron divided by the time interval, and multiplied by the detector gain, or $\langle I_{G-R}^2 \rangle^{1/2} = q(2N)^{1/2}G/t$. Taking the square and using equation (3.16),

$$\langle I_{G-R}^2 \rangle = \left(\frac{2q}{t}\right)\left(\frac{qNG}{t}\right)G = \left(\frac{2q}{t}\right)\langle I_{ph} \rangle G, \quad (3.36)$$

where $\langle I_{ph} \rangle$ is the detector current averaged over the noise fluctuations. If the detector has a significant number of thermally generated charge carriers, they contribute their own G–R noise component in addition to the one in equation (3.36).

It is noteworthy that the intrinsic $\sqrt{N}$ noise of the photogenerated charge carriers has been degraded by $\sqrt{2}$ because the recombination of these charge carriers occurs in the high resistance detector volume. Alternate detector types discussed in the following chapters avoid this noise increase because for them recombination occurs in relatively high conductivity regions of the detector.

G–R noise is typically measured with an instrument that is sensitive to signals over some range of frequencies. It would be convenient if this response were sharply defined by lower and upper frequencies, f_1 and f_2, with constant response between f_1 and f_2 and no response otherwise. We could then define the frequency bandwidth as $\Delta f = f_2 - f_1$. Unfortunately, it is impossible to build circuitry with the properties just described, and, even if it were possible, the full realization of this concept would require measurements of infinite duration. It is therefore necessary to introduce an equivalent noise bandwidth, which is computed as the sharply defined hypothetical 'square' bandwidth $(f_2 - f_1)$ through which the noise power would be the same as the noise power through the frequency response of the actual system, where both have the same maximum gain. Since the power in the electrical signal is proportional to the square of its amplitude, the bandwidth is defined in terms of the square of the response function. Thus,

$$\Delta f = \int_0^\infty |G(\xi)|^2 \, \mathrm{d}\xi, \tag{3.37}$$

where $G(\xi)$ is the electrical response (for example, current or voltage) of the system as a function of frequency ξ, normalized to unity response at the frequency of maximum response.

For a system with exponential response, $G \sim e^{-t/\tau}$, it can be shown that

$$\Delta f = \mathrm{d}f = \frac{1}{4\tau}. \tag{3.38}$$

Similarly, if a measurement is made by integrating the electrical signal over a time interval T_{int},

$$\mathrm{d}f = \frac{1}{2T_{int}}. \tag{3.39}$$

Expressions such as equations (3.38) and (3.39) are often used to convert noise measurements taken under one set of conditions to equivalent measurements under different conditions.

The case represented by equation (3.39) is applicable to the assumptions we made in deriving G–R noise. Substituting equations (3.16) and (3.39) into the expression for $I_{\mathrm{G-R}}$ (equation (3.36)),

$$\langle I^2_{\mathrm{G-R}} \rangle = 4q^2 \varphi \eta G^2 \, df. \tag{3.40}$$

We continue to assume that the detector is cold enough that thermally generated conductivity is negligible. Otherwise, the thermally generated charge carriers contribute to $I_{\mathrm{G-R}}$ and increase the noise.[1]

There are also noise sources that originate in the detector itself in the absence of external signals. Johnson, or Nyquist, noise is a fundamental form of thermodynamic noise that arises because of thermal motions of charge carriers in any resistive circuit element, such as our photoconductor. Refer to Figure 3.8 for the following discussion. The circuit illustrated has one degree of freedom, V_{N}. From thermodynamics, if the system is in equilibrium, it has an average energy of $kT/2$ associated with each degree of freedom, or in our case with V_{N}. Since the energy on a capacitor is $CV^2/2$, we obtain

$$\frac{1}{2} C \langle V^2_{\mathrm{N}} \rangle = \frac{1}{2} kT. \tag{3.41}$$

The capacitor is a storage device, so the randomly fluctuating potential energy on it must have a corresponding random kinetic energy component, which is the Johnson noise current I_{J}. Again, from thermodynamic considerations we expect this energy to be of the form $\langle P \rangle t = kT/2$, where we have expressed the energy as the power, P, times the response time of the circuit, $t = RC$. From elementary circuit theory, the maximum power that can be delivered to a device connected across the terminals of the circuit in Figure 3.8 is

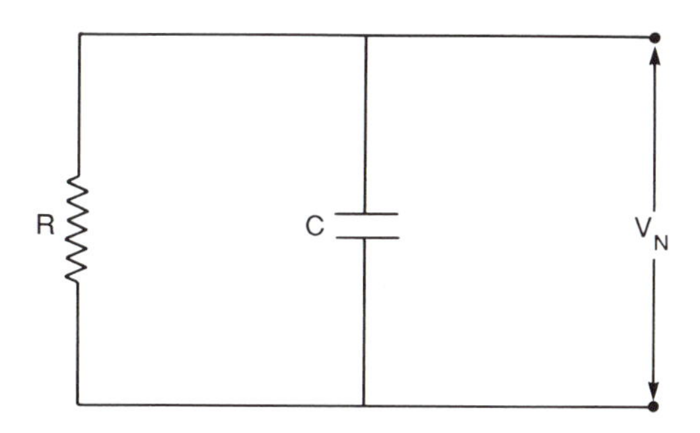

Figure 3.8. Circuit for calculation of Johnson and kTC noise.

$\langle P \rangle = \langle I^2 \rangle R/2$. For a circuit with exponential response (for example, Figure 3.8), the relationship between response time and electrical bandwidth is given by equation (3.38). We therefore obtain

$$\langle I_J^2 \rangle = \frac{4kT\,df}{R}. \tag{3.42}$$

While we are studying the circuit in Figure 3.8, it is convenient to derive the charge on the capacitor. This form of charge noise will be of interest when we discuss circuits that read out the signals from photoconductors and other detectors; it is called kTC or reset noise. From equation (3.41), it is easily shown that

$$\langle Q_N^2 \rangle = kTC, \tag{3.43}$$

where we have made use of the relationship between charge and voltage for a capacitor, $V = Q/C$.

It should be clear from the above discussion that Johnson noise and kTC noise are manifestations of the same thing. On a microscopic scale, they arise because of the random currents generated by the Brownian motion of the charge carriers in a resistor. In addition to these fundamental noise mechanisms, most electronic devices have increased noise at low frequencies. This trait makes AC amplifiers much easier to design than DC amplifiers. This kind of noise is sometimes blamed on the details of electrical conduction at the contacts to the semiconductor material or on surface trapping sites. It increases with increasing current through the device, and the noise power often goes inversely as the frequency; hence it is called 'one over f noise'. Thus, we have

$$\langle I_{1/f}^2 \rangle = \frac{KI^a\,df}{f^b}, \tag{3.44}$$

where K is a normalization constant and $a \approx 2$ and $b \approx 1$. At sufficiently high frequencies, the noise is dominated by mechanisms different from those that produce $1/f$ noise and the value of b tends to decrease toward zero. When b is zero, the resulting frequency independent noise is called white noise.

Equations (3.40), (3.42), and (3.44) illustrate that the rms noise currents, $\langle I_N^2 \rangle^{1/2}$, are proportional to the square root of the frequency bandwidth, df. The cause of this behavior is illustrated by equation (3.39). Imagine that the signal current is produced by a series of independent, Poisson-distributed events. The number of such events received will increase in proportion to the time of integration, so the accuracy of measurement will improve as the square root of the time, that is, the relative error of measurement goes as the inverse square root of integration time. From equation (3.39), it is equivalent to say the relative error of measurement goes as $(df)^{1/2}$. If the signal varies at a frequency that is within the range defined by df, it is roughly independent of df and the signal-to-noise ratio improves as $(df)^{1/2}$. Consequently, there is a strong incentive to use electronic filtering to reject frequencies that do not carry significant signal. Such electronic filtering can also be effective in rejecting spurious noise sources that are concentrated at certain frequencies, such as pickup from AC power lines or microphonic electrical signals from mechanical vibrations within the detector system.

The noise mechanisms described above are assumed to be independent, so they are combined quadratically to estimate the total noise of the system:

$$\langle I_{\mathrm{N}}^2\rangle = \langle I_{\mathrm{G-R}}^2\rangle + \langle I_{\mathrm{J}}^2\rangle + \langle I_{1/f}^2\rangle. \tag{3.45}$$

For these mechanisms, the noise excursions can be taken to obey Poisson statistics. Detectors may be subject to other noise mechanisms that are not distributed in this manner. Examples include current spikes that are produced spontaneously in the detector when it has a momentary breakdown or when it is struck by an energetic charged particle such as a cosmic ray. If at all possible, such events should be identified by their departure from Poisson statistics and removed from the detector output either electronically or in later reduction steps. If they remain, they can both dominate the noise of the system and make accurate noise estimation difficult because the normal statistical techniques assume Poissonian noise behavior.

Equation (3.45) has an alternative interpretation that emphasizes the difference between noise and signal. Because noise power goes as $\langle I_{\mathrm{N}}^2\rangle$, we see that noises add in terms of powers. This behavior holds very generally and occurs because noise currents vary randomly in phase. For example, when the *noise* is to be obtained over a range of frequencies, the *powers* from adjacent frequency intervals are combined, that is, currents are added quadratically. On the other hand, *signal* currents over a range of frequency are correlated in phase and therefore add *linearly*.

3.2.6 Thermal excitation

Throughout this discussion, we have insisted that the detector be operated at a sufficiently low temperature to suppress thermal excitation. Some of the advantages (besides mathematical simplicity) have been pointed out, and one would expect any low light level detector to be run in this mode. Liquid cryogen dewars that provide a convenient means to cool detectors are discussed briefly in Section 9.3.6. Nonetheless, it is necessary to consider thermal excitation briefly to determine the temperature at which it can be frozen out. The discussion will also demonstrate the steep dependence of thermally excited dark current on detector temperature. In cases where the dark current is appreciable and the detector temperature is not perfectly constant, this dependence can make calibration difficult.

Classically, the electrons in a semiconductor obey Maxwell–Boltzmann statistics, and the relative numbers in two energy levels are related as follows:

$$\frac{n_2}{n_1} = e^{-(E_2 - E_1)/kT} = e^{-E_g/kT}. \tag{3.46}$$

More properly, the electrons follow Fermi–Dirac statistics. In this case, the distribution of the electrons over the allowed energy states is governed by the Fermi–Dirac distribution function

$$f(E) = \frac{1}{1 + e^{(E - E_F)/kT}}, \tag{3.47}$$

where E_F is the energy of the Fermi level. This distribution gives, as a function of temperature, the probability that an available state at energy E will be occupied by an electron. Note that $f(E_F) = 0.5$. Moreover, when $T = 0$, $f(E < E_F) = 1$, and $f(E > E_F) = 0$. Thus, at absolute zero, the electrons fill all of the states up to the Fermi level and none above it. At all temperatures, the Fermi level separates the states that are probably occupied ($E < E_F$) from those that are probably empty ($E > E_F$). The relationship of $f(E)$ to the bandgap diagram is shown in Figure 3.9, which illustrates the increasing probability for the electrons to be excited into the conduction band as the temperature is increased.

The concentration of electrons in the conduction band is

$$n_0 = \int_{E_c}^{\infty} f(E) N(E) \, dE, \tag{3.48}$$

where $N(E)dE$ is the density of states (conventional units are cm^{-3}) in the energy interval dE. This integral can be simplified by introducing the effective density of states, N_c, located at the conduction band edge, E_c, such that

$$n_0 = N_c f(E_c). \tag{3.49}$$

This approximation works well because, at modest temperatures, the conduction electrons occupy only the bottom few states in the conduction band; thus, the integrand in equation (3.48) is significantly larger than zero only for energies near E_c, and the integral is well represented by an appropriate

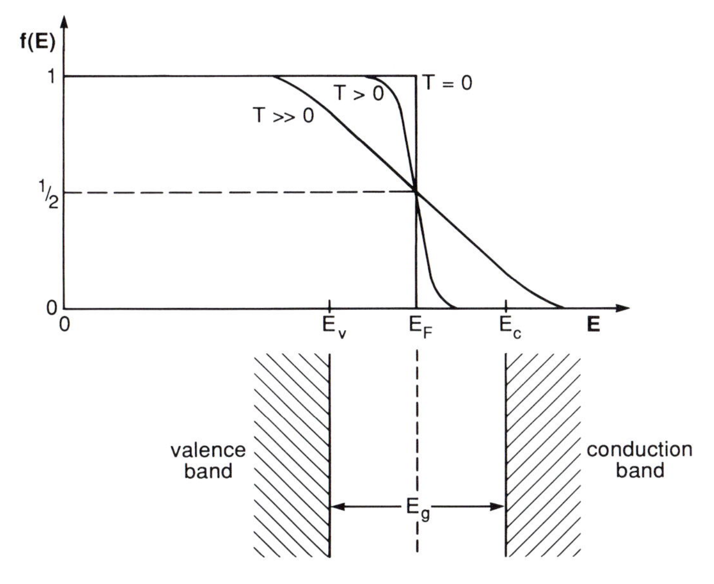

Figure 3.9. Electron probability distribution $f(E)$ as a function of temperature T, compared with an energy band diagram.

average. It is usually the case that $E_c - E_F \gg kT$ (for example, $kT = 0.026\,\text{eV}$ at $T = 300\,\text{K}$). As a result, equation (3.47) can be simplified to

$$f(E_c) = e^{-(E_c - E_F)/kT}. \tag{3.50}$$

It can be shown (see, for example, Streetman, 1990, Appendix IV) that

$$N_c = 2\left[\frac{2\pi m_n^* kT}{h^2}\right]^{3/2}, \tag{3.51}$$

where m_n^* is the effective mass of a conduction electron, allowing for large scale effects of the crystal. Similarly, m_p^* is the effective mass of a conduction hole. The values of the effective masses vary slightly with temperature and also depend on the material. For silicon, $m_n^* \approx 1.1 m_e$ and $m_p^* \approx 0.56 m_e$, where m_e is the mass of the electron. For germanium, $m_n^* \approx 0.55 m_e$ and $m_p^* \approx 0.37 m_e$. Combining equations (3.49), (3.50), and (3.51), we obtain

$$n_0 = 2\left[\frac{2\pi m_n^* kT}{h^2}\right]^{3/2} e^{-(E_c - E_F)/kT}. \tag{3.52}$$

The applicability of equation (3.52) is subject to the validity of the approximations leading to equations (3.49) and (3.50). Given that most intrinsic semiconductors have relatively large bandgaps and that good detector performance requires temperatures at or below 300 K, equation (3.52) will generally be valid in the following discussions. Analogously, for holes we have

$$N_v = 2\left[\frac{2\pi m_p^* kT}{h^2}\right]^{3/2} \tag{3.53}$$

and

$$p_0 = 2\left[\frac{2\pi m_p^* kT}{h^2}\right]^{3/2} e^{-(E_F - E_v)/kT}. \tag{3.54}$$

For intrinsic semiconductors, $E_c - E_F \sim E_F - E_v \sim E_g/2$; equality will be assumed for these relations unless otherwise noted.

As an example, applying equation (3.52) to silicon at 300 K yields $n_0 \sim 1.38 \times 10^{10}\ \mathrm{cm}^{-3}$. At 77 K, we get $n_0 \sim 1.8 \times 10^{-18}\ \mathrm{cm}^{-3}$. These estimates imply that the thermal charge carriers can be frozen out extremely thoroughly without resorting to exceptionally low temperatures (77 K is the temperature of liquid nitrogen, which is readily available). In Chapter 4, however, we will see that this result is no longer valid when we allow for the inevitable impurities in the material.

3.3 Performance specification

To complete the discussion of intrinsic photoconductors, we will discuss some general parameters used to specify their performance and that of other types of detector. The most useful general description of a detector incorporates its signal-to-noise ratio under some prescribed set of conditions. One such specification is the noise equivalent power (NEP), normally specified in units of $\mathrm{W\,Hz^{-1/2}}$.[2] It is defined as the signal power that gives an rms signal to noise of one in a system that has an electronic bandpass of 1 Hz. Conditions that should be specified along with performance measures such as NEP include the wavelength or frequency of the incident photons, the power incident on the detector, and the electronic frequency at which the signal and noise are measured. Beware that the more sensitive the detector, the smaller the NEP!

By the definition of NEP, the signal-to-noise ratio in an observation is

$$\frac{S}{N} = \frac{P_S}{\mathrm{NEP}(\mathrm{d}f)^{1/2}}, \tag{3.55}$$

where P_S is the signal power incident on the detector and $\mathrm{d}f$ is the integration electronic bandwidth. Measurements are frequently conducted for a given time interval, T_{int}, in which case $\mathrm{d}f$ can be determined from equation (3.39), and equation (3.55) can be written

$$\frac{S}{N} = \frac{P_S(2T_{int})^{1/2}}{\text{NEP}}. \tag{3.56}$$

Rearranging equation (3.56) expresses the NEP as a function of the signal-to-noise ratio on a calibrated signal:

$$\text{NEP} = \frac{P_S(2T_{int})^{1/2}}{(S/N)}. \tag{3.57}$$

This expression is convenient, since S/N is the quantity that is usually measured in a practical situation.

S/N as employed in the calculation of NEP refers to the conceptually simplest case of a detector exposed for a specified time interval to a source of signal power. The S/N actually achieved will depend on specific measurement strategies. For example, it is often necessary to divide the measurement time between the source and background flux so the measurements include information about the strength and variability of the background. To be specific, assume that the signal from the source is much weaker than the signal from the background with which it is detected. It is then usually appropriate to divide the measurement time equally between source plus background and background alone. Since the total integration time is divided in two halves, the S/N ratio for each half is reduced by $\sqrt{2}$ (see equation (3.56)). In addition, since the final result is obtained by subtracting the background signal from the signal received on background plus source, a further reduction by $\sqrt{2}$ occurs (we assume in these calculations that the errors are independent). Therefore, the realizable ratio of S/N is only half what one would estimate from a naive application of equation (3.56).

Another useful expression in terms of detector specifications is

$$\text{NEP} = \frac{I_N(\text{A Hz}^{-1/2})}{S(\text{A W}^{-1})}, \tag{3.58}$$

where S is the current responsivity and I_N is the current noise defined in equation (3.45). A similar relation holds for voltages.

A concept related to NEP is the noise equivalent flux density, NEFD, which is the incident flux density that gives unity signal to noise in unity bandwidth. It is given by

$$\text{NEFD} = \frac{E_S(2T_{int})^{1/2}}{(S/N)}, \tag{3.59}$$

where E_s is the flux density of the source for which the signal-to-noise ratio is measured. NEFD is usually quoted at the system level; that is, it describes the signal-to-noise ratio achieved by a full optics/detector system using a practical measurement strategy.

In our bigger-equals-better world, not everyone is comfortable with a figure of merit that exhibits 'upside down' behavior, and an alternative performance measurement called the detectivity, $D = 1/\mathrm{NEP}$, has been introduced. By combining equations (3.1) and (3.42), we see for a transparent contact geometry that the Johnson-noise-limited detector noise goes as the square root of the detector area (all other things being equal). This result suggests that the performance of detectors of different sizes can be compared by multiplying D by the square root of the area, yielding the figure of merit

$$D^* = \frac{A^{1/2}}{\mathrm{NEP}}, \tag{3.60}$$

(in units of $\mathrm{cm\,Hz^{1/2}\,W^{-1}}$), where A is the detector area. The scaling with area, however, is appropriate only for certain types of limiting noise and then only over a limited range of areas. In particular, the very high performance levels predicted by this expression for extremely small detectors are hardly ever realized in practice. D^* must therefore be used with caution in estimating sensitivities.

For the simple cases we are considering, the total noise current is given by equation (3.45). Two general cases are of interest. When the detector is limited by the statistics of the incoming photon stream, it is said to be photon noise limited, and $I_{\mathrm{G-R}}$ dominates. Then

$$\mathrm{NEP_{G-R}}(\mathrm{W\,Hz^{-1/2}}) = \frac{I_{\mathrm{G-R}}(\mathrm{A\,Hz^{-1/2}})}{S(\mathrm{A\,W^{-1}})} = \frac{2hc}{\lambda}\left(\frac{\varphi}{\eta}\right)^{1/2}, \tag{3.61}$$

where we have substituted for the responsivity from equation (3.13) and for $I_{\mathrm{G-R}}$ from equation (3.40). In this case, the NEP can be reduced (improved) only by increasing the quantum efficiency, η. In the infrared, the dominant photon stream received by the detector is often background rather than signal, and a background-limited detector is known as a background-limited infrared photodetector, or BLIP. Equation (3.61) defines the NEP for BLIP operation. Note that equation (3.61) contains the detector quantum efficiency, so two detectors with differing η may achieve differing BLIP performance levels under identical conditions.

When the detector is limited by its internal thermodynamic noise, it is said to be Johnson noise limited, and the NEP becomes

$$\mathrm{NEP_J}(\mathrm{W\,Hz^{-1/2}}) = \frac{I_\mathrm{J}}{S} = \frac{2hc(kT)^{1/2}}{\eta\lambda q G R^{1/2}}, \tag{3.62}$$

from equations (3.13) and (3.42). Here, the NEP can be improved by increasing the quantum efficiency, η, the photoconductive gain, G, or the detector resistance, R, or decreasing the operating temperature, T.

NEPs based on differing noise sources can be added quadratically to calculate the net NEP. For example, in the general case where a detector is limited by a combination of G–R and Johnson noise, we will obtain

$$\mathrm{NEP} = (\mathrm{NEP}^2_{\mathrm{G\text{–}R}} + \mathrm{NEP}^2_{\mathrm{J}})^{1/2}. \tag{3.63}$$

Within a given detector type, one can sometimes be confident that responsivities and quantum efficiencies are identical from one device to another. Comparing equations (3.60) and (3.62), it then follows that a useful figure of merit for detector performance in the Johnson-noise-limited regime is the product of the resistance and the area of the detector, frequently referred to as the R_0A product.

An important caution is necessary. There are measurement conditions under which the use of NEP (and the related figures of merit D and D^*) can be seriously misleading. Examples are the operation both of integrating amplifiers, as discussed in Chapter 6, and of heterodyne detectors, as discussed in Chapter 10. Under these circumstances, it is better to construct a figure of merit that is based on the signal to noise achieved under the appropriate conditions, rather than to try to adapt NEP or some other general figure of merit. This point will be illustrated in the chapters just mentioned.

Additional discussions of performance specification can be found in Wolfe and Zissis (1978), Dereniak and Crowe (1984), Vincent (1990), and Wyatt (1991).

3.4 Example: design of a photoconductor

Consider an intrinsic silicon photoconductor operating at 1 μm and constructed as shown in Figure 3.1. Let it be 1 mm^2, and operate it at 300 K. Assume the detector breaks down when its bias voltage, V_b, exceeds 60 mV. Determine: (a) a reasonable detector thickness for good quantum efficiency; (b) the responsivity; (c) the dark resistance; (d) the time response; (e) the Johnson noise; and (f) the NEP, D, and D^*.

(a) Since the transverse contacts make it impossible to coat the back of the detector with a reflective metal, the minimum thickness for reasonably good quantum efficiency is one absorption length. From Figure 3.5, at 1 μm this thickness corresponds to about 80 μm. Allowing for reflection loss at the front face (the refractive index of silicon is $n \sim 3.4$), the quantum efficiency is 44% (see equation (2.5)).

(b) To maintain safely stable operation, we choose to set $V_b = 40$ mV. The photoconductive gain can be calculated from equation (3.12). For simplicity, we will ignore the hole component of photoconductivity. From Table 3.1, we obtain $\tau = 10^{-4}$ s and $\mu = 1350\,\text{cm}^2\,\text{V}^{-1}\,\text{s}^{-1}$. The electric field $\mathscr{E}_x = V_b/\ell = (0.04\,\text{V})/(0.1\,\text{cm}) = 0.4\,\text{V cm}^{-1}$. These parameters yield $G = 0.54$. Substituting for G and η in equation (3.13), the responsivity will be $S = 0.19\,\text{A W}^{-1}$.

(c) For the sake of simplicity, we will again ignore the contribution of holes in computing the dark resistance of the detector. The concentration of conduction electrons can be obtained from equation (3.52) with $E_g = 1.11$ eV, $m_n^* = 1.00 \times 10^{-30}$ kg, and $T = 300$ K. We estimate $n_0 = 1.38 \times 10^{16}\,\text{m}^{-3}$. From equation (3.5), the conductivity of the detector material is $2.98 \times 10^{-6}\,(\Omega\,\text{cm})^{-1}$. Substituting into equation (3.1), we find that the detector dark resistance is $4.2 \times 10^7\,\Omega$.

(d) The time response of the detector is determined by the longest of three time constants. The first is the recombination time, which in this case is 10^{-4} s. The second is the RC time constant. The capacitance can be calculated from equation (3.19) with the dielectric constant from Table 3.1; the result is $C = 8.4 \times 10^{-15}$ F, leading to $\tau_{RC} = 3.5 \times 10^{-7}$ s. Since we have not specified the circuit in which the detector is operated, the dielectric relaxation time constant will be the same as τ_{RC}. Comparing these three time response limitations, we find that the recombination time is the important limitation.

(e) The Johnson noise current is given by equation (3.42) with $T = 300$ K and $R = 4.2 \times 10^7\,\Omega$. It is $1.99 \times 10^{-14}\,\text{d}f^{1/2}$ A.

(f) The performance of the detector can be characterized by calculating the signal-to-noise ratio it would achieve in a given integration time on a given signal power. For example, in a 0.5 s integration, from equation (3.39), $\text{d}f = 1$ Hz and the Johnson noise current $I_J = 1.99 \times 10^{-14}$ A. An input photon signal of 1.05×10^{-13} W gives a signal current of 2×10^{-14} A, since the detector responsivity is $0.19\,\text{A W}^{-1}$. Therefore, the ratio of signal to noise on this strength of signal is 1 in 0.5 s of integration. By definition, the detector $\text{NEP} = 1.05 \times 10^{-13}\,\text{W Hz}^{-1/2}$, the detectivity $D = 9.5 \times 10^{12}\,\text{Hz}^{1/2}\,\text{W}^{-1}$, and $D^* = 0.1\,\text{cm} \times D = 9.5 \times 10^{11}\,\text{cm Hz}^{1/2}\,\text{W}^{-1}$.

3.5 Problems

3.1 Use equation (3.25) to prove equation (3.38).

3.2 A circular photoconductor of diameter 0.5 mm operating at a wavelength of 1 μm has a spectral bandwidth of 1% (0.01 μm). Suppose it views a circular blackbody source of diameter 1 mm at a distance of 1 m and at a temperature of 1500 K. If the detector puts out a signal of 5×10^{-13} A, what is its responsivity and ηG product?

3.3 Suppose the output of the detector in Problem 3.2 is sampled in a series of well defined one-second intervals. With the view of the source (and other photons) blocked off, it is found that the rms noise current is 2.00×10^{-16} A, which we take to be excess electronics noise. Viewing the source, the rms noise is 4.58×10^{-16} A. Determine the detective quantum efficiency (in the absence of electronics noise) and use it to estimate the photoconductive gain.

3.4 Show that the quantum efficiency of a photoconductor that has reflectivity r at both the front and back faces, absorption coefficient a, and length ℓ from the front to the back is

$$\eta = \frac{(1-r)(1-e^{a\ell})}{1-re^{a\ell}}.$$

Hint: recall the Maclaurin series

$$\frac{1}{1-x} = \sum_{0}^{\infty} x^n.$$

Comment on the applicability of this result to a detector with a perfectly reflective back face (for example, one with a transparent front contact and a metallic back face).

3.5 Compare the performance of a germanium photoconductor in the same application as the silicon one in the example. Assume the detector has transparent front contact, sensitive area 1 mm^2, and an absorbing back contact. Assume breakdown occurs at 1 V cm^{-1} and that the detector is biased to half this value. How can the operating conditions be adjusted to make the germanium more competitive and therefore to take advantage of the high absorption efficiency of germanium near 1 μm?

3.6 Consider two resistors at different temperatures. Show that the Johnson noise of the combination when they are connected in series (see Figure 3.10(a)) is

$$\langle V_J^2 \rangle = 4k\, df(R_1 T_1 + R_2 T_2),$$

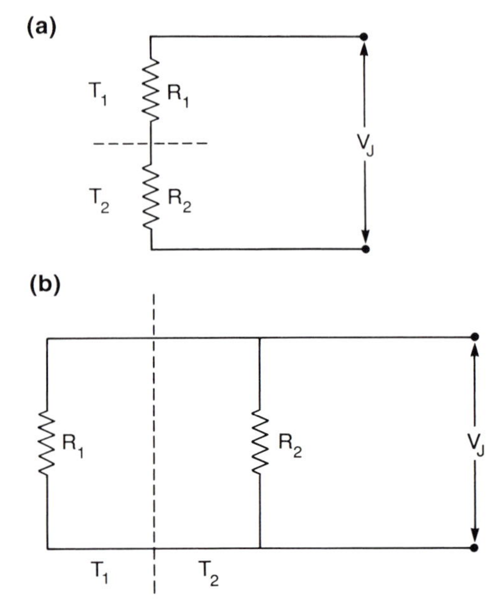

Figure 3.10. Circuits for Problem 3.6.

but when they are connected in parallel (see Figure 3.10(b)), it is

$$\langle V_J^2 \rangle = 4k\,\mathrm{d}fR_1R_2\left[\frac{R_1T_2+R_2T_1}{(R_1+R_2)^2}\right].$$

3.7 Consider the circuit of Figure 3.11, and assume that the detector is a photoconductor with a resistance of $\sim 1\ \mathrm{M}\Omega$ that is exposed to a small, varying signal. Assuming that the system is limited by Johnson noise, derive an expression for the signal-to-noise ratio as a function of the load and detector resistances and temperatures (use the results of Problem 3.6). Determine the value of R_L where S/N is maximum for the case of a detector operating at 4 K, 77 K, and 300 K, always with the load resistor at 300 K. How does the signal-to-noise ratio vary with detector operating temperature?

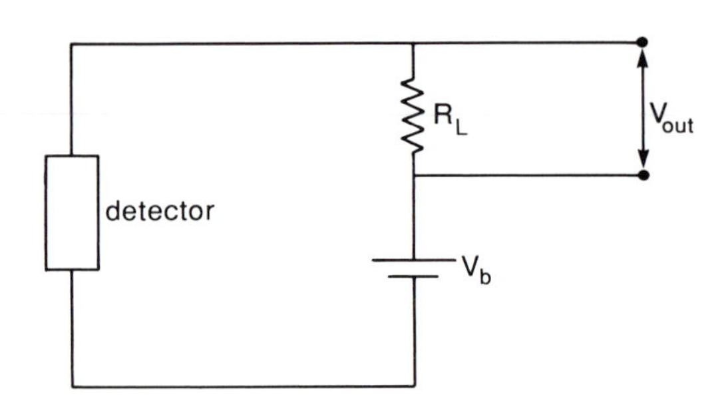

Figure 3.11. Circuit for Problem 3.7.

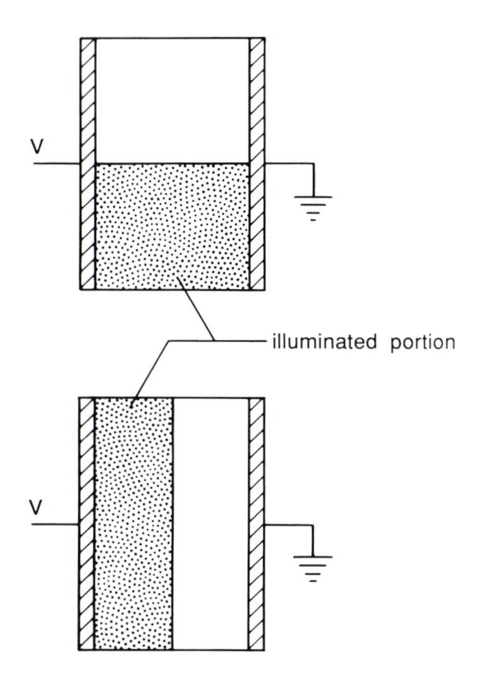

Figure 3.12. Illustration for Problem 3.8.

3.8 Consider a detector with transverse contacts that is illuminated with a bright source of light that only shines on half the sensitive area of the detector (with the remainder of the detector surface in darkness). Discuss whether the detector will respond differently depending on whether the illuminated section extends across the contacts or is parallel to them (see Figure 3.12). Is there an analogous situation for transparent contact detectors?

Notes

1 In fact, a number of other simplifying assumptions have been made in the derivation of G–R noise. One of the most significant is that the relaxation time for a charge carrier has been taken to be independent of the density of free carriers. This assumption is not generally true (see equation (3.33)). A more rigorous treatment of G–R noise (and other noise mechanisms) can be found in van Vliet (1967), van der Ziel (1976), or Boyd (1983).

2 In some references, the NEP is not normalized to unity bandpass but is quoted as the signal (in watts) that gives unity S/N with the bandpass quoted separately. We have selected the frequency-normalized convention because it makes NEP a single-parameter figure of merit that is convenient for direct

comparison of detector behavior. A different measurement, the 'blackbody NEP', uses the signal to noise achieved by the detector with no filters when viewing a blackbody of some temperature. Because such measurements are difficult to interpret, we will not use this parameter either.

4

Extrinsic photoconductors

As described in the preceding chapter, the spectral response of intrinsic photoconductors is limited to photons that have energies equal to or exceeding the bandgap energy of the detector material. For the high quality semiconductors silicon and germanium, these energies correspond to maximum wavelengths of 1.1 μm and 1.8 μm, respectively. A number of semiconductor compounds have smaller bandgaps. They are often used as intrinsic photoconductors with rapid time response; however, they are generally unable to achieve, at very low light levels, the extremely high impedances needed to overcome Johnson noise adequately. In addition, the performance of intrinsic photoconductors degrades rapidly as the wavelength extends beyond about 15 μm due, for example, to poor stability of the materials, difficulties in achieving high uniformity in material properties, and problems in making good electrical contacts to them. Detector operation far into the infrared must therefore be based on some other mechanism. The addition of impurities to a semiconductor allows conductivity to be induced by freeing the impurity-based charge carriers. This extrinsic process requires smaller energy increments than does intrinsic photoconduction, enabling response at long infrared wavelengths. As a result, extrinsic photoconductors are a useful variation for infrared detectors. Because the lower excitation energies also allow for large thermally excited dark currents, these detectors must be operated at low temperatures. They are readily made in two-dimensional arrays containing thousands of pixels and can provide background limited performance out to 200 μm even at extremely low light levels.

4.1 Basics

4.1.1 Operation

As discussed in Chapter 1, impurities strongly modify the properties of a semiconductor. Consider the structure of silicon, which conveniently lends

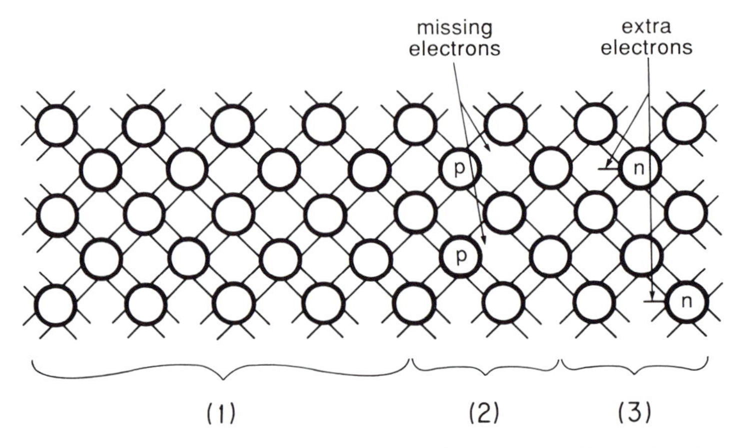

Figure 4.1. Comparison of n-type and p-type doping in silicon. Region (1) shows undoped material, region (2) is doped p-type, and region (3) is n-type.

itself to representation in two dimensions, as shown in Figure 4.1. (The true structure of silicon is three dimensional with tetrahedral symmetry.) The silicon atoms are represented here by open circles connected by bonds consisting of pairs of electrons; each electron is indicated by a single line. In the intrinsic section of the structure, the uniform pattern of line pairs shows that all of the atoms share pairs of electrons to complete their outer electron shells. If an impurity is added from an element in column III of the periodic table (for example, B, Al, Ga, In), it has one too few electrons to complete the crystal structure. Under modest excitation, either by thermal excitation or by absorption of a photon, it can steal an electron from its neighbor, which becomes a positive ion. This process produces a hole in the valence band, which can be passed from one silicon atom to another to conduct electricity. The impurity is called an acceptor, and the material is referred to as p-type because this type of doping absorbs electrons and its dominant charge carriers are positive (see Section 1.3). If the impurity is from column V (for example, P, As, Sb), it has sufficient electrons to accommodate the sharing with one electron left over. This 'extra' electron is relatively easily detached to enter the conduction band. The impurity is called a donor, and the material is called n-type because this type of doping tends to generate negative charge carriers. In either case, if the electrical behavior is dominated by the effects introduced by the impurities, the material is called extrinsic.

Figure 4.1 and the paragraph above give an overly concrete picture of the semiconductor crystal. Quantum mechanically, the impurity-induced holes and conduction electrons must be treated as collective states of the crystal that are described by probability distribution functions that extend over many atoms. A satisfactory theory of the excitation energies required for extrinsic photoconductors requires a quantum mechanical derivation, which is beyond the scope of our discussion.

If the impurity concentration is high enough to provide good absorption efficiency, extrinsic semiconductors can be useful detectors. Photoconductive detectors based on extrinsic materials operate in a similar fashion to the intrinsic photoconductors discussed in Chapter 3, except for differences noted below. Detailed reviews of the characteristics of such detectors have been written by Bratt (1977) and Sclar (1984).

Notice that the electrical conductivity of extrinsic material differs in a fundamental way from that of intrinsic material. In the latter case, charge carriers are created as electron/hole pairs consisting of a conduction electron and a hole in the valence band from the inter-atomic bond it has vacated. In the former case, individual charge carriers are created whose complementary charge resides in an ionized atom which remains bonded into the crystal structure; this complementary charge is therefore immobile and cannot carry currents. At high temperatures, the intrinsic mechanism dominates in both types of material. At modest and low temperatures, the carrier contributed by the dominant impurity is the most numerous and is called the majority charge carrier. The complementary type, or minority charge carrier, coexists at a lower concentration. It is contributed either by the intrinsic conductivity mechanism or by ionization of impurities of complementary type and lower concentration. The notation *semiconductor:dopant* is used to indicate the semiconductor and majority impurity; for example, Si:As designates silicon with arsenic as the majority impurity.

The dopants do not interfere with the intrinsic absorption process in the material. In fact, because the number of semiconductor atoms is always far larger than the number of impurity atoms, intrinsic absorption always prevails over extrinsic absorption in wavelength regions where the intrinsic mechanism is effective. The response from the intrinsic absorption must be anticipated in any successful use of extrinsic photoconductors, for example by using optical filters that strongly block the wavelengths that can excite intrinsic conductivity.

The basic operation of extrinsic photoconductors is similar to that of intrinsic photoconductors except that the excitation energy, E_i, must be substituted for the bandgap energy (E_g). In first-order theory, the response drops to zero at the wavelength corresponding to this energy, $\lambda_c = hc/E_i$. As with intrinsic photoconductors, a number of effects act to round off the dependence of response on wavelength near this cutoff. The absorption

Table 4.1. *Properties of some extrinsic photoconductors*

		Ge		Si	
Impurity	Type	Cutoff wavelength λ_c (μm)	Photo-ionization cross-section σ_i (cm^2)	Cutoff wavelength λ_c (μm)	Photo-ionization cross-section σ_i (cm^2)
Al	p			18.5[a]	8×10^{-16}[b]
B	p	119[b]		28[a]	1.4×10^{-15}[b]
Be	p	52[b]	1.0×10^{-14}[c]	8.3[a]	5×10^{-18}[a]
Ga	p	115[b]	1.0×10^{-14}[c]	17.2[a]	5×10^{-16}[b]
In	p	111[b]		7.9[a]	3.3×10^{-17}[b]
As	n	98[b]	1.1×10^{-14}[b]	23[a]	2.2×10^{-15}[a]
Cu	p	31[b]	1.0×10^{-15}[b]	5.2[a]	5×10^{-18}[a]
P	n	103[b]	1.5×10^{-14}[b]	27[a]	1.7×10^{-15}[b]
Sb	n	129[b]	1.6×10^{-14}[b]	29[a]	6.2×10^{-15}[a]

[a] Sclar (1984).
[b] Bratt (1977).
[c] Wang *et al.* (1986).
Parameters for additional materials are tabulated in Landolt-Börnstein (1989).

coefficient is

$$a(\lambda)=\sigma_i(\lambda)N_I, \tag{4.1}$$

where $\sigma_i(\lambda)$ is the photoionization cross-section and N_I is the neutral impurity concentration. Table 4.1 gives sample values of the cutoff wavelengths and photoionization cross-sections (at the wavelength of peak response). Detector quantum efficiencies can be calculated as in equation (3.18), given N_I.

It would appear desirable to increase N_I without limit to maximize the quantum efficiency, but upper limits to the impurity concentration arise from two sources. First, there is a limit to the solubility of the impurity atoms in the semiconductor crystal. These limits range from $\approx 10^{18}$ to 10^{21} cm^{-3} for commonly used dopants. Secondly, before the solubility limit is reached, the electrical properties of the crystal usually undergo unwanted changes in the form of conductivity modes that cannot be adequately controlled either by operating the detector at low temperature to freeze them out or by other means.

One example of these unwanted conductivity modes is hopping. When impurity atoms are sufficiently close together, the electron wave function from one impurity atom extends at a non-negligible value to a neighboring impurity atom. Under these conditions, conduction can occur directly from one impurity atom to another without supplying the energy necessary to raise an electron into the conduction band.

Given these limitations, typical acceptable impurity concentrations are around 10^{15} to $10^{16}\,cm^{-3}$ for silicon and somewhat lower for germanium. Using the photoionization cross-sections in Table 4.1 and equation (4.1), it can be seen that the absorption coefficients are some three orders of magnitude less than those for direct absorption in intrinsic photoconductors. As a result, the active volumes in extrinsic photoconductors must be large to get adequate quantum efficiency, with dimensions on the order of a millimeter in the direction along the incoming photon beam. These detectors are sometimes termed 'bulk photoconductors' to distinguish them from detector types where the absorption takes place in a much smaller volume (see, for example, Section 4.3.2).

4.1.2 Construction

Extrinsic photoconductors are usually made of silicon or germanium using techniques drawn from the semiconductor industry (Bratt, 1977). High purity crystals can be grown from a small seed crystal whose end is dipped in supercooled melted material. It is advantageous to set up the growth process in such a way that the crystal is pulled out of the melt without touching the walls of the vessel (the Czochralski method); this technique both avoids stresses in the crystal that can develop if it solidifies in contact with the crucible walls and also prevents the escape of impurities from the walls into the crystal. The material can be doped by adding impurities to the melt from which the crystal is grown. Alternatively, the purity of the material can be improved after growth by zone refining the crystal. In this process, a heater melts a small region of the crystal; the heater is slowly driven along the length of the crystal to allow each part to melt and regrow. Because impurities disrupt the crystal structure, they tend to accumulate in the liquid and to remain there when the crystal grows again behind the molten section. Repeated application of this process tends to collect the impurities near the end toward which the heater is driven; this end is cut off the crystal and discarded. The dopant impurities can be added in a late stage of the zone refining by placing the impurity in the melted zone. Doping of grown crystals can also be accomplished by exposing a crystal to impurity vapors or by bringing its surface into contact with a sample of the impurity material; in both of these cases, the impurity atoms are distributed into the crystal by diffusion at high temperature. Floating zone growth combines features of Czochralski growth and zone refining. The crystal is grown onto a seed from a melted portion of an ingot that is suspended to avoid contact with the walls of the crucible.

To make detectors, a thin wafer of material is cut from the crystal with a precision saw and then polished. The two sides of this wafer will form the contacts of the detectors. To smooth the variations in the electric fields near the contacts, heavy doping is used in the semiconductor where it joins the actual contact. The doping can be applied by implanting impurity atoms of similar type to the majority impurity (for example, boron for a p-type crystal) from a high energy beam directly in a thin layer under the surface of the semiconductor wafer (this process is called ion implantation) and then annealing the crystal damage by heating. A transparent contact can be made by adding a second layer of ion-implanted impurities at lower energy than the first so it does not penetrate so far into the crystal and at a higher impurity density so it has larger electrical conductivity. For an opaque contact, metal is evaporated over the ion-implanted layers. Individual detectors or arrays of detectors are cut from the wafer with a precision saw. The surface damage left by the saw can be removed by etching the detectors.

For completeness, we discuss some additional techniques used to produce the semiconductor detectors and electronic devices that we will discuss in following chapters. When high temperatures must be avoided, crystals can be obtained by *epitaxial growth.* For example, a mixture of semiconductor and some other material may have a lower melting temperature than the semiconductor alone; introduction of a seed crystal into a such a melt can lead to growth of the crystal without incorporating significant amounts of the second material into the crystal (a process called liquid-phase epitaxy). Similarly, it may be possible to grow the crystal by placing a seed crystal in contact with a gas whose molecules contain the crystal material in combination with other atoms (vapor-phase epitaxy). Still another possibility is to expose the seed crystal to a molecular or atomic beam that creates new growth (molecular-beam epitaxy). Epitaxy is a particularly useful technique for growing a thin crystalline layer on a pre-existing crystal substrate without melting or otherwise disturbing the substrate; this new layer may have different doping from the substrate or may even be a different material with matching crystal structure

Protective or insulating layers or both can be grown easily on silicon because it naturally forms a rugged and stable oxide. Growth of this oxide layer occurs when the silicon is exposed to air, but it is usually accelerated by heating the material in an oven. In the case of germanium and many other semiconductors, protective layers must be added by more complex processing steps. The ease of growing insulating layers is central to the use of silicon in the production of complex integrated circuits

Circuit elements are grown on silicon by repeated application of the techniques described above. Complex structures can be created by coating the material with a *photoresist*, exposing it to ultraviolet light through a *mask*, developing the photoresist, and then etching or processing in some other way. The regions of the photoresist that are exposed to the ultraviolet are removed by development, allowing the processing access to the underlying material in specific areas; the remaining areas are protected by the resist. When the remainder of the resist is removed, an image of the mask remains. This new structure can be used as the basis for additional processing steps.

These basic construction techniques are used to make the semiconductor devices we discuss in Chapters 3–7 and 9–11 of this book. They are applied to a broad variety of semiconductor materials and with many different dopants and geometries to tailor electronic devices for a vast range of applications. Further discussion of these issues can be found in Sze (1985) and Streetman (1990).

4.2 Limitations

We return to extrinsic photoconductors to explore their performance attributes in more detail.

4.2.1 Frequency response

All other things being equal, the carrier lifetime in an extrinsic semiconductor is roughly proportional to the inverse of the concentration of the *ionized* majority impurities with which the free carriers recombine. Neutral minority and majority impurity atoms ionize each other by exchanging charge carriers – that is, the n-type atoms contribute their unbonded electrons to the holes in the p-type atoms. Therefore, the recombination time is inversely proportional to the concentration of *minority* impurities. Particularly with silicon detectors, where the recombination times are relatively short to begin with (see, for example, Table 3.1), adequately fast time response for virtually any application can be achieved with a sufficient concentration of minority impurities. Consequently, the time response is usually dominated by the RC time constant or by dielectric relaxation, both of which were described in Section 3.2.4. With germanium detectors, where the impurity concentrations must be kept relatively low, the frequency response in demanding applications may be limited by the recombination time.

4.2.2 Noise

The basic noise mechanisms in extrinsic photoconductors are identical to those described for intrinsic ones. See Section 3.2.5.

4.2.3 Thermal excitation

For n-type material, the higher concentration of conduction electrons (as compared to holes) implies that the electron state probability function must be shifted toward the conduction band. Examining the form of this function (equation (3.47)), we see that E_F must lie above the middle of the energy gap. Similarly, for p-type material E_F lies below the middle of the energy gap. In either case, the electron state probability function shows there will be an increase in the probability of free charge carriers in the material. As a result, thermal excitation is greatly expedited in extrinsic photoconductors; that is, the presence of impurities makes it easier from the standpoint of energetics to excite significant conductivity.

In deriving an expression for thermal excitation, we will first consider the concentration of free negative carriers, n, in n-type material. Let the concentrations of donors and acceptors be N_D and N_A, respectively. Let N_D^n be the concentration of neutral donors and N_D^i the concentration of ionized donors, so $N_D = N_D^n + N_D^i$. The simplest case, applicable at relatively high temperatures, would assume that all the acceptors have captured electrons from the donors to complete their crystal bonds, and that all the donors are ionized by escape of their 'extra' electrons ($N_D^n = 0$, or $N_D = N_D^i$). Then $n = N_D - N_A$. Under these conditions, however, the typical concentrations of free charge carriers are too high for sensitive extrinsic detector operation.

To derive n for lower temperatures, we still assume that the acceptors have captured electrons from donors, so $N_D^i = N_A + n$ is the concentration of ionized donors available for recombination, and $N_D^n = N_D - N_A - n$ is the concentration of neutral donors from which free electrons can be generated. Thus, the generation rate is proportional to $(N_D - N_A - n)$ and the recombination rate to $n(N_A + n)$. Adopting constants of proportionality A and B,

$$\frac{\mathrm{d}n}{\mathrm{d}t} = A(N_D - N_A - n) - Bn(N_A + n). \tag{4.2}$$

In equilibrium, $\mathrm{d}n/\mathrm{d}t = 0$ so

$$\frac{A}{B} = \frac{n(N_A + n)}{N_D - N_A - n}. \tag{4.3}$$

However, A/B is also equivalent to the product of the ratio of available densities of states for generation and recombination with $e^{-E_i/kT}$ (the exponential factor takes into account the electron probability distribution). Analogously to the derivation of equation (3.52), it can be shown that

$$\frac{A}{B} = \left[\frac{2}{\delta}\right]\left[\frac{2\pi m_n^* kT}{h^2}\right]^{3/2} e^{-E_i/kT}, \tag{4.4}$$

where δ is the ground state degeneracy of the donor impurity (it is usually of the order of unity). Substituting for A/B from equation (4.3) and assuming $N_D \gg n$,

$$n = \left[\frac{N_D - N_A}{N_A + n}\right]\left[\frac{2}{\delta}\right]\left[\frac{2\pi m_n^* kT}{h^2}\right]^{3/2} e^{-E_i/kT}. \tag{4.5}$$

Taking note of the comments following the derivation of equation (3.52), we observe that equation (4.5) will be valid for $E_i \gg kT$; this condition is not met near room temperature for impurities with small values of E_i, but is generally true at the low operating temperatures of extrinsic photoconductors.

A similar discussion holds for p-type material if p is substituted for n, N_D for N_A, and N_A for N_D:

$$p = \left[\frac{N_A - N_D}{N_D + p}\right]\left[\frac{2}{\delta}\right]\left[\frac{2\pi m_p^* kT}{h^2}\right]^{3/2} e^{-E_i/kT}. \tag{4.6}$$

m_n^* and m_p^* are defined just prior to equation (3.52)

The conductivity of the material is $\sigma_n = qn\mu_n$ (or $\sigma_p = qp\mu_p$ for p-type) as before (see equation (3.5)), where μ_n and μ_p are functions both of the level of doping and of T. Measurements of the mobility of detector grade Si:P and Si:As are shown in Figure 3.3.

A potential problem both with intrinsic semiconductors and with extrinsic ones with relatively large excitation energies is revealed by equations (4.5) and (4.6). Even with the incredible purity levels that can be achieved, some unwanted residual impurities remain in the best material. A particularly difficult one to remove from silicon is boron, which has a very low excitation energy; from Table 4.1 and the relation for λ_c, $E_i = 0.045$ eV. A sample of material of mediocre purity may have a boron concentration of 10^{13} cm^{-3}; if we assume that $N_D = 0$, $\delta = 1$, and $T = 77$ K (the temperature of liquid nitrogen), we can show from equation (4.6) that virtually all these impurity atoms will be ionized and that the concentration of thermally excited holes will be $\approx 10^{13}$ cm^{-3}. The conductivity will be large, resulting in an unacceptable level of Johnson noise when operated as a low light level photoconductor.

4.2.4 Compensation

To counteract the problem of unwanted thermal conductivity due to impurities with small values of E_i, these dopants are 'compensated' by adding carefully controlled amounts of the opposite kind of impurity – for example, n-type if the unwanted impurity is p-type. The result is readily seen from equations (4.5) and (4.6) by setting $N_D = N_A$. Although some degree of compensation is frequently critical in obtaining good detector performance, heavily compensated material is usually inferior to material that starts as pure as possible and needs to be only lightly compensated. A high concentration of compensating impurities usually results in the rapid capture of photoexcited charge carriers and reduces the photoconductive gain. It is also difficult to control the compensation process precisely enough to be totally effective. Problems can arise because of nonuniformities in the sample or because of imprecision in the exact amount of compensating material. Finally, hopping conductivity may be enhanced because the atoms of the compensating impurity ionize some of the majority impurity atoms, making available empty sites into which carriers can hop.

The term compensation is also frequently used in a more general sense to refer to the effects of impurities of opposite type to the dominant one, regardless of how they are introduced or whether their effects are beneficial.

4.2.5 Radiation effects

All solid state detectors are very sensitive to energetic particles because these particles are heavily ionizing and hence create large numbers of free charge carriers when they pass through the sensitive volume of the detector. As we have seen, extrinsic detectors must have large volumes to achieve reasonable quantum efficiencies. Consequently, in a high radiation environment, they are struck frequently and generate a high rate of spurious signals. In well behaved detectors, the duration of such signals is limited by the frequency response of the amplifier. A 'radiation hit' then appears as a large voltage spike followed by a relatively quick recovery (that is, limited by the amplifier response time) to quiescent operating conditions.

Unfortunately, detectors operating at low backgrounds and under high radiation conditions can exhibit other effects. The high energy particles create extensive damage in the crystal bonds; the resulting lattice defects can act much as impurities would by introducing additional energy levels that can behave as acceptors and donors. The increase in pseudo-minority-impurities can increase the recombination time and result in an increase in the detector responsivity. Unfortunately, the behavior is not stable, so the noise also increases, typically much faster than the responsivity. Under low background operating conditions this condition can decay very slowly. Thus, the signal-to-noise ratio can be seriously decreased well after the energetic radiation has been removed. In addition, the slow decay of the responsivity back to its normal value makes the calibration of the detectors time dependent and hence degrades the accuracy of any measurements made with them. These problems are severe with extrinsic germanium detectors, but they also occur in silicon detectors, particularly those that operate at relatively long wavelengths.

The radiation damage can be erased by first heating the detector sufficiently (~6 K for germanium, ~20 K for silicon) to re-establish thermal equilibrium, and then cooling it to normal operating temperature (~2 K for germanium, ~6 K for silicon). Alternatively, the bias voltage can be raised above the detector breakdown level, flooding the active region momentarily with conduction electrons, or the detector can be flooded with a high flux of infrared photons.

4.2.6 Hook anomaly and spiking

At low background levels, extrinsic detectors can exhibit a variety of undesirable response characteristics, including 'hook' (so named because of the appearance of the electronic output waveform), spontaneous spiking, and dielectric relaxation (already discussed in Chapter 3). The first two effects are illustrated in Figure 4.2. They are thought to arise from voltage barriers set up at the detector contacts by space charge effects. Hook response is associated with the quantum mechanical tunneling of charge carriers through the voltage

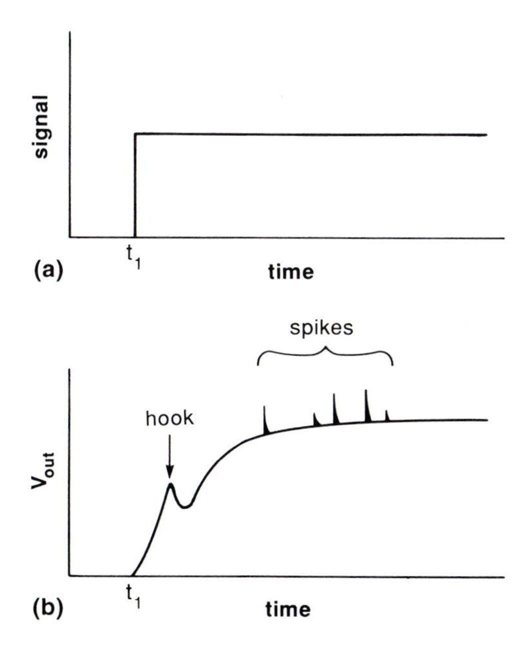

Figure 4.2. Response of a photoconductor to a step input signal (a), showing hook characteristic and voltage spikes (b).

barrier at the contact. Spikes are produced when charges are accelerated in the electric fields near the contacts and acquire sufficient energy to impact-ionize impurities in the material, creating a mini-avalanche of charge carriers.

Care in the construction of the detector's electrical contacts can minimize hook response and spiking. Spiking and hook response are also reduced at low photoconductive gain, for example, by operating the detector at reduced bias voltage. A detailed discussion is given by Sclar (1984).

4.3 Variants

4.3.1 *Stressed detectors*

The long wavelength cutoff of a p-type photoconductor can be modified by physically stressing the crystal. This behavior should at least seem plausible because conduction in p-type material occurs through the breaking and remaking of inter-atomic bonds (that is, migration of a hole). An external force places stress on the inter-atomic bonds, so it would be plausible that less additional energy is required to break them.

A particularly dramatic effect can be achieved with diamond lattice crystals stressed along a particular (the [100]) crystal axis. The stress can produce a significant reduction in the acceptor binding energy and can extend the response of Ge:Ga photoconductors from a maximum wavelength of 115 μm (unstressed) to beyond 200 μm (Kazanskii, Richards, and Haller, 1977; Haller, Hueschen, and Richards, 1979).

In making practical use of this effect, it is essential to apply and maintain very uniform and controlled pressure to the detector so that the entire detector volume is placed under high stress without exceeding its breaking strength at any point. Detectors stressed in this manner show a wavelength dependent response similar to that of conventional detectors, with responsivity proportional to wavelength until the wavelength is close to the cutoff.

4.3.2 *Blocked impurity band (BIB) detectors*

We have seen that the goals of low electrical conductivity and efficient absorption of infrared photons place contradictory requirements on the doping level in extrinsic photoconductors. One solution to this dilemma is to use separate detector layers to optimize the electrical and optical properties separately. A device built with this philosophy is the blocked impurity band (BIB) detector[1]; it is shown schematically in Figure 4.3 and described in detail by Szmulowicz and Madarsz (1987).

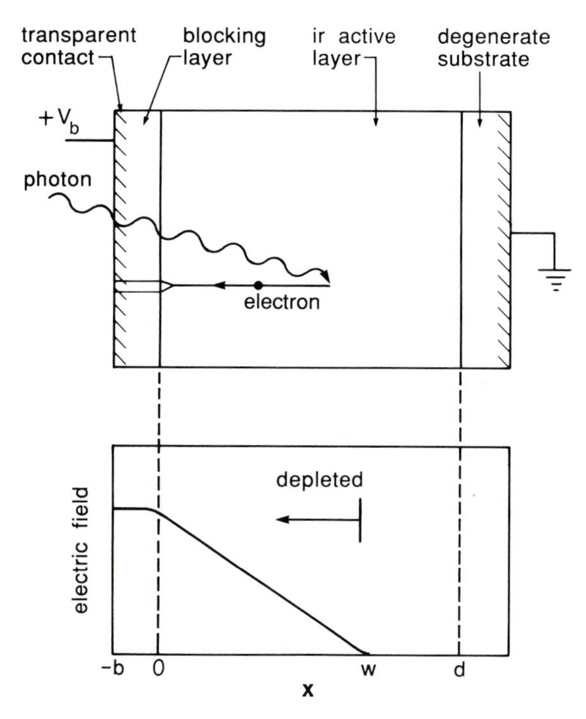

Figure 4.3. Cross-section of a blocked impurity band detector.

In a BIB detector, the absorbing, infrared-active layer is doped heavily to a level where impurity band conductivity through hopping would be completely unacceptable in a conventional extrinsic detector. In the following discussion, we will assume this layer is n-type; in fact, this assumption coincides with the most common type of BIB detectors, which are made of arsenic-doped silicon. We will also assume that there is a low, but not negligible, concentration of a p-type dopant in the infrared-active layer. An additional thin, high purity layer, called the blocking layer, is grown over the front of the absorbing layer. This layer provides the large electrical resistance required for low light level operation. One electrical contact is made to the blocking layer, and the second contact is made to the back of the active layer. The details of the arrangement of the contacts depend on whether the detector is to be front illuminated through the first contact and blocking layer or back illuminated through the second contact. In the former case, a transparent contact is implanted into the blocking layer and the second contact is made by growing the detector on an extremely heavily doped, electrically conducting (degenerate) substrate. In the latter case, a thin degenerate but transparent contact layer is grown underneath the active layer on a high purity, transparent substrate. These two geometries are illustrated in Figure 4.4.

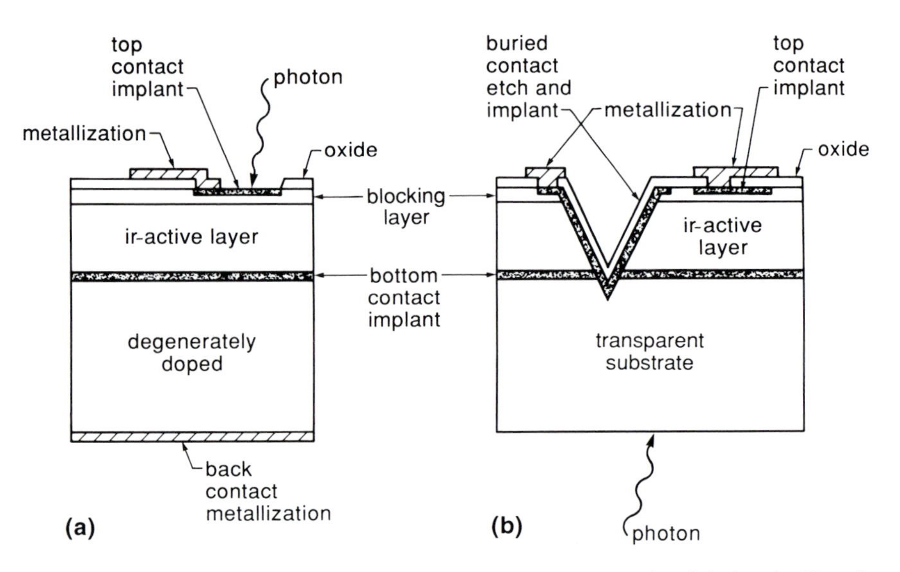

Figure 4.4. Front illuminated BIB detector (a), compared with back illuminated detector (b).

A valuable characteristic of BIB detectors is that the infrared-active layer is so heavily doped that it can be made quite thin without compromising the quantum efficiency, which can be 40–80% (the latter value with antireflection coating) at the peak of the detector response. Since the surface area of the active volume is small, these detectors can operate efficiently in environments where there is a high density of energetic charged particles. Another advantage of BIB detectors is that the infrared-active layers of these detectors need not have extremely high impedance, and hence dielectric relaxation effects are reduced.

Because of the heavy doping of the infrared-active layer, the impurity band increases in width around its nominal energy level and the energy gap decreases between the conduction band (or valence band for p-type material) and the nearest portion of the impurity band. Consequently, the minimum photon energy required to excite photoconduction is lower for a BIB detector than for a conventional bulk photoconductor with the same dopant, and the spectral response extends to longer wavelengths than that of the conventional detector. This effect is illustrated in Figure 4.5. Because of the high impurity concentrations that can be achieved without degrading the dark current, a BIB detector can also be tailored to provide better quantum efficiency toward short wavelengths than is possible with a conventional detector. Together, these effects provide high sensitivity operation over a broader spectral range than with other detectors and can therefore simplify system design.

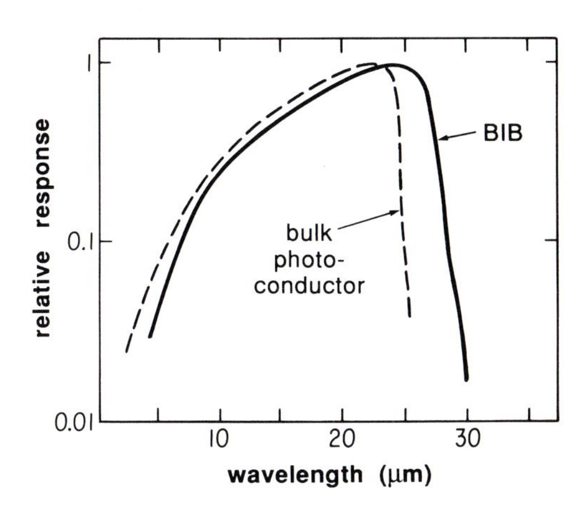

Figure 4.5. Comparison of spectral responses for a bulk Si:As photoconductor and a Si:As BIB.

In addition, the heavily doped material under one contact of BIB detectors reduces the possibility of quantum mechanical tunneling and thus hook response, while the intrinsic material under the other contact, with its low impurity concentration, reduces impact-ionization and thus spiking relative to bulk photoconductors under low backgrounds. Since the recombination in BIB detectors occurs in relatively low resistance material and is not a random process distributed through the high impedance section of the detector, there is only a single random event associated with the detection of a single photon. Consequently, the rms G–R noise current is reduced by a factor of $\sqrt{2}$ compared with conventional photoconductors.

For correct operation, a positive bias (relative to that on the active layer) must be placed on the blocking layer of an n-type BIB detector. If a negative bias were placed on the blocking layer, electrons from the contact would drift through the blocking and active layers to be collected on the opposite contact, producing a large dark current. Thus, the device is asymmetric electrically.

Under an appropriate level of positive bias, the detector operates as illustrated in Figure 4.6. Electrons in the impurity band under modest thermal excitation can hop toward the blocking layer, but their progress is stopped there unless they have sufficient energy to rise into the conduction band. When a photon is absorbed in the infrared-active layer, it raises an electron from the impurity band into the conduction band, where it is also attracted to the blocking layer. However, the conduction bands of the active and blocking layers are continuous, so the electron can pass through the blocking layer unimpeded to the contact. At the same time, the holes produced by absorption of a photon migrate to the opposite contact and are collected.

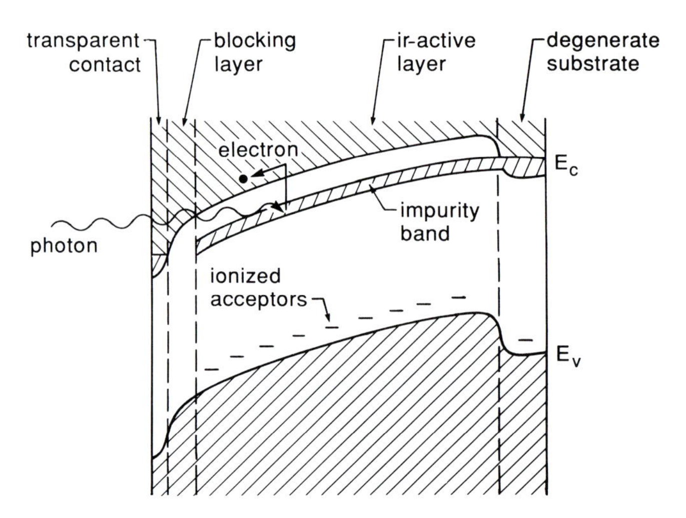

Figure 4.6. Band diagram for Si:As BIB detector.

For this detection process to be efficient, there must be an electric field in the infrared-active layer sufficient to drive the photon-produced charge carriers to the blocking layer. Because of the high impedance of the blocking layer, most of the electric field in the detector is developed across it. Any thermally produced charge carriers must be driven out of the active layer to raise its impedance sufficiently that a significant field also develops across it. The positive bias on the intrinsic layer contact drives free negative charge carriers out of the active layer by attracting them to the active/intrinsic layer interface, from where they can traverse the blocking layer and be collected at the electrode. Similarly, it drives any ionized donor sites toward the opposite contact. Consequently, the portion of the infrared-active region near the interface with the intrinsic layer is *depleted* of charge carriers. Because of the field across this high impedance depletion region, any photogenerated charge carriers within it will drift rapidly to the electrodes. Beyond the depletion region toward the negative electrode, the field is greatly reduced and the collection of free charge carriers is inefficient; consequently, absorbed photons produce little response. Since to first order only the depleted portion of the infrared-active layer is effective in detection, the quantum efficiency depends on the width of the depletion region, which in turn depends on the bias voltage and the density of negative space charge associated with the acceptors.

The width of the depletion region can be determined by calculating where in the detector the electric field becomes zero. To do so, it is necessary to solve Poisson's equation. The p-type impurities in the absorbing layer will be compensated by the arsenic, leaving a concentration of ionized acceptor atoms. These atoms will contribute a negative space charge; thus, Poisson's equation in the infrared-active layer is

$$\frac{d\mathscr{E}_x}{dx} = \frac{\rho}{\kappa_0 \varepsilon_0} = -\frac{qN_A}{\kappa_0 \varepsilon_0}, \tag{4.7}$$

where κ_0 is the dielectric constant and N_A is the density of ionized acceptors. We let x run from the blocking layer/infrared-active layer interface into the detector, assume a blocking layer devoid of impurities (see Figure 4.3), and recall that $dV/dx = -\mathscr{E}_x$. The solution for the thickness of the depletion region is

$$w = \left[\frac{2\kappa_0 \varepsilon_0}{qN_A}|V_b| + t_B^2\right]^{1/2} - t_B, \tag{4.8}$$

where t_B is the thickness of the blocking layer and V_b is the bias voltage. To avoid the problems in reduced response from overcompensated material, it is desirable to keep N_A of the order of 10^{13} cm^{-3} (or less). Adopting this value, $V_b = 4$ V, and $t_B = 4\ \mu$m, we derive $w = 19.2\ \mu$m.

The quantum efficiency is obtained by substituting w for d_1 into an expression like equation (2.3) (or equation (2.5) if reflective losses are to be taken into account). A typical donor concentration is $N_D = 5 \times 10^{17}$ cm^{-3}; comparing with the absorption cross-section in Table 4.1, it can be seen that $\sigma_i N_D w = 2.12$, and the absorptive quantum efficiency will be about 88% for a single pass, or as high as 98% for a double pass in a detector with a reflective back surface. If the thickness of the properly doped infrared-active layer is t_{IR}, the quantum efficiency will increase with increasing V_b until $w \geqslant t_{IR}$. The critical bias level at which $w = t_{IR}$ is termed V_{bC}, which can be shown from equation (4.8) to be

$$V_{bC} = \frac{qN_A}{2\kappa_0 \varepsilon_0}(t_{IR}^2 + 2t_{IR}t_B). \tag{4.9}$$

If w is the width of the depletion region, the electric field strength in this region is

$$\mathscr{E}_x = \frac{qN_A}{\kappa_0 \varepsilon_0}(w - x). \tag{4.10}$$

For our typical values ($N_A \sim 10^{13}$ cm^{-3}, $w = 19.2\ \mu$m), the field strength near the blocking layer is large; for example, for $x = 1.2\ \mu$m, $w - x = 18\ \mu$m and the field is $\sim 2.8 \times 10^3$ V cm^{-1}, far larger than for the bulk photoconductors we have discussed previously. The field increases linearly across the depletion region, reaching a maximum at the interface with the blocking layer (see Figure 4.3).

The mean free path of the electrons in this large field region is $\sim 0.2\,\mu m$. An electron that is accelerated across this path by a field of $2.8 \times 10^3\ V\,cm^{-1}$ will acquire an energy of 0.056 eV. The excitation energy for Si:As is 0.054 eV (corresponding to a cutoff wavelength of 23 μm); therefore, 0.054 eV of energy will be adequate to ionize the neutral arsenic impurity atoms. When photoelectrons in the region of large field impact-ionize neutral arsenic atoms, they excite additional electrons from the impurity band into the conduction band. As the field draws them toward the blocking layer, these secondary electrons may also acquire enough energy to ionize additional arsenic atoms. The process can lead to a single photon producing a cascade of electrons entering the blocking layer. Unless the bias voltage is very large, the multiplication stops in the blocking layer because impact-ionization in intrinsic material requires electron energies equal to the intrinsic bandgap, that is, about 20 times greater than those required in the doped layer. As a result, the charge multiplication is confined and controlled by the setting of the bias voltage and stable photoconductive gains >1 are produced.

The amount of gain increases rapidly with increasing bias voltage up to breakdown. This behavior arises from the strong dependence of the impact-ionization coefficient for the arsenic atoms, α, on the electric field strength:

$$\alpha = \sigma N_D e^{-E_c/E}, \tag{4.11}$$

where the cross-section $\sigma = 1.6 \times 10^{-13}\ cm^2$, N_D is the arsenic concentration, and the critical electric field E_c is found to be $\sim 7000\ V\,cm^{-1}$ from fits to experimental data (Szmulowicz and Madarsz, 1987). If the width of the gain region is u, the multiplication factor assuming a uniform field in the gain region is

$$M = e^{\alpha u}. \tag{4.12}$$

The field is not uniform, as shown in equation (4.10), so a rigorous solution requires that M be calculated with allowance for the dependence of field on x. It is frequently convenient to do such calculations by numerical simulation (La Violette and Stapelbroek, 1989). Continuing with our example, however, by taking the field to be uniform and evaluating for the position $w - x = 18\ \mu m$, we find $M \sim 2.2$.

Within the detector, the increase of the field near the infrared-active layer/blocking layer interface means that the amplification of the signal by this process occurs preferentially near the blocking layer, leading to designation of this portion of the infrared-active layer as the gain region (although there is no distinct boundary in the detector material; the dimensions of the gain region are set by the bias voltage). The gain process is relatively noisy because of the statistics of the interactions that determine the gain. Additional variations in gain can occur because the gain region is not sharply defined. Let the factor β be the amount by which the noise is increased relative to that for a noiseless gain mechanism. That is, β is the gain dispersion in the device, $\beta = \langle G^2 \rangle / \langle G \rangle^2$. Then the detective quantum efficiency is degraded by this factor:

$$\mathrm{DQE} = \frac{\eta}{\beta}. \tag{4.13}$$

Because of the lack of recombination noise, the photon-limited noise current is reduced by a factor of $\sqrt{2}$ compared with equation (3.36) and is termed shot noise:

$$\langle I_S^2 \rangle = 2q^2 \varphi \frac{\eta}{\beta} (\beta G)^2 \, \mathrm{d}f. \tag{4.14}$$

The responsivity remains as in equation (3.13).

In practical applications, it is frequently advantageous to operate BIB detectors with gains of $\sim$5–10 to overcome amplifier noise (see, for example, Herter *et al.*, 1989). At these gains, one expects $\beta \leqslant 2$ (Szmulowicz and Madarsz, 1987); hence, the penalty in detective quantum efficiency is modest.

4.3.3 Solid state photomultiplier

The above discussion of the separation of the infrared-active region of a BIB detector into an absorbing region and a gain region, and of the influence on the detector performance of gain dispersion arising in part from the lack of a sharp boundary between these two regions, suggests the further step of growing the detectors with a distinct and optimized gain region. The result of this concept is the solid state photomultiplier (SSPM) (Petroff, Stapelbroek, and Kleinhans, 1987). The performance of the SSPM is described by Hays *et al.* (1989) as follows. The spectral response is similar to that of a conventional Si:As BIB detector. However, a single detected photon produces an output pulse containing $\sim$40 000 electrons, which is easily distinguished from the electrical noise. These pulses have intrinsic widths of a few nanoseconds, although they are typically broadened by the output electronics to about a microsecond. Under optimum operating conditions, the dark pulse rate is $\leqslant 1000\,\mathrm{s}^{-1}$; lower dark pulse rates can be obtained at reduced operating temperatures, at the expense of reduced quantum efficiency.

To illustrate the application of SSPMs, consider an operating wavelength of 25 μm and assume the SSPM has a dark count rate of $1000\,s^{-1}$ and a quantum efficiency of 0.3. The dark count in 0.5 seconds is then 500 and the corresponding noise is $\sqrt{(500)} = 22.4$ counts. The input photon signal required to produce an output of 22.4 counts is 1.2×10^{-18} W, or using the relation between integration time and bandwidth in equation (3.39), the NEP $= 1.2 \times 10^{-18}\,W\,Hz^{-1/2}$. This level of performance can be achieved with time resolutions of $\leqslant 1 \times 10^{-6}$ s. Compare these performance characteristics with the silicon photoconductor discussed in the example in Chapter 3; NEP $= 1.05 \times 10^{-13}\,W\,Hz^{-1/2}$ with $\tau_{RC} = 3.5 \times 10^{-7}$ s. Following the dependencies in equation (3.62), if this detector were cooled to ~ 7 K (similar to the operating temperature of the SSPM) and its resistance were increased to reduce its Johnson-noise-limited NEP to $1.2 \times 10^{-18}\,W\,Hz^{-1/2}$, then its RC time constant would be ~ 60 s. Similar results would be obtained with any conventional silicon photoconductor. Thus, the SSPM provides time resolution some 10^8 times better than can be achieved with conventional photoconductors with similar NEPs.

Successful operation as an SSPM requires optimization of a number of parameters, including not only the construction of the detector but also its bias voltage and operating temperature. The internal structure of the detector is similar to a Si:As BIB except that a well defined gain region is grown between the blocking layer and infrared-absorbing layer with an acceptor concentration of $0.5 - 1 \times 10^{14}\,cm^{-3}$ and a thickness of about 4 μm. Typically, the infrared-absorbing region has a lower acceptor concentration. When the detector is properly biased, the gain region goes into depletion and a strong electric field is developed across it as with the BIB detector. From equation (4.10), it can be seen that increasing N_A in the gain region allows the generation of a larger field than with a normal BIB detector and therefore increases the amount of avalanching. At a carefully optimized operating temperature, a controlled dark current is produced by field-assisted thermal ionization (the Poole–Frenkel effect) where the field is near maximum in the gain region. The origin of this current can be understood by noting that the strong bending of the energy level bands in regions of strong electric field will expedite a thermally excited charge carrier traversing from the impurity to the conduction band. An equivalent current must flow through the infrared-active layer, where by Ohm's law it generates an electric field that pushes any photo-excited electrons into the gain region. Once there, these electrons avalanche to produce the output pulse of the device (this process is discussed by La Violette and Stapelbroek, 1989). The dark current necessary for correct biasing of the infrared-absorbing region does not contribute to the detector noise (as it would in a conventional BIB detector with modest gain) because it does not

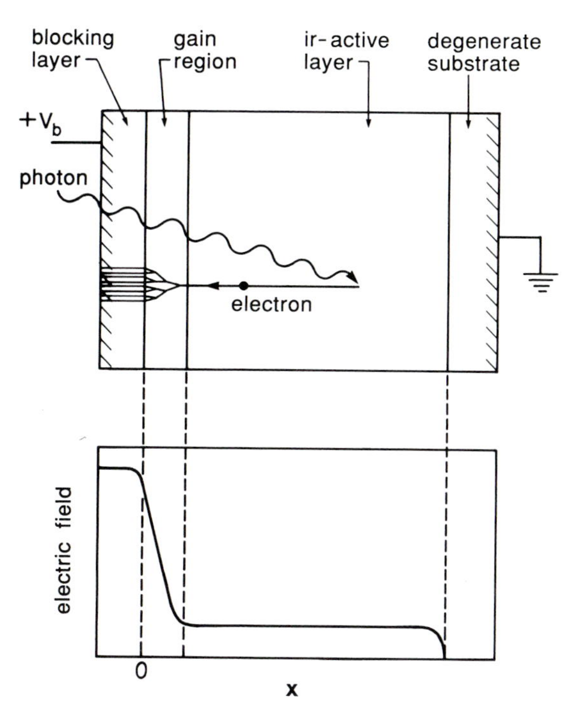

Figure 4.7. Cross-section of a solid state photomultiplier.

produce pulses that could be confused with photon signals. Figure 4.7 illustrates the construction and operation of the SSPM.

4.4 Problems

4.1 Design an arsenic-doped silicon photoconductor to operate near 20 μm at a temperature of 10 K. Assume transparent contacts, a sensitive area 1 mm^2, an arsenic concentration of 10^{16} cm^{-3}, a mobility of 6×10^4 cm^2 V^{-1} s^{-1} below 40 K, a recombination time of 3×10^{-9} s, and a maximum possible electric field before breakdown of 200 V cm^{-1}. Compute the following: (a) the detector geometry that gives a thickness of one absorption length (quantum efficiency of 52% including reflection effects), and (assuming this geometry) (b) the photoconductive gain, (c) the responsivity, (d) the detector resistance (assume $\delta = 1$), (e) the low frequency NEP with no signal, and (f) the time response.

4.2 Comment on the expected performance of a Ge:Sb BIB detector. Assume the blocking layer is 4 μm thick, that the infrared-active layer is one absorption length thick, that the detector operates stably at a maximum bias of 30 mV, that the maximum permissible antimony concentration is $2 \times 10^{16}\ \mathrm{cm}^{-3}$, and that the mean free path for conduction electrons is similar to that for silicon. Determine an optimum acceptor density, estimate the detector quantum efficiency, and discuss the ability of the detector to provide gain.

4.3 Consider two photoconductors of equal volume and cutoff wavelength, one intrinsic and the other extrinsic (n-type). Assume the intrinsic detector has no impurities and that the minority impurity concentration in the extrinsic detector is zero. Compare dark currents as a function of temperature. Assume $\delta = 1$ for the extrinsic detector, $m_n^* = 1.1 m_e$ for both materials, and $N_D = 3 \times 10^{15}\ \mathrm{cm}^{-3}$ for the extrinsic detector. Discuss your result.

4.4 Assume the Si:As BIB detector used as an example in the text with $N_D = 5 \times 10^{17}\ \mathrm{cm}^{-3}$, a blocking layer $4\ \mu\mathrm{m}$ thick, and operating at a bias voltage of 4 V. Assume the detector back surface is nonreflective. Consider the four cases $N_A = 1 \times 10^{13}$, 3×10^{13}, 1×10^{14}, and $3 \times 10^{14}\ \mathrm{cm}^{-3}$. Evaluate the quantum efficiencies, gains, and relative responsivities achieved with these acceptor concentrations. Assume in all cases that the width of the gain region is $1.2\ \mu\mathrm{m}$.

Note

1 Strictly speaking, blocked impurity band and BIB are trademarks of Rockwell International, where this detector type was invented. Therefore, a generic name – impurity band conduction (IBC) – is sometimes used; in addition, other manufacturers sometimes apply their own tradenames.

5
Photodiodes

A photodiode is based on a junction between two oppositely doped zones in a sample of semiconductor. These adjacent zones create a region depleted of charge carriers, producing a high impedance. In silicon and germanium, this arrangement permits construction of detectors that operate at high sensitivity even at room temperature. In semiconductors whose bandgaps permit intrinsic operation in the 1–15 μm region, a junction is often necessary to achieve good performance at any temperature. Because these detectors operate through intrinsic rather than extrinsic absorption, they can achieve high quantum efficiency in small volumes. The spectral response of photodiodes is currently restricted to wavelengths shorter than about 15 μm because of the lack of high quality intrinsic semiconductors with extremely small bandgaps. Standard techniques of semiconductor device fabrication allow photodiodes to be constructed in arrays with many thousands of pixels. Photodiodes are usually the detectors of choice for 1–6 μm and are often useful not only at longer infrared wavelengths but also in the visible and near ultraviolet.

5.1 Basic operation

As an illustration of the usefulness of photodiodes, we first consider the problems that arise in constructing an intrinsic photoconductor from InSb for the 1–5 μm region. From Table 3.1, we get the material parameters $\tau \approx 10^{-7}$ s and $\mu \approx 10^5$ cm^2 V^{-1} s^{-1}. We compute the photoconductive gain to be $G \sim 10^{-2}\ V/\ell^2$ (from equation (3.12) – and with ℓ in centimeters). The breakdown voltage for InSb is low, so the only way to achieve a reasonably high photoconductive gain is to make ℓ small. From equation (3.1), the detector resistance is $\ell/\sigma wd$, where $\sigma = qn_0\mu$. Because of the large electron mobility in InSb ($\sim$100 times that in silicon), it is impossible to achieve a large resistance with a small value of ℓ (while keeping the other detector dimensions fixed). That is, for this material, it is impossible to achieve simultaneously a high photoconductive gain and a large resistance. If G is small, the signals are small, and the detector system is likely to be limited by amplifier noise. If the resistance is small, the system will be limited by Johnson noise (equation (3.42)).

This dilemma can be overcome by fabricating a diode in the InSb (Rieke, DeVaux, and Tuzzolino, 1959). Diodes are made by growing oppositely doped regions adjacent to each other in a single piece of material or by implanting impurity ions of opposite type to the dominant doping of the material with an ion accelerator. The n-type material has a surplus (and the p-type a deficiency) of electrons compared to what are needed for the crystal bonds. As a result, if thermal excitation is adequate to free them, electrons in the vicinity of the junction between the two types of doping diffuse from the n-type into the p-type material where they combine with holes, producing a space charge region with a net negative charge in the p-type material and a net positive charge in the n-type material they left behind. This process is illustrated in Figure 5.1. The region where the charge has diffused from the n-type to the p-type material has nearly all complete bonds and a depletion of potential charge carriers. The high resistance of this depletion region overcomes the dilemma we faced previously in trying to make a photoconductor of intrinsic InSb.

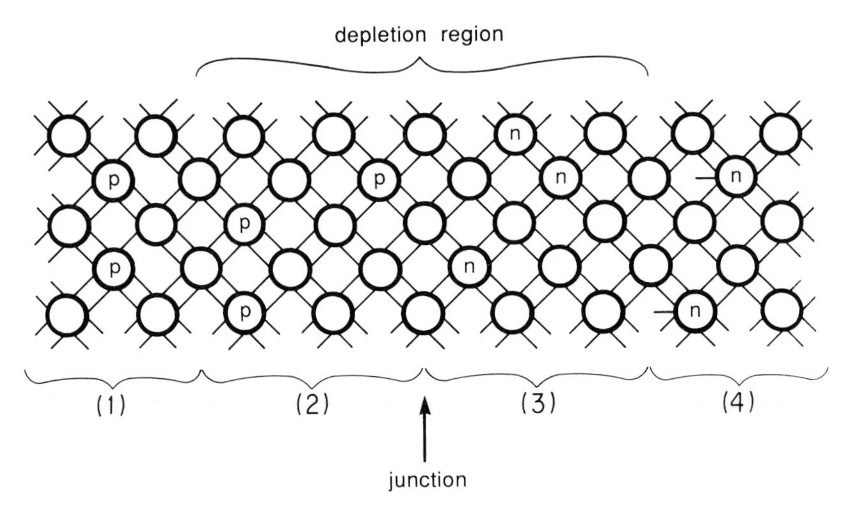

Figure 5.1. Microscopic view of a junction. Region (1) contains neutral p-type material; region (4) contains neutral n-type material; the electrons associated with the impurity n-type atoms in region (3) have diffused into region (2) where they have filled the holes due to the p-type atoms. Regions (2) and (3) have net space charges and are depleted of free charge carriers.

At extremely low temperatures, too few conduction electrons are freed to establish the depletion region. Both the diode junction and the conductivity in the semiconductor material are said to be 'frozen out'. This case will not be considered further.

The diffusion of charges is self-limiting because it results in a voltage being set up across the junction; this voltage opposes the diffusion of additional electrons into the p-type material. The voltage at which equilibrium has been reached is called the contact potential, V_0. The material on either side of the depletion region has a relatively small electrical resistance because of its doping; consequently, there is virtually no potential across it, and the potential and energy level diagrams for the diode are as shown in Figure 5.2.

The size of V_0 is determined by the Fermi levels on the two sides of the junction. Recall that $f(E_F) = 0.5$ (equation (3.47)); the Fermi level in the n-type material prior to contact is at a higher energy than in the p-type (see Section 4.2.3). Electrons will flow between the materials until their Fermi levels are the same. Although this behavior might be expected from the definition of the Fermi level, it can also be proven from the formalism we are about to develop (see Problem 5.4). The difference in the Fermi levels before contact is equal to qV_0, as is the difference in the conduction band levels after contact.

We will derive the electrical properties of the diode in detail later, but a

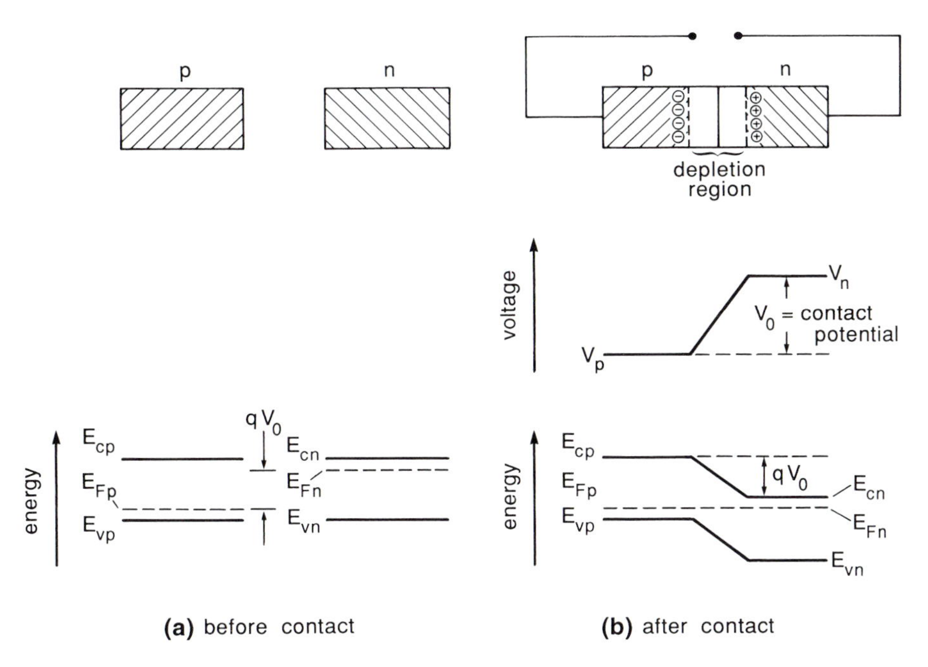

Figure 5.2. Development of a contact potential across a junction.

qualitative description is useful now. If an external bias is connected such that it adds to the contact potential (that is, the positive voltage is connected to the n-type material), we say that the diode is reverse-biased. Under this condition, the potential across the depletion region is increased by the external voltage, which increases the size of the depletion region and thus the resistance of the junction.

If the reverse bias is increased, eventually the junction will break down and become highly conducting. At modest reverse biases, breakdown can occur through tunneling; the reverse bias brings E_{cn} (the conduction band in the n-type material) below E_{vp} (the valence band in the p-type material; see Figure 5.3), making it energetically favorable for an electron to penetrate the depletion region without first moving into the conduction band of the p-type material, E_{cp}. If the depletion region is narrow enough, the electron wave function can extend across it. As a result, there is a finite probability that the electron will appear on the other side of the depletion region; if it does, it is said to have tunneled through the junction. At high reverse biases, breakdown occurs by avalanching. In this case, the strong electric field can accelerate an electron from the p-type region sufficiently strongly that the electron creates additional conduction electrons when it collides with atoms in the depletion region. This cascade of conduction electrons carries a large current.

When the junction is forward-biased, the sign of the applied potential is reversed so it decreases the bias across the depletion region. If the bias voltage

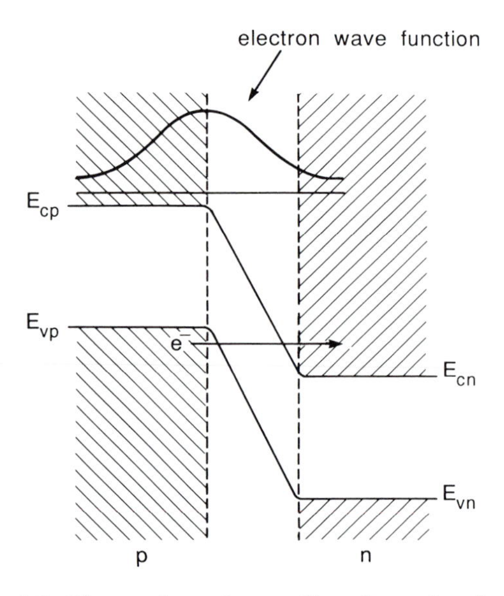

Figure 5.3. Illustration of tunneling through a junction.

is larger than V_0, the junction becomes strongly conducting. The overall behavior of the diode is summarized by the I–V curve (Figure 5.4).

If free charge carriers are generated and recombine in either the n-type or p-type regions they produce little net modulation of the current through the device because the relatively low resistances of these regions allow equilibrium to be re-established quickly. However, charge carriers produced in or near a reverse-biased or unbiased junction can be driven across it by the junction field and then can recombine on the other side, thus carrying a net current. Charge carriers can be produced thermally or by photoexcitation; as always, we assume that the detector is cold enough that we can ignore the former. Photoexcitation in the p-type material is illustrated in Figure 5.5; a photon is absorbed and excites an electron/hole pair. The hole is eventually collected at the negative electrode or recombines. The electron diffuses through the material; if it enters the depletion region, the junction field drives it across, creating a photocurrent. The same process occurs if the n-type material is illuminated, except the roles of the electron and hole are reversed.

As long as a photodiode is designed to allow efficient diffusion of the photoexcited charge carriers into the junction, virtually every absorbed photon will contribute to the photocurrent. Thus,

$$I_{ph} = -\varphi q \eta, \tag{5.1}$$

where I_{ph} is the photocurrent, η is the quantum efficiency, φ is the incident photon rate (s^{-1}), and the negative sign is consistent with the coordinate system used later in this chapter. Compared with the expression in equation (3.16), equation (5.1) can be interpreted as stating that the photoconductive

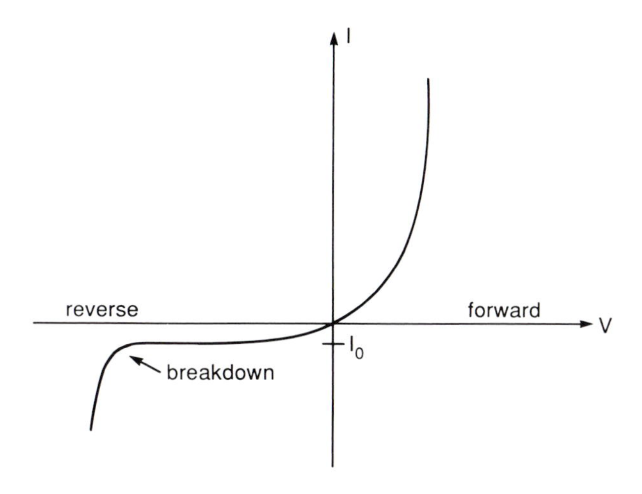

Figure 5.4. Current–voltage curve for a diode. I_0 is the saturation current.

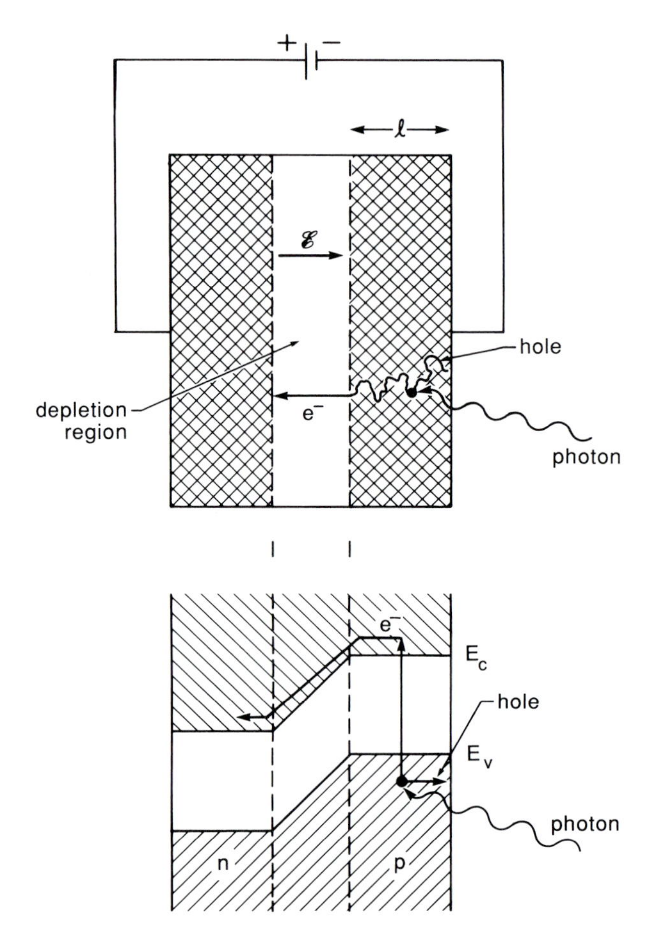

Figure 5.5. Illustration of the detection process in a photodiode. The photogenerated electron diffuses into the depletion region, where the field sweeps it across to the n-type region.

gain of photodiodes is $G = 1$. The power in the incident photon beam is given by equation (3.11). Thus, the responsivity

$$S = \frac{I_{ph}}{P_{ph}} = \frac{\eta \lambda q}{hc}, \tag{5.2}$$

so long as $\lambda \leqslant hc/E_g$. This response has a wavelength dependence similar to that for photoconductors (for example, see Figure 3.4). Noise currents are as described in Section 3.2.5, except the G–R noise is reduced by a factor of $\sqrt{2}$ because recombination of the charge carriers occurs in the heavily doped regions of the detectors, not in the depletion region. As discussed in the preceding chapter, in this case the noise is called shot noise; for a diode in the photon noise limited regime it is

$$\langle I_S^2 \rangle = 2q^2 \varphi \eta \, df \tag{5.3}$$

(see equation (3.40)), and the NEP is therefore reduced by a factor of $\sqrt{2}$ from that in equation (3.61),

$$\mathrm{NEP}(\mathrm{W\,Hz^{-1/2}}) = \frac{hc}{\lambda}\left(\frac{2\varphi}{\eta}\right)^{1/2}. \tag{5.4}$$

The NEP in the Johnson-noise-limited regime is given by equation (3.62) with $G = 1$.

For reasons which will be explained soon, photodiodes, though constructed with extrinsic material, work only through intrinsic absorption; therefore, the possible spectral responses correspond to the intrinsic bandgaps of the relevant semiconductors. Typical photodiode materials with cutoff wavelengths at room temperature are Si ($\lambda_c = 1.1\ \mu$m), Ge ($\lambda_c = 1.8\ \mu$m), InAs ($\lambda_c = 3.4\ \mu$m), InSb ($\lambda_c = 6.8\ \mu$m), and HgCdTe. The variable bandgap of HgCdTe can only be exploited out to a cutoff wavelength of about 15 μm, due to the fabrication difficulties mentioned in the introduction to Chapter 4. The bandgaps and hence cutoff wavelengths of these materials change slightly with operating temperature as discussed in Section 3.2.1; the cutoffs of InAs and InSb also become 5–10% shorter at 77 K than at room temperature. Additional materials used in high speed receiver diodes in fiber optics communications include GaInAs ($\lambda_c = 1.65\ \mu$m) and AlGaAsSb (with λ_c adjustable from 0.75 to 1.7 μm by changing the relative amounts of Al and Ga).

Additional discussion of the operation of photodiodes can be found in Reine, Sood, and Tredwell (1981) and Rogalski and Piotrowski (1988).

5.2 Quantitative description

From the above description, there are a number of attributes of interest in photodiodes. Once charge carriers are produced in the diode material, they must make their way to the junction by *diffusion*, a process that is also important in understanding the electrical behavior of the diode. Absorption and diffusion combine to control the *quantum efficiency*. The electrical properties of the junction will be described in terms of two important performance aspects: the *impedance* of the junction and the *photoresponse* of the detector. The frequency response of the detector will be controlled by its *capacitance*. These characteristics and their implications for detector performance are discussed in turn in this section.

5.2.1 *Diffusion*

Diffusion refers to the tendency of thermal motions to spread a population of particles uniformly over the accessible volume in the absence of confining forces. The operation of a photodiode depends on diffusion of charge carriers into its junction region. After the charge carriers have been created, perhaps near the surface of the diode and away from the junction, diffusion spreads them through the material; those that reach the junction are swept across it by the junction field so that a gradient is maintained against which further diffusion occurs. An efficient photodiode is designed so that virtually all the charge carriers diffuse into the junction.

To understand this process, we need to describe it quantitatively. Refer to Figure 5.6 for the following discussion. Consider a column of semiconductor containing charge carriers at density $n(x)$. To be somewhat more specific, let the carriers be electrons, and let their density decrease with increasing x. (The same general arguments are valid for hole diffusion.) Suppose there is an electric field in the material, $\mathscr{E}_x$, in the direction of increasing x. The electrons thus experience an electrostatic force (per unit volume)

$$F_e = -q\mathscr{E}_x n(x), \tag{5.5}$$

which drives them in the direction of decreasing x; that is, it will tend to increase the already existing density concentration in this direction. However, the random thermal motions of the electrons will tend to spread them out against the effect of the electric field; for example, if the field were removed, we would expect the electrons on average to be spread uniformly through

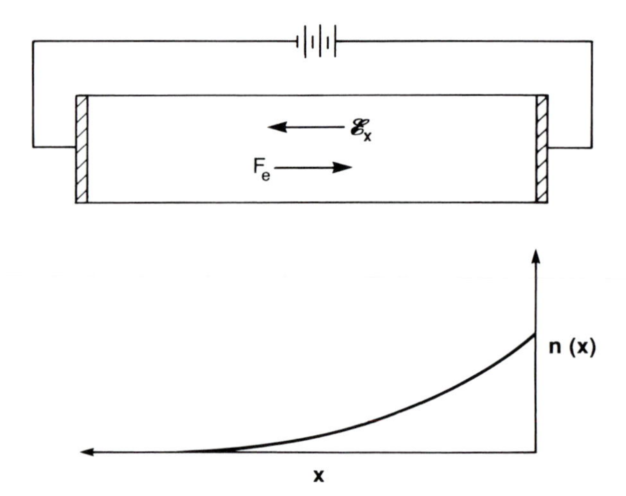

Figure 5.6. Distribution of electrons under joint influence of electric field and diffusion.

the entire accessible volume. The osmotic pressure that drives this diffusion is

$$P_0(x)=\left(\frac{N}{V}\right)kT=n(x)kT. \tag{5.6}$$

The corresponding osmotic force (again per unit volume) is

$$F_0(x)=-\frac{\mathrm{d}P_0(x)}{\mathrm{d}x}=-kT\frac{\mathrm{d}n(x)}{\mathrm{d}x}. \tag{5.7}$$

At equilibrium, $F_0(x)+F_E(x)=0$, or

$$\frac{\mathrm{d}n(x)}{\mathrm{d}x}+\frac{q\mathscr{E}_x n(x)}{kT}=0. \tag{5.8}$$

The solution of equation (5.8) is

$$n(x)=n(0)\,e^{-q\mathscr{E}_x x/kT}. \tag{5.9}$$

The diffusion coefficient, D (conventionally given in units of $\mathrm{cm^2\,s^{-1}}$; $D\,(\mathrm{m^2\,s^{-1}})=10^{-4}\,D\,(\mathrm{cm^2\,s^{-1}})$), is defined to give the osmotic flux of particles (number $\mathrm{cm^{-2}\,s^{-1}}$) as

$$S_0=-D_n\frac{\mathrm{d}n(x)}{\mathrm{d}x}. \tag{5.10}$$

Using equation (3.6), the flux of particles driven by the electrostatic field is

$$S_E=\langle v_x\rangle n(x)=-\mu_n\mathscr{E}_x n(x). \tag{5.11}$$

Under equilibrium, the net flux is zero, that is, $S_0+S_E=0$, or

$$\frac{\mathrm{d}n(x)}{\mathrm{d}x}+\frac{\mu_n\mathscr{E}_x n(x)}{D_n}=0. \tag{5.12}$$

Comparing equations (5.8) and (5.12), we see that

$$\frac{D_n}{\mu_n}=\frac{kT}{q}, \tag{5.13}$$

connecting the diffusion coefficient with the mobility. This equation is known as the Einstein relation. A similar equation for holes relates D_p to μ_p.

Now consider a volume of semiconductor material of length $\mathrm{d}x$ with a flux f_1 of electrons entering at x_1 and a flux f_2 leaving at x_2. The rate of change in the number of conduction electrons in the volume is

$$\frac{f_2-f_1}{\mathrm{d}x}=D_n\frac{\mathrm{d}^2n(x)}{\mathrm{d}x^2}. \tag{5.14}$$

If the electrons have a lifetime τ_n before recombination, they will disappear at a rate $n(x)/\tau_n$. They can also be created by thermal- or photo-excitation, which we will represent by the source term g. Then, by continuity, we have

$$\frac{dn}{dt} = D_n \frac{d^2 n}{dx^2} - \frac{n}{\tau_n} + g. \tag{5.15}$$

At equilibrium, $dn/dt = 0$, and for the moment we set $g = 0$, so equation (5.15) becomes

$$\frac{d^2 n}{dx^2} = \frac{n}{D_n \tau_n}. \tag{5.16}$$

The solution of equation (5.16) with the boundary condition that recombination should reduce $n(x)$ to zero for very large x is

$$n(x) = n(0)\, e^{-x/L_n}, \tag{5.17}$$

where we have defined the diffusion length to be

$$L_n = (D_n \tau_n)^{1/2}. \tag{5.18}$$

The diffusion coefficient and diffusion length play important roles in both the optical and electrical properties of photodiodes, as will be discussed in the following two sections.

5.2.2 *Quantum efficiency*

To be detected, the photoexcited charge carriers in a photodiode must reach the high resistance region of the detector. That is, they must diffuse into the depletion region out of the relatively low resistance p- or n-type neutral material on either side of the junction. From equation (5.17), this requirement suggests that the layer of neutral material overlying the junction should be no more than a diffusion length thick. The resulting relationship between L and photodiode quantum efficiency leads to important constraints on detector design. For example, following the discussion in Section 3.1, at low temperatures we take μ to be due to neutral impurity scattering and therefore to be independent of T; from the Einstein relation (5.13), D will go as T. Also from Section 3.1, we can take the recombination time, τ, to change approximately in proportion to $T^{1/2}$. The diffusion length then goes as $T^{3/4}$. Obtaining good quantum efficiency with diodes operated at low temperatures requires a compromise between thinning the overlying absorbing layer for good charge collection and making it thick enough for good absorption. Equation (5.18) also demonstrates why it is not feasible to extend the response of a photodiode through extrinsic absorption. The thickness of the overlying absorber must be $\ell \leqslant (D\tau)^{1/2} = C_1 N_I^{-1/2}$, where C_1 is a constant and N_I is the impurity concentration in the material (taking τ to go roughly as N_I^{-1}). At the same time, obtaining good absorption requires that $\ell \geqslant 1/a(\lambda)$, where the absorption coefficient $a(\lambda) = \sigma_i N_I$, or $\ell \geqslant C_2 N_I^{-1}$, where C_2 is a constant. It is generally not possible to meet these two conditions on ℓ simultaneously with extrinsic absorption.

Following Holloway (1986), the above arguments regarding the photodiode quantum efficiency can be made more quantitative by considering the diffusion of charge carriers into the depletion region according to equation (5.15), which can be rewritten (still assuming the steady state case, $dn/dt = 0$):

$$\frac{d^2 n}{dx^2} - \frac{n}{L^2} + \frac{g}{D} = 0. \tag{5.19}$$

We will use g to represent a uniform planar charge source and will adopt the approximation of an infinitely extended diode junction to avoid having to consider edge effects. If g is the number of photogenerated charge carriers per unit time and per unit volume (assuming that thermal excitation is reduced by adequate cooling), then equation (5.19) represents the diffusion of the signal charge carriers from their generation sites to the junction. We let x run from 0 at the junction to c at the diode surface. The general solution of equation (5.19) is

$$n(x) = A \cosh\left(\frac{x}{L}\right) + B \sinh\left(\frac{x}{L}\right) + \frac{gL^2}{D}. \tag{5.20}$$

Equation (5.20) can be simplified if we specify appropriate boundary conditions. We will assume the charge carriers are absorbed with 100% efficiency when they diffuse to the junction,

$$n = 0 \quad \text{at} \quad x = 0. \tag{5.21}$$

We also assume that the absorption by the detector material is very efficient so the photons are absorbed very near the surface of the diode. This assumption means that $g \sim 0$ everywhere but in this thin surface layer. (The situation for inefficient absorption is treated as a limiting case in Problem 5.2.) Applying condition (5.21), we also set $A = 0$. The flux of charge carriers at $x = c$ can be taken to be

$$S_{\text{in}} = \left(D \frac{dn}{dx}\right)_{(x=c)} = b\varphi, \tag{5.22}$$

where φ is the photon flux and b is the fraction of incident photons available for absorption to produce charge carriers (thus accounting for loss mechanisms such as reflection from the surface). The first part of the expression above is identical to equation (5.10) except for a sign change. This change arises because the direction of increasing x has been reversed in the present discussion to allow easier application of the boundary conditions. Equation (5.22) allows us to determine B in (5.20), yielding

$$n(x)=\frac{S_{\text{in}}L\sinh\left(\frac{x}{L}\right)}{D\cosh\left(\frac{c}{L}\right)}. \tag{5.23}$$

The quantum efficiency is the flux of charge carriers into the junction divided by the flux of input photons, or (from equations (5.10) and (5.23)),

$$\eta=\frac{D\left(\frac{\mathrm{d}n}{\mathrm{d}x}\right)_{(x=0)}}{\varphi}=b\,\mathrm{sech}\left(\frac{c}{L}\right)=\frac{2b}{e^{c/L}+e^{-c/L}}. \tag{5.24}$$

For example, when the material overlying the junction has a thickness of one diffusion length, $c=L$ and $\eta=0.65b$, but if $c=2L$, $\eta=0.27b$. Our intuitive guess at the beginning of this section that we should have no more than one absorption length of overlying material proves to be reasonable.

Recalling that L can scale as $T^{3/4}$, equation (5.24) indicates that the quantum efficiency of a photodiode will be low if it is operated at a temperature far below the range for which it was designed. Fortunately, it is found that many diodes operate with higher quantum efficiency at very low temperatures than would be implied by equation (5.24). In many practical cases, the quantum efficiency falls by a factor of two to five with decreasing temperature in accordance with (5.24) but then becomes roughly independent of temperature down to freezeout.

5.2.3 Current and impedance

The considerations described above define the photoresponse of a diode; we will next describe its electrical behavior. A fundamental description of this behavior is given by the diode equation, which describes the relationship between the voltage applied across the diode and the current that flows through it. The current comprises contributions from holes and electrons, but we will for simplicity consider only the holes at the moment.

Refer to Figure 5.7. We assume that the p-type material contains an adequate

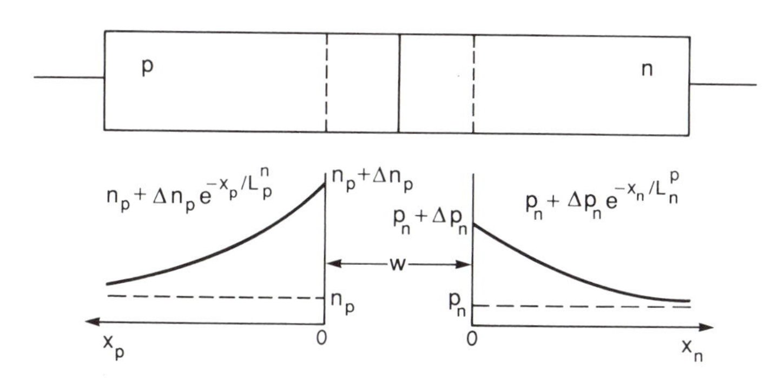

Figure 5.7. Carrier densities in a diode.

supply of conduction holes to supply any current of interest. (When the diode is strongly forward-biased this assumption may be invalid, but this condition is not of interest to us.) To carry a current, these holes must diffuse across the depletion region against the field set up by the contact potential; this situation was discussed in Section 5.2.1. In the present case, we can apply equation (5.9) for the positive charge carriers. We place the boundary of the p-type region at $x=0$ and denote the density of free holes at this point as p_p; the density of free holes at the edge of the n-type region is represented by p_n. The terms p_p and p_n are commonly referred to as majority and minority charge carrier concentrations, respectively. These terms emphasize the difference in location of otherwise identical charge carriers; majority charge carriers are in a region of semiconductor where the doping favors generation of similar carriers, and minority carriers are in semiconductor that is doped to favor generation of carriers of opposite type.

Applying equation (5.9), we find that the ratio of hole concentration at the boundary of the depletion region in the p-type material to that at the boundary in the n-type material is

$$\frac{p_p}{p_n}=e^{qV_0/kT}. \tag{5.25}$$

By the same argument, if a bias V_b is applied to the junction, we obtain

$$\frac{p_p^b}{p_n^b}=e^{q(V_0-V_b)/kT}, \tag{5.26}$$

where the b superscript denotes the carrier concentrations with the bias voltage present. To first order, the bias voltage has little effect on the majority carrier, so $p_p^b=p_p$. Then, using equation (5.25), we get

$$p_n^b=p_n\, e^{qV_b/kT}, \tag{5.27}$$

and

$$\Delta p_n(0) = p_n^b - p_n = p_n(e^{qV_b/kT} - 1), \tag{5.28}$$

where $\Delta p_n(0)$ is the change in the conduction hole concentration at the boundary of the depletion region that results from applying the bias voltage.

To generate a current, the charge represented by equation (5.28) must diffuse through the remaining n-type material in the diode and reach the contact. From equation (5.17) (and as illustrated in Figure 5.7),

$$\Delta p_n(x_n) = \Delta p_n(0)\, e^{-x_n/L_n^p}, \tag{5.29}$$

where L_n^p is the diffusion length for holes in the n-type material. From equation (5.10), the osmotic flux of holes in the n-type materialis $-D_n^p\, d(\Delta p_n)/dx$, which produces a current of

$$I_p(x_n) = -qAD_n^p \frac{d[\Delta p_n(x_n)]}{dx_n}, \tag{5.30}$$

where A is the junction area. Using equations (5.28), (5.29), and (5.30),

$$I_p(x_n) = \frac{qAD_n^p}{L_n^p} \Delta p_n(x_n),$$

and

$$I_p(0) = \frac{qAD_n^p}{L_n^p} p_n(e^{qV_b/kT} - 1). \tag{5.31}$$

Similarly for the electron component of the current

$$I_n(x_p = 0) = -\frac{qAD_p^n}{L_p^n} n_p(e^{qV_b/kT} - 1) \tag{5.32}$$

Remembering (Figure 5.7) that x_n and x_p run in opposite directions, the total current through the junction is $I = I_p - I_n$, or

$$I = \left[qA\left(\frac{D_n^p}{L_n^p} p_n + \frac{D_p^n}{L_p^n} n_p \right) \right](e^{qV_b/kT} - 1) = I_0(e^{qV_b/kT} - 1), \tag{5.33}$$

where I_0 (in brackets) is the saturation current. Equation (5.33) is called the diode equation and gives an approximate quantitative description of the behavior illustrated in Figure 5.4 (except for breakdown).

To make use of this expression, we need to know how to evaluate the various terms in I_0. The diffusion coefficients can be taken from equation (5.13) and the diffusion lengths from equation (5.18). The minority carrier concentrations, that is p_n and n_p, cannot be obtained directly from equations (3.52) and (3.54). They can be derived, however, by first observing that the product of the charge carrier concentrations, n_0p_0, can be written using equations (3.52) and(3.54) as

$$n_0p_0 = N_c e^{-E_g/2kT} N_v e^{-E_g/2kT} = n_ip_i = n_i^2, \tag{5.34}$$

where n_i and p_i are the conduction electron and hole concentrations in the intrinsic material, as indicated by the subscript i. The first equality in equation (5.34) is important because it shows that the dependencies on the Fermi energy in equations (3.52) and (3.54) drop out in the product n_0p_0; that is, this product is independent of any doping of the material that might shift E_F. Therefore, we can associate n_0p_0 with the product of majority and minority carrier densities, that is, n_np_n or n_pp_p. The last equality in equation (5.34) is true because conduction electrons and holes are created in pairs in intrinsic material, allowing us to set $n_i = p_i$. The carrier concentrations in intrinsic material are given by equations (3.52) and (3.54), again with $E_c - E_F = |E_v - E_F| = E_g/2$; they are therefore readily calculated.

Equation (5.34) is referred to as the carrier product equation and is a powerful tool in deriving carrier concentrations. It shows how the minority carrier concentration is suppressed due to the capture of majority carriers by the minority impurity atoms. In particular, the minority carrier concentrations in equation (5.33) can be derived from the majority carrier concentrations by substituting n_np_n or n_pp_p for n_0p_0, i.e.,

$$p_n = \frac{n_i^2}{n_n} \quad \text{and} \quad n_p = \frac{n_i^2}{p_p}. \tag{5.35}$$

Let the dopant concentration in the n-type material be N_D (cm^{-3}) and that in the p-type material N_A (cm^{-3}). Assume the diode is operated at a high enough temperature that all the impurity atoms are ionized. Hence, ignoring compensation by trace impurities, the concentration of majority charge carriers equals the concentration of dopants; for example, $n_n \sim N_D$. In this case we have, from equations (3.52), (5.34), and (5.35)

$$p_n = \frac{(N_c e^{-E_g/2kT})^2}{N_D}. \tag{5.36}$$

A similar expression can be derived for n_p:

$$n_p = \frac{(N_v e^{-E_g/2kT})^2}{N_A}. \tag{5.37}$$

It will also be useful to be able to estimate the contact potential, V_0. From equation (5.25),

$$V_0 = \left(\frac{kT}{q}\right) \ln\left(\frac{p_p}{p_n}\right), \tag{5.38}$$

and we have already seen how to obtain all the quantities that are needed to evaluate this expression.

As always, an overriding requirement for high sensitivity operation is to cool the detector sufficiently to stop thermally generated carriers from passing through it at a significant rate. With the photodiode, we have an advantage over photoconductors because majority carriers generated outside the depletion region have little effect on the junction current and hence on the effective resistance of the diode; note in equation (5.33) that I_0 depends only on the minority carrier concentrations p_n and n_p.

In principle, the diode resistance (V/I) can be increased by back-biasing the detector close to breakdown. This strategy frequently produces excess current-related noise, and the best overall performance for low light levels is usually obtained near zero bias. A figure of merit for the detector is then the zero bias resistance; it is obtained by rearranging equation (5.33) to solve for V_b and then taking the derivative with respect to I:

$$\left(\frac{dV_b}{dI}\right)_{(V_b=0)} = \frac{d}{dI}\left[\left(\frac{kT}{q}\right) \ln\left(1+\frac{I}{I_0}\right)\right]_{(I=0)} = \frac{kT}{qI_0} = R_0. \tag{5.39}$$

As seen in equation (5.33), I_0 depends on diffusion coefficients and lengths and on minority carrier concentrations. D and L go as modest powers of the temperature, as can be seen from equations (5.13), (5.18), and the temperature dependencies of the mobility and recombination time that were discussed in Chapter 3. The behavior of I_0 is dominated by the exponential temperature dependencies of p_n and n_p and (see equations (5.33), (5.36), and (5.37)). Thus R_0 for an ideal diode increases exponentially with cooling. For real diodes, this improvement occurs only to some limiting value of R_0. For example, tunneling of charge carriers across the junction has relatively little temperature dependence and is likely to place an upper limit on R_0 at very low temperatures.

In addition to the dependence on temperature already discussed, I_0 and hence R_0 depend on the level of doping in the diode. This dependence arises from the dependence of p_n and n_p on doping level, as shown in equations (5.36) and (5.37). In addition, a second dependence on doping level arises from the diffusion lengths in equation (5.33). Near room temperature, D is roughly independent of impurity concentration, but L is proportional to $N_I^{-1/2}$ through the dependence of recombination time on impurities. We can therefore show that

$$I_0 \approx C_1(N_D)^{-1/2} + C_2(N_A)^{-1/2}, \tag{5.40}$$

where C_1 and C_2 are constants. At low temperatures where μ is due to neutral impurity scattering, D goes inversely as a modest power of N_I, increasing the effect of impurity concentration in reducing I_0 compared with equation (5.40). Thus, R_0 increases at least as fast as the square root of the doping concentration in the diode. All other things being equal, better diode performance should be achieved with higher doping to increase R_0.

Our goal has been to achieve a first-order understanding of diode behavior, and it is now time to confess to a number of simplifying assumptions that were made along the way. We have taken the signal frequency to be low enough that the diode could be assumed to be in equilibrium, thus dropping out all time-dependent terms. In the operating regime of interest for high sensitivity detectors, we are justified in considering only diffusion currents, so we have not included terms involving charge carrier drift in the electrostatic field.

With far less justification, we have ignored the generation and recombination of charge carriers in the depletion region. For example, when the diode is moderately forward-biased, some electrons entering from the n-type region recombine with holes entering from the p-type region, which is equivalent to a transfer of a positive charge across the depletion region (because the hole density in the n-type region has been increased and that in the p-type region decreased). When the diode is moderately reverse-biased, the increased field across the depletion region stops this recombination mechanism, but electron/hole pairs can be generated in the depletion region and swept out of it by the junction field. In both of these cases, since there is an additional current in the diode, the equivalent saturation current is larger than the I_0 given in equation (5.33). In the reverse-biased case there may be no clearly defined saturation current; instead, as the width of the depletion region increases with back bias, so does the volume in which generation of charge carriers can occur and hence the current across the junction. In the forward-biased case, the depletion width decreases with increasing bias, tending to flatten out the inflection in the diode I–V curve as it switches into conduction. An approximate allowance for this latter behavior can be made by taking the I–V curve to be given by

$$I = I_s\left(e^{qV_b/mkT} - 1\right), \tag{5.41}$$

where I_s and m are fitted empirically to the measured curve. The parameter m is called the ideality factor.

5.2.4 Response

In the preceding discussion, we derived the electrical properties of an ideal diode in the absence of light. When light falls on this diode, any resulting charge carriers that penetrate to the depletion region will be driven across it and will produce a current. In other words, the photocurrent through the junction is as given in equation (5.1), where the quantum efficiency of the detector can be calculated as shown in equation (5.24) for the case of high absorption efficiency. The total current through an ideal photodiode is given by the sum of the expressions in equations (5.33) and (5.1):

$$I = -\frac{\eta q P}{h\nu} + I_0(e^{qV_b/kT} - 1), \tag{5.42}$$

where we have converted φ to $P/h\nu$. Figure 5.8 shows the I–V curve for a diode exposed to various levels of radiation.

Figure 5.8 also suggests a variety of photodiode operation modes. For example, the detector can be run by monitoring the output voltage at a fixed current. A simple method might be to use a high input impedance voltmeter so $I \sim 0$. As Figure 5.8 suggests and Problem 5.1 asks you to prove, the detector response may be significantly nonlinear when the device is operated in this way. In the photoconductive mode, the diode is placed in a circuit that holds the voltage across the detector constant (and negative to back-bias

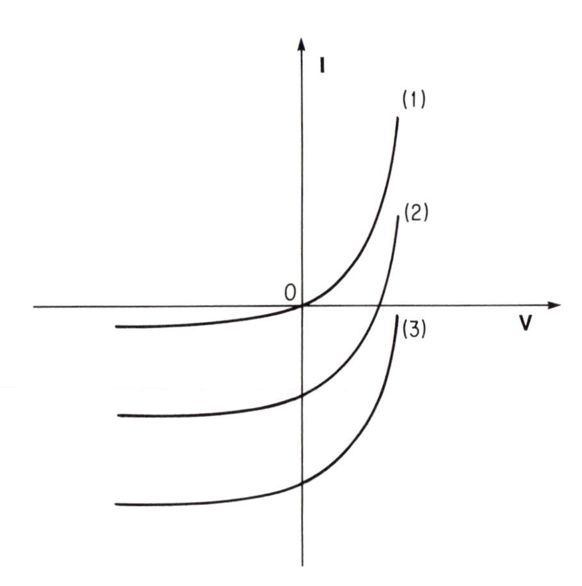

Figure 5.8. Response of a diode to illumination. The illumination increases for curves (1) to (3), starting from zero for curve (1).

it), and the current through the circuit is measured as an indication of the illumination level. As indicated by equation (5.42), the output current in this mode is linear with input power. A particularly attractive operating method is to hold the voltage across the detector at zero; certain types of excess voltage noise are then suppressed (Hall *et al.*, 1975). In Chapter 6, we will illustrate the ability of a simple circuit called a transimpedance amplifier to operate photodiodes in this way.

5.2.5 Capacitance

A photodiode has relatively high capacitance because the distribution of positive and negative charge across its junction forms a parallel plate capacitor with small separation between the plates. The capacitance of the photodiode can control its frequency response and often determines the limiting noise of the amplifier used to read out its signal. Here we show how to estimate the capacitance of the junction and what parameters control it.

Consider an 'abrupt junction' at $x=0$, with N_A the concentration of ionized acceptors for $x<0$ and N_D the concentration of ionized donors for $x>0$. Let l_p be the depth of the depletion region into the p-type region and l_n the depth into the n-type region. Applying Poisson's equation in one dimension, we have for the junction

$$\frac{d^2V}{dx^2}=\begin{cases} qN_A/\varepsilon, & -l_p<x<0 \\ -qN_D/\varepsilon, & 0<x<l_n \\ 0 & \text{otherwise.} \end{cases} \tag{5.43}$$

We adopt the boundary condition $dV/dx=0$ for $x\leqslant -l_p$ and $x\geqslant l_n$, and $V(l_n)-V(-l_p)=V_0-V_b$, where V_0 is the contact potential and V_b is the applied potential. The solution to equation (5.43) is

$$V(x)=\begin{cases} \dfrac{qN_A}{2\varepsilon}(x^2+2l_px), & -l_p<x<0 \\ \dfrac{-qN_D}{2\varepsilon}(x^2-2l_nx), & 0<x<l_n. \end{cases} \tag{5.44}$$

Since the electric field in the neutral regions of the diode material is zero, the space charge on either side of the junction must be equal and opposite:

$$N_Al_p=N_Dl_n. \tag{5.45}$$

If we express V_0-V_b from (5.44) (and the discussion immediately preceding it) and use equation (5.45), we can solve for lengths of the depletion region into the p-type and n-type material as:

$$l_p = \left[\frac{2\varepsilon N_D(V_0 - V_b)}{qN_A(N_A + N_D)}\right]^{1/2}$$
$$l_n = \left[\frac{2\varepsilon N_A(V_0 - V_b)}{qN_D(N_A + N_D)}\right]^{1/2}. \tag{5.46}$$

The width of the depletion region is

$$w = l_p + l_n = \left[\frac{2\varepsilon(N_A + N_D)(V_0 - V_b)}{qN_A N_D}\right]^{1/2}. \tag{5.47}$$

The junction capacitance is

$$C_J = \varepsilon\frac{A}{w} = \kappa_0\varepsilon_0\frac{A}{w}, \tag{5.48}$$

where the dielectric constant κ_0 is given in Table 3.1, ε_0 is the permittivity of free space, and A is the junction area. Note that C_J goes as the square root of the impurity concentration.

In Chapter 6, we will see that achieving low noise with practical readouts requires that the detector capacitance be kept small, which is achieved in a photodiode by minimizing the impurity concentrations. We already know that a detector with a large resistance is needed to minimize Johnson noise; from the preceding discussion, higher R_0 is achieved with larger impurity concentrations. Thus, the two requirements on the impurity concentrations are in conflict. In practical detectors, some compromise must be made.

The abrupt junction is only one example of a doping density profile that may apply to a diode. Other profiles yield differing relations between capacitance and impurity concentration and bias voltage. Measurements of the dependence of capacitance on voltage can be used to deduce the contact potential and the profile of the impurity concentration (see Problems 5.5 and 5.6).

5.3 Photodiode variations

5.3.1 PIN diode

As can be seen from equation (5.48), the frequency response of a photodiode can be improved by widening the depletion region. If the depletion region is wide enough that most of the photon absorption occurs there, the photoexcited charge carriers will drift under the influence of the junction field rather than having to diffuse into the junction. We have seen that the width of the depletion region can be increased by reducing the doping that creates the junction (see equation (5.47)), but with a concomitant undesirable decrease in the detector resistance. Often a better solution is to interpose a high resistivity intrinsic (or nearly intrinsic) layer between the p- and n-sides of the diode; hence, these detectors are called PIN diodes. The time response for a PIN diode is $w/|\langle v_x \rangle|$, where

$$|\langle v_x \rangle| = \mu \mathscr{E}_x = \frac{\mu(V_0 + V_b)}{w}, \tag{5.49}$$

or

$$\tau_{PIN} = \frac{w^2}{\mu(V_0 + V_b)}. \tag{5.50}$$

The intrinsic region must be thin enough that the charge carriers drift across the junction and are collected before they recombine; in other words, τ_{PIN} in equation (5.50) should be less than the recombination time τ_{rec}. For example, taking parameters for silicon from Table 3.1 and assuming $V_0 + V_b > 1$ V, we can allow $w > 1$ mm. Comparing this width to the absorption coefficient for silicon (Figure 3.3), the intrinsic layer can be thin enough for the charge carriers to drift across it and still be thick enough for high absorption efficiency nearly to the bandgap energy. The very high breakdown voltage of PIN diodes also allows them to be used with large biases (~ 100 V), in which case equation (5.50) shows they will have very fast time response.

5.3.2 Avalanche diode

If a photodiode is operated at a back-bias so large that it is just short of breakdown, photoexcited charge carriers can be sufficiently accelerated in the depletion region that they produce additional carriers by avalanching. The process is similar to that discussed toward the end of Section 4.3.2 with regard to blocked impurity band (BIB) detectors, with two exceptions. First, the field must be sufficiently large to excite charge carriers across the intrinsic bandgap (rather than just from an impurity level into the conduction band). Field strengths some 20 times greater are needed than for the impurity-based avalanching in the BIB, that is, fields $> 1 \times 10^5$ V cm^{-1}. Secondly, unlike the BIB, the avalanching involves both holes and electrons. As a result, the increase in noise, that is the β factor discussed for the BIBs, is larger for a given level of multiplication in the diode than in the BIB. Despite the increase in noise, the large current amplification can be a useful feature if the detector is to be used with a relatively high noise readout. Avalanche diodes are particularly useful where very high frequency response is required; time response of a few nanoseconds is attainable. At comparable response times, a conventional PIN diode system may be limited by noise from the output amplifier.

Where illumination levels are modest, the rapid response of an avalanche diode can even allow pulse counting of single photon events, so long as quench circuitry is used to suppress continuous breakdown and the detector is cooled sufficiently to control dark current. Pulse rise times of a few nanoseconds are achieved in this manner with a dead time of a few hundred nanoseconds after each detection to allow for quenching.

To provide rapid response, these diodes are designed as PIN devices with very thin front doped layers so as many photons as possible are absorbed in the depletion region, across which the resulting charge carriers are driven rapidly as in equation (5.50). The electric field in the depletion region can be derived as in equation (4.10) and Figure 4.3. The field exceeds the avalanche threshold in only a small region of the depletion region; the multiplication is therefore similar for nearly all charge carriers and can be controlled by appropriate adjustment of the bias voltage.

Both avalanche and PIN diodes are used where a simple, rugged, compact, and inexpensive detector is required. For fast time response at extremely low light levels and with minimal dark counts, photoemissive devices (see Chapter 8) are generally preferred if their higher complexity and fragility, larger size, and expense can be tolerated. However, continued improvements in photodiodes make these detectors optimum for an increasingly large variety of applications. For example, high performance avalanche diodes are now available with sensitive areas of $\sim 200\,\mathrm{mm}^2$, providing detector areas competitive with small photomultipliers. Avalanche photodiodes are discussed in depth by Stillman and Wolfe (1977), Capasso (1985), Kaneda (1985), and Pearsall and Pollack (1985).

5.3.3 Schottky diode

A junction between a semiconductor and a metal produces an asymmetric potential barrier that acts as a diode. These *Schottky diodes* are discussed both here as direct photon detectors and in Chapter 11 as key electrical elements in millimeter-wave receivers. Their operation can be understood from Figure 5.9, which shows a p-type silicon semiconductor joined to the metal PtSi. Electrons flow across the junction from the metal into the semiconductor until the Fermi level is level; at this point, the flow has been halted by the electric field caused by the additional negative charge in the semiconductor. As a result, an asymmetric potential barrier is set up for

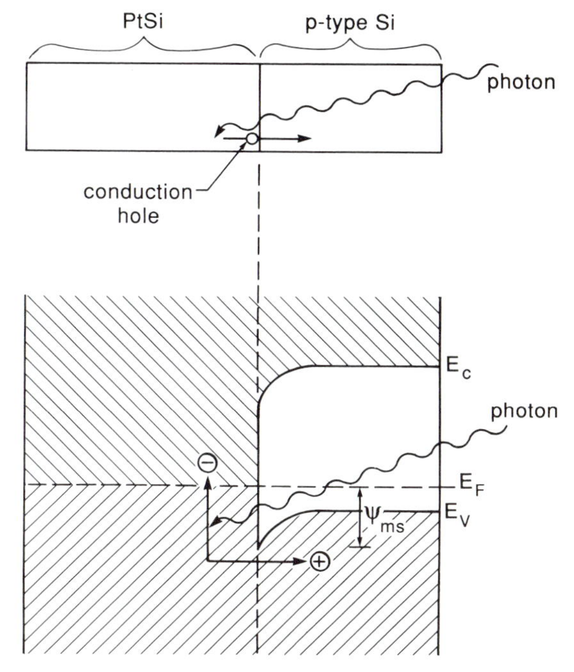

Figure 5.9. Energy band diagram to illustrate photoemission in a Schottky diode detector.

conduction holes reaching the junction; those in the valence band of the semiconductor must surmount a low potential barrier to reach the valence band of the metal, while those in the metal have a relatively higher barrier against entry into the valence band in the semiconductor.

Schottky barrier detectors can be constructed on silicon using a variety of metals, each of which has a characteristic barrier height, Ψ_{ms}, and corresponding cutoff wavelength for photoresponse. Ψ_{ms} is determined by the contact potential and can be significantly less than the bandgap in the semiconductor. Therefore, photoresponse is obtained to much longer wavelengths than those that correspond to the bandgap (in fact, the intrinsic absorption in silicon sets the short wavelength cutoff of the diode because photons absorbed in the silicon layer never reach the metal). Common examples of Schottky barrier detectors are Pd_2Si ($\Psi_{ms} = 0.35$ eV, $\lambda_c = 3.5\ \mu m$) and PtSi ($\Psi_{ms} = 0.22$ eV, $\lambda_c = 5.6\ \mu m$). To extend the performance of this type of detector to longer wavelengths, IrSi ($\Psi_{ms} = 0.15$ eV, $\lambda_c = 8\ \mu m$) has been explored.

We will base our discussion on platinum silicide (PtSi), which is a particularly popular version of this detector type. Schottky barriers of this material are made by evaporating a thin layer of platinum onto a piece of p-type silicon, followed by a heat treatment. Additional processing steps add a dielectric layer of SiO_2 and an aluminum reflector. The completed detector is shown in Figure 5.10. In this device, the metal films in which the photons must be absorbed tend to have relatively high reflectivity; in thick PtSi Schottky barrier detectors, only about 10% of the photons penetrate into the metal to be absorbed.

The operation of this detector is described by an energy diagram for *holes* as in Figure 5.11 (just the inverse of our usual energy level diagrams which take the viewpoint of the electrons). The large energy difference between the Fermi level and the conduction band in the dielectric, along with the behavior

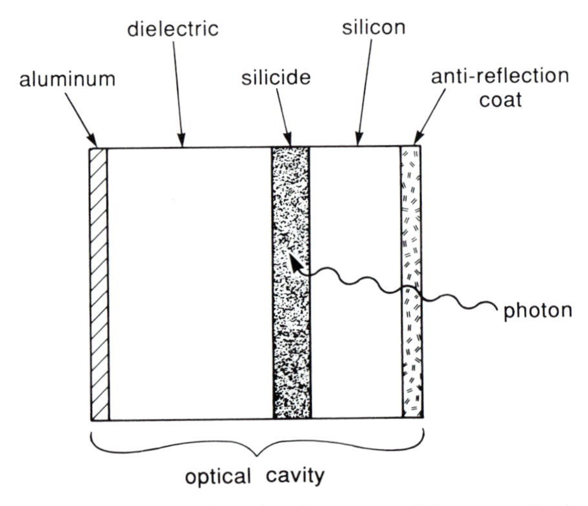

Figure 5.10. Design of a Schottky barrier detector with an optical cavity to enhance quantum efficiency.

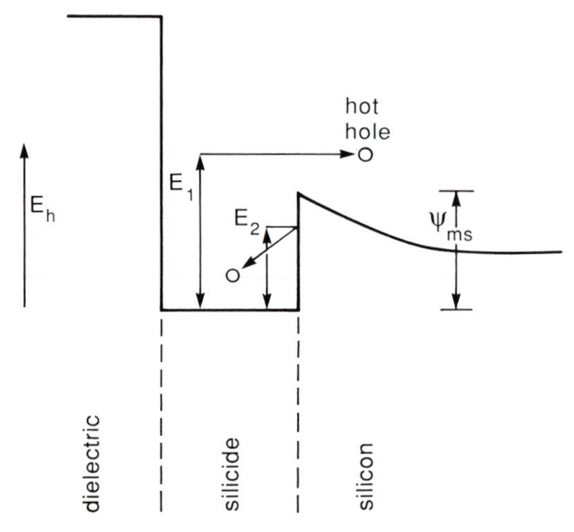

Figure 5.11. Illustration of the emission of a hot hole over the Schottky barrier.

of the semiconductor–metal interface, results in an asymmetric potential well for the holes. If a hot hole has energy greater than Ψ_{ms}, and if its momentum is in a direction that will carry it over the barrier, then it can escape into the semiconductor. If it does so, it causes a current to flow across the junction that can be sensed, thereby resulting in detection of the photon that created the hole. The hole in Figure 5.11 with energy E_1 is an example; by contrast, the hole with E_2 has insufficient energy to cross the barrier and re-thermalizes in the silicide material.

The efficiency of this process is governed by the solid angle over which the hot holes will reach the junction with sufficient energy to pass over the barrier. If the detector temperature is sufficiently low, we can assume that there is a sharp threshold energy of Ψ_{ms} for escape over the barrier (this assumption is called the zero temperature approximation). The escape probability of hot holes under this approximation can be computed assuming that the holes have isotropic velocities when they are created and that they will travel without loss of energy until they encounter the barrier. Noting that the momentum is proportional to $E^{1/2}$, where E is the energy, the solid angle defined by this condition is contained within θ in Figure 5.12. Under our assumptions, the escape probability, $P(E)$, is this solid angle divided by the solid angle of a sphere, 4π. Making use of equation (1.9),

$$P(E) = \frac{4\pi \sin^2\left(\frac{\theta}{2}\right)}{4\pi} = \frac{(1-\cos\theta)}{2} = \frac{1-\left(\frac{\Psi_{ms}}{E}\right)^{1/2}}{2}. \tag{5.51}$$

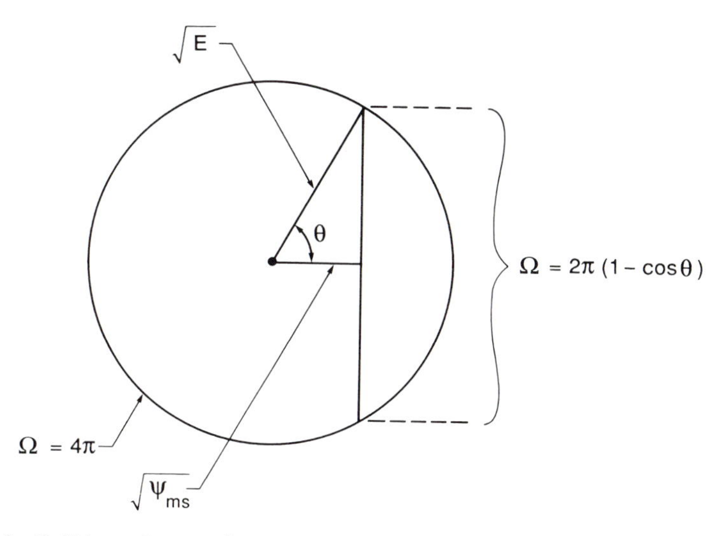

Figure 5.12. Solid angle requirements for emission of hot holes over the Schottky barrier.

For a given photon energy $h\nu$, there is a maximum angle θ_0 above which there will be no escape of the hole and therefore no detection. From equation (5.51),

$$\theta_0 = \cos^{-1}\left(\frac{\Psi_{ms}}{h\nu}\right)^{1/2}. \tag{5.52}$$

The density of states for holes is dN/dE; since the photon energies are well below the Fermi energy, we can take dN/dE to be constant over the energy range of interest. The total number of excited states accessible for holes upon absorption of a photon of energy $h\nu$ is then

$$N_T = \int_0^{h\nu} \frac{dN}{dE} dE = \frac{dN}{dE} h\nu. \tag{5.53}$$

The number of states which result in the hole escaping across the energy barrier at the junction is

$$N = \int_{\Psi_{ms}}^{h\nu} \frac{dN}{dE} P(E)\, dE = \frac{1}{2}\frac{dN}{dE} h\nu\left[1 - \left(\frac{\Psi_{ms}}{h\nu}\right)^{1/2}\right]^2. \tag{5.54}$$

Assuming that a photon is absorbed in the silicide, the probability that it produces a signal is then proportional to the number of states for which the hole it produces escapes over the barrier divided by the total number of states accessible to the hole, or

$$\eta_{int} = \frac{N}{N_T} = \frac{1}{2}\left[1 - \left(\frac{\Psi_{ms}}{h\nu}\right)^{1/2}\right]^2. \tag{5.55}$$

The quantum efficiency of the detector is the probability that a photon of frequency ν and incident on the detector will be absorbed in the silicide layer, $\eta_{ext}(\nu)$, times the probability that an absorbed photon will produce a hole that will escape over the barrier at the junction, η_{int}, that is,

$$\eta = \frac{\eta_{ext}(\nu)}{2}\left[1 - \left(\frac{\Psi_{ms}}{h\nu}\right)^{1/2}\right]^2. \tag{5.56}$$

As mentioned above, typically $\eta_{ext}(\nu) \leqslant 0.1$, so the quantum efficiency of these detectors is relatively low and drops monotonically as the wavelength approaches the long wavelength cutoff.

The quantum efficiency can be increased by three optimizations. First, the silicide layer can be made very thin, reducing the reflection losses and increasing the portion of the photons that penetrate this layer. Secondly, by placing an aluminum mirror behind the silicide layer, photons that pass through it altogether will be reflected and may be absorbed on a second pass. The space between the mirror and silicide layer can be tuned in thickness so that interference effects increase the absorption probability for photons of certain wavelengths (and decrease it for photons of other wavelengths). Thirdly, the dielectric layer behind the silicide acts as a reflector for hot holes, so a portion of those which are emitted away from the junction are diverted and pass over the potential barrier. Elabd and Kosonocky (1982) discuss these optimizations and other aspects of these detectors from both theoretical and experimental viewpoints and show that well-designed detectors can have quantum efficiencies at certain wavelengths that are about a factor of 20 greater than predicted by equation (5.56). These values remain generally lower than for well-optimized semiconductor junction diodes.

To be useful at very low light levels, the Schottky barrier detector must be cooled sufficiently that there is an insignificant number of thermally excited charge carriers with enough energy to flow over the diode potential barrier. Thermally excited currents would also arise if carriers could tunnel through the junction; as will be shown in Chapter 11, so long as the barrier width is $\gg 1 \times 10^{-9}$ m, tunneling currents will be small. This condition is relatively easy to meet.

These detectors have a number of advantages. They are relatively simple to construct, and the simple processing allows them to be made with a very high degree of uniformity. When made in arrays, the excellent uniformity reduces the extent of data processing necessary to obtain usable results. Since Schottky barrier detectors are built on silicon, it is possible to build readout transistors using traditional silicon processing on the same piece of material, again providing for simple and large format array construction.

5.4 Quantum well detectors

With the exception of Schottky junctions, the photodiodes we have discussed are made using the same semiconductor material throughout but with doping to modify its properties in different ways on opposite sides of a junction. Another type of junction can be based on joining different semiconductor materials; in this case, not only can we make use of doping of the materials to adjust the position of the bands on either side of the junction, but the bandgap itself can change across the junction. Such a junction is called a heterojunction.

To make a successful heterojunction, it is required that the crystal properties of the two materials match closely. Although there are other material pairs that can also be used, a particularly attractive set is GaAs and $Al_xGa_{1-x}As$.

The aluminum atoms can replace gallium atoms in GaAs with virtually no effect on the crystal lattice structure, but the bandgap increases with increasing amounts of aluminum. At a GaAs to $Al_xGa_{1-x}As$ junction, there is a discontinuity of the conduction band edge by approximately x eV and of the valence band edge by approximately $0.15x$ eV. Thin layers of these materials can be deposited in a highly controlled fashion using molecular beam epitaxy (see Chapter 4).

Consider Figure 5.13. It shows the edge of the conduction band for a structure with a thin layer of GaAs grown between two pieces of GaAlAs (we do not need to be specific about the relative amount of Al and Ga because the discussion will be qualitative). If conduction electrons are created in the GaAlAs, some of them will wander into the GaAs, will lose energy there, and will no longer have sufficient energy to re-enter the GaAlAs. The electrons are therefore trapped in a potential well. From quantum mechanics, we know that the electrons can only assume certain discrete energies when they are in such a well; hence the name quantum well. Energy levels are sketched in the quantum well in Figure 5.13. If we have a multilayered structure with thin layers of GaAlAs alternated with thin ones of GaAs (see Figure 5.14(a)), we obtain multiple quantum wells such as in Figure 5.14(b).

Multiple quantum well devices can be used for infrared detectors. If the walls of the wells (that is, the layers of GaAlAs) are made sufficiently thin, then electrons can pass from one well to another by quantum mechanical tunneling through the wall between them. We will show in a different context (Chapter 11) that efficient tunneling occurs through potential barriers of the order of 10^{-9} m in thickness. In the present case, a bias is placed across the device as in Figure 5.14(b); tunneling then becomes highly favored if the energy

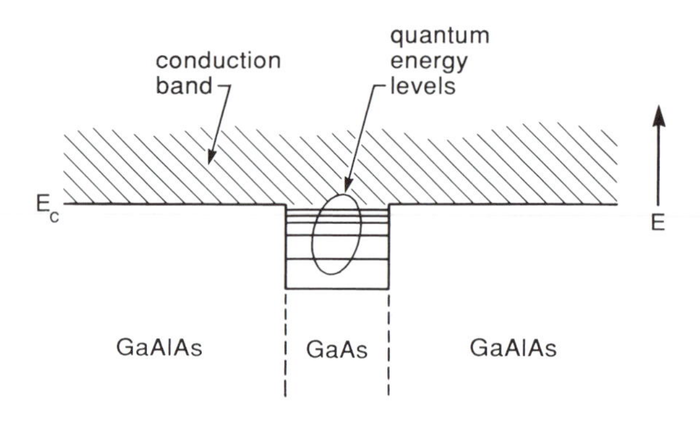

Figure 5.13. The formation of a quantum well in a layer of GaAs between layers of GaAlAs.

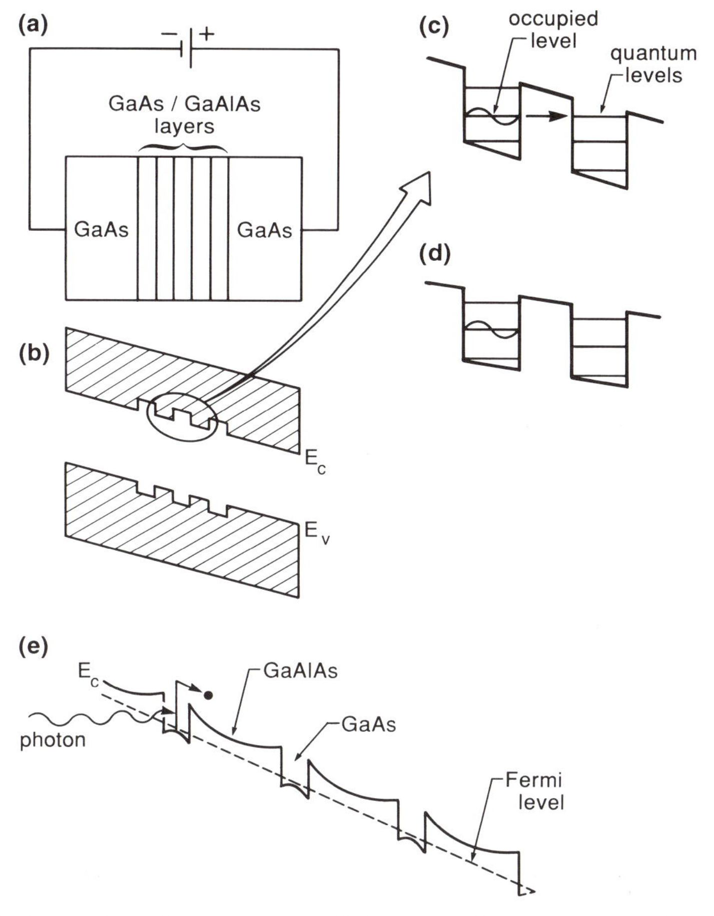

Figure 5.14. Operation of a quantum well detector. The structure of the detector and the bandgap diagram are shown schematically in (a) and (b). The transition of electrons between adjacent wells is shown in (c) and (d). Panel (e) shows the use of doping to lift the Fermi level into the bottoms of the wells to provide a high density of charge carriers for photon absorption.

state of the electron in its well is aligned with an open energy state in the adjacent well. For example, an absorbed photon can excite an electron to an upper energy level within a well and can cause it to tunnel to a matching energy level in the next well. This situation is illustrated in Figure 5.14(c) and contrasted with Figure 5.14(d) where the energy levels are not aligned. The electrons freed in this manner then cascade through the series of quantum wells and conduct an electrical current through the device.

In the above operating mode, the photon absorption occurs through a transition between two sharply defined quantum levels; as a result, the spectral response of the detector is relatively narrow. Since this behavior is usually undesirable, recent work on quantum well detectors has concentrated on a second operating mode in which an absorbed photon lifts an electron out of the well and the detector bias carries it over the potential wall (see Figure 5.14(e)). Detectors operating in this fashion have responses over wavelength ranges that are 15 to 35% of the center wavelength of the spectral response.

As shown in Figure 5.14(e), the material for these detectors is heavily doped with a donor-type impurity to lift the Fermi level into the bottom of the quantum well so that the GaAs regions have a relatively high concentration of conduction electrons, providing free electrons for photon interactions. Thermally excited conductivity is inhibited by the potential barrier formed by the GaAlAs.

This type of quantum well detector is somewhat analogous to Schottky barrier detectors, and the first-order physics of operation follows from the discussion in Section 5.3.3. There are three important differences compared with Schottky diode detectors. The first is that the height of the potential barrier and hence the long wavelength response can be adjusted readily in the quantum well devices by controlling the amount of Al in the AlGaAs layers. Current development emphasizes the 10 μm region where there is no high performance Schottky device. The second difference is that absorption in the quantum well is by free carriers in a semiconductor, not a metal. The free carrier density and hence the absorption is therefore much lower than for a Schottky barrier detector. A third, and more subtle, difference is that the quantum mechanical selection rules for the electron transition require that the photons be introduced into the detector in a direction along the quantum well layers; conventional front illumination of the detector is not usable. With side illumination, quantum efficiencies of 20% can be obtained; detectors with grating-like faces can achieve similar values in a pseudo-front-illuminated geometry.

Quantum well detectors do not offer the same performance at the level of individual devices as can be obtained with direct bandgap intrinsic semiconductors (for example, HgCdTe). A more detailed performance comparison can be found in Kinch and Yariv (1989), and Levine (1990). However, there are many parameters that can be adjusted in quantum well devices and their performance is steadily improving as these parameters are optimized (for example, Janousek *et al.*, 1990 and Levine *et al.*, 1990). As they are developed further, these detectors may provide system-level advantages similar to those of PtSi that cause them to be favored in certain applications – good uniformity, low cost, and the ability to fabricate the detectors monolithically with readouts on a GaAs substrate.

5.5 Example

We consider the photodiode equivalent of the silicon photoconductor designed in Chapter 3. That is, the diode should be operated at 300 K, it should be 1 mm^2, and it should be operated at 1 μm. We dope the material with arsenic at 10^{15} cm^{-3} to make the n-type side and use a similar concentration of boron to make the p-type side. Compute I_0, R_0, and C_J, and estimate the quantum efficiency at 1 μm.

(a) I_0: From the discussion following equation (3.53), the conduction electron concentration in the intrinsic material is 1.38×10^{10} cm^{-3}. Equations (5.36) and (5.37) then give us $p_n = 1.9 \times 10^{11}$ m^{-3} and $n_p = 2.5 \times 10^{10}$ m^{-3}. From Table 3.1, $\tau \sim 10^{-4}$s for impurity concentrations $\sim 10^{12}$ cm^{-3}. Assuming τ varies as N_I^{-1}, we get $\tau \sim 10^{-7}$ s. Taking mobilities from this same table and using equations (5.13) and (5.18), the diffusion coefficient for electrons in the p-type material is $D_p^n = 3.5 \times 10^{-3}$ m^2 s^{-1}, and for holes in the n-type material is $D_n^p = 1.24 \times 10^{-3}$ m^2 s^{-1}. These values yield $L_p^n = 1.87 \times 10^{-5}$ m and $L_n^p = 1.11 \times 10^{-5}$ m. From equation (5.33), I_0 is then $qA(2.6 \times 10^{13}$ m^{-2} s$^{-1}) = 4.1 \times 10^{-12}$ A.

(b) R_0: From equation (5.39), $R_0 = 6.3 \times 10^9\ \Omega$

(c) C_J: From equation (5.38), $V_0 = 0.58$ V. From Table 3.1, the dielectric constant for silicon is 11.8. Setting $V_b = 0$, we can use equation (5.47) to compute the width of the depletion region to be 1.23×10^{-6} m. Substituting into equation (5.48), the junction capacitance is found to be 85 pF.

(d) Quantum efficiency: Given a diffusion length of 19 μm for electrons and the absorption length of ~ 80 μm at a wavelength of 1 μm, it is clear that the quantum efficiency will be low, of order 10% (see Problem 5.2). For example, if the absorbing layer is set to one diffusion length, then $\tanh(c/L) = 0.76$. From reflection loss, $b = 0.70$. The generation rate times the absorbing layer thickness is 0.21 (assuming exponential absorption with absorption length 80 μm). These values yield $\eta = 11\%$.

5.6 Problems

5.1 Show that the voltage across a photodiode held to zero current (that is, measured with a very high impedance voltmeter) is

$$V = \frac{kT}{q} \ln\left(1 + \frac{\eta q P}{h\nu I_0}\right).$$

Compute the local voltage responsivity of the diode, dV/dP. For a diode with $I_0 = 10^{-6}$ A, $\eta = 0.5$, and $T = 300$ K, compare the local voltage responsivity for $P = 0$ and $P = 10^{-6}$ W and 10^{-5} W at a wavelength of 0.9 μm.

5.2 Assume that we are operating a photodiode at a wavelength where the absorption is so low that we can assume that the photoexcited charge carriers are produced uniformly throughout the material overlying the junction. The corresponding boundary condition at the surface of the diode is

$$\left(\frac{dn}{dx}\right)_{(x=c)}=0.$$

Show that the quantum efficiency is

$$\eta=\frac{b\gamma c\tanh(c/L)}{(c/L)},$$

where γ is interpreted to be the number of charge carriers generated per unit path per photon in the detector material.

5.3 Assume a photodiode with quantum efficiency η is operated at zero bias, receives φ photons s^{-1}, and has no $1/f$ or other excess noise. Derive a constraint on I_0 as a function of φ and η that will result in the diode operating with DQE$\geqslant 0.5\,\eta$.

5.4 Use equation (5.12) to demonstrate that the Fermi level must be constant across a junction in equilibrium.

5.5 Consider a junction with $N_A \gg N_D$ and $N_D(x)=Bx^\beta$. The depletion region will then extend from $x\sim 0$ to some distance $x=w$ within the n-type region. Show that the junction capacitance is

$$C_J=A\left[\frac{qB\varepsilon^{\beta+1}}{(\beta+2)(V_0-V_b)}\right]^{1/(\beta+2)}.$$

5.6 A one-sided, ideal silicon diode with $N_A \gg N_D$ is square and 100 μm on a side. Its capacitance as a function of bias voltage is measured as in the following table:

V_b(V)	C (pF)
0.500	2.72
0.000	2.00
−1.000	1.49
−2.000	1.26
−3.000	1.12
−4.000	1.025
−5.000	0.95

Determine the contact potential and the doping profile for the diode.

5.7 Redesign the silicon diode of the example for operation at 4 K (assume that freezeout of the junction is of no concern and that the recombination time goes as $T^{1/2}$ and the diffusion coefficient as T). Assume that the doping method gives reliable results only for $n \geqslant 6 \times 10^{13}\,\mathrm{cm}^{-3}$. Comment on any improvements and limitations this low operating temperature produces.

6

Amplifiers and readouts

To be useful, the output signal from any of the electronic detectors we have discussed must be processed by external electronics. Conventional electronics, however, are not very well suited for an infinitesimal current emerging from a device with virtually infinite impedance. Nonetheless, highly optimized circuit elements have been developed to receive this type of signal and amplify it. Most of these devices are based on very high input impedance, low noise amplifiers that can be built with field effect transistors (FETs). FETs are used in a variety of circuits that are constructed to give the desired frequency response and to accommodate the electrical properties and operating temperature of the detector, among other considerations. In the most sensitive circuits, signals of only a few electrons can be sensed reliably.

6.1 Building blocks

There are two basic kinds of transistor out of which amplifiers for the detector outputs could be built: bipolar junction transistors (BJTs) and field effect transistors (FETs). BJTs are generally unsuitable for directly receiving the signal from the detectors we have been discussing because they have relatively modest input impedances. FETs are used to build first stage electronics for virtually all high sensitivity detectors.

There are two basic classes of FET: the junction field effect transistor (JFET) and the metal–oxide–semiconductor field effect transistor (MOSFET). Although they operate by rather different means, a common terminology is used to discuss their performance. Both FET types have an electrically conducting channel through which current flows from the source terminal to the drain terminal. The current is controlled by an electric field that is established by the voltage on a third terminal, the gate.

In the JFET, two diodes are grown back-to-back, and the current is conducted along the common side of the junctions. Figure 6.1 shows a schematic n-channel JFET along with its simplified circuit diagram; a p-channel device reverses the positions and roles of the p- and n-doped materials. The operation of the JFET depends on the very high impedance that is obtained in the depletion region of a diode, a phenomenon that was discussed in the preceding chapter. As the gate voltage is increased to raise the back-bias on the junctions, the depletion regions will grow as shown by the progression of dashed contours in Figure 6.1. When the gate voltage is large enough to cause the depletion regions to join, the current flow is strongly impeded and the transistor is said to be pinched off. If the gate voltage is reduced slightly below this level, a high degree of control can be exercised over the current.

In an enhancement mode n-channel MOSFET, two diodes are formed by implanting an n-type dopant into a p-type substrate, as shown in Figure 6.2. A p-channel MOSFET reverses the positions and roles of the p- and n-doped materials. The metal gate electrode is evaporated onto an insulating layer of SiO_2. Normally, the device is pinched off because the two diodes are back-to-back and there is no continuous path of n-type material between the source and the drain. If, however, a positive voltage is put on the gate, it tends to attract negative charge carriers to the underside of the insulator. If the voltage is strong enough, an n-type channel forms, and current is able to flow through the continuous n-type path from the source to the drain. The size of the channel, and therefore the current, are controlled by the size of the voltage (as indicated by the dot–dash lines). There are a number of other

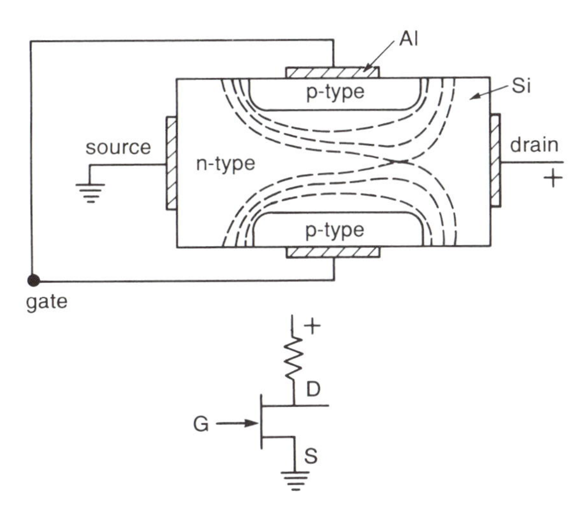

Figure 6.1. Principle of operation of an n-channel junction field effect transistor.

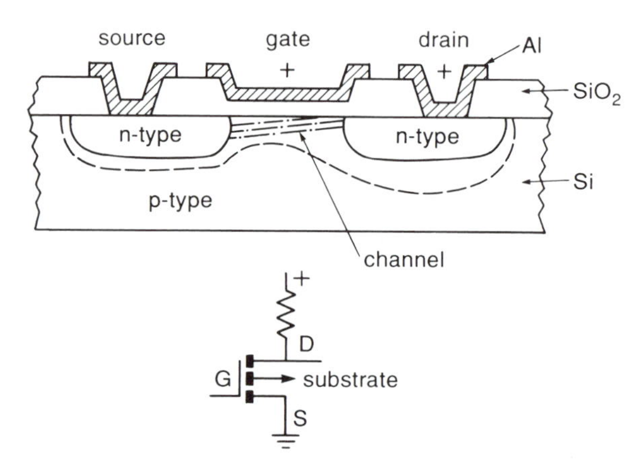

Figure 6.2. Principle of operation of a n-channel enhancement mode metal–oxide–semiconductor transistor.

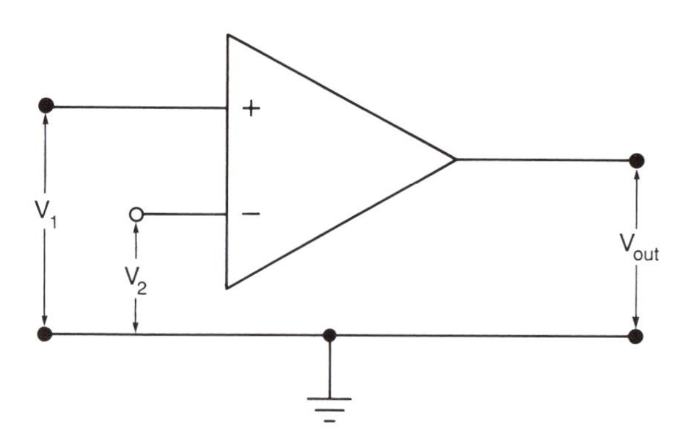

Figure 6.3. Representation of an operational amplifier.

types of MOSFET that are variations on this theme. For more detailed discussion of FETs, consult Sze (1985) or Streetman (1990).

After the detector signal has been amplified by the first stage of FETs, it is usually passed on to additional amplifiers. A generic form is the operational amplifier, or op-amp. This designation is left over from the early days of electronic computers, when analog computers were built in which amplifiers were used to simulate various mathematical operations. Although a modern op-amp is a complex integrated circuit, in most situations it can be treated as a single circuit element: a high voltage gain, direct current (DC) amplifier with high input impedance. It is convenient to denote these amplifiers by a single symbol (that is, we will not try to show the individual electronic components) as in Figure 6.3. The output voltage, V_{out}, is proportional to the difference between the two input voltages, V_1 and V_2. Note that the '−'

(inverting) and '+' (noninverting) symbols on the input terminals designate only that the output voltage changes respectively either in the opposite or in the same direction as the input voltage applied to that terminal. They indicate nothing about the relative voltage difference across the two input terminals. In the uses described below, high input impedance and low noise are achieved by using FETs as the first electronic stage, either as part of the op-amp or as a separate input stage that can be cooled along with the detector and placed very close to it to minimize the length of wiring that must carry high impedance signals. In Figure 6.3, the input FETs have been subsumed into the symbol for the op-amp. Op-amps and other electronic circuit issues are discussed in many places, for example by Horowitz and Hill (1989).

6.2 Load resistor and amplifier

The simplest way to measure the signal from a detector is to place the detector in a voltage divider with a resistor, pass a current through it, and measure the voltage developed across it (see Figure 6.4). Notice that the voltmeter impedance is in parallel with the detector impedance; care must be taken that reduced signals and excess noise do not result from use of a low impedance voltmeter. Fortunately, FET input transistors can provide very high input impedances. This arrangement can be satisfactory in situations where modest sensitivity is required.

When we want very high sensitivities with photoconductors and photodiodes without the expense and complication of operating them at very low temperatures, we require (see equation (3.62)) a high detector impedance to reduce the Johnson-noise-limited NEP. In the circuit of Figure 6.4, it is necessary that $R_L \geqslant R_d$ to get a reasonably high signal ($R_d \gg R_L$ would lead to an output voltage barely varying from the bias voltage, V). The large

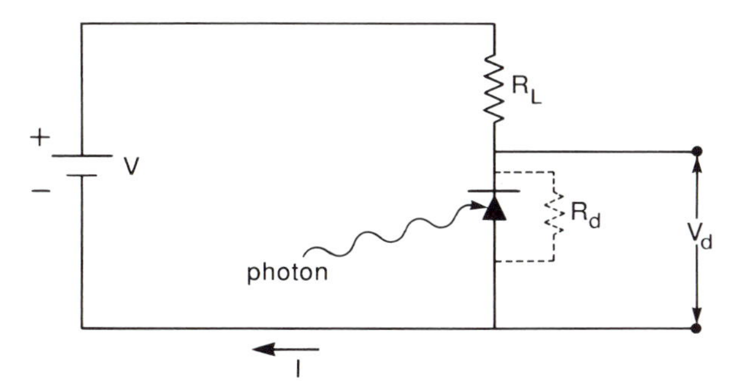

Figure 6.4. Simple readout circuit using a load resistor and sensing the voltage across the detector directly.

resistances required can lead to undesirably slow response; as an example, for a photodiode with $R_0 > 10^{12}\,\Omega$ and capacitance > 1 pF, the time response will be > 1 s. A second difficulty is that this circuit does not let us establish and maintain the detector bias independent of the signal. This situation is obviously a problem for a photodiode, where we wish to set the operating point at the optimal position on the $I - V$ curve (Section 5.2.4); it can also cause difficulties with other detectors. Finally, for both photodiodes and photoconductors, the detector nonlinearities are passed through the circuit without correction.

In Chapter 9 we will discuss bolometers. These detectors must be held at extremely low temperatures to suppress noise mechanisms that cannot be controlled by adjusting the electrical properties of the detector. The low temperature controls Johnson noise without requiring ultra-high detector impedances, and the shortcomings just discussed for the circuit in Figure 6.4 do not apply. In fact, we will show how this circuit has an interesting advantage in suppressing Johnson noise when used with a bolometer.

6.3 Transimpedance amplifier (TIA)

6.3.1 Basic operation

The shortcomings of the circuit described above for use with high impedance detectors are dealt with simply and elegantly by the transimpedance amplifier, or TIA, circuit shown in Figure 6.5. It is illustrated there with a photodiode; a similar circuit is also frequently used with photoconductors.

The output of the amplifier in Figure 6.5 is proportional to the difference of the input voltages: $V_{out} \propto V_+ - V_-$. In the TIA, the detector is connected to the inverting ($-$) input terminal of the op-amp, and the output of the op-amp is connected (via R_f) to the same input terminal. As a result, the output

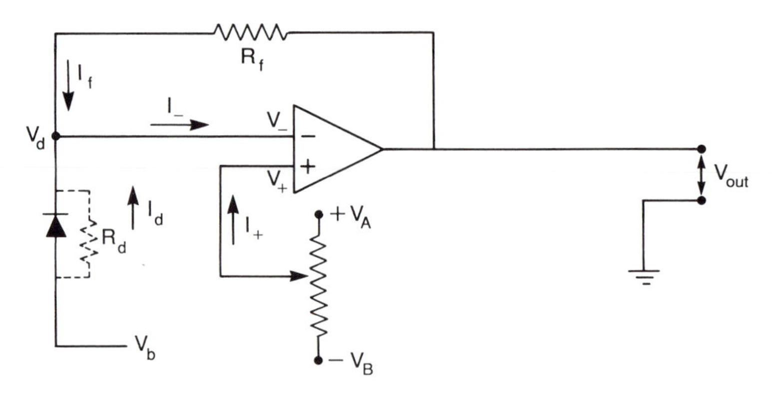

Figure 6.5. Transimpedance amplifier.

current opposes the detector current (negative feedback). Consequently, the op-amp drives itself to minimize the voltage difference between its two inputs, $V_+ - V_- \approx 0$. The amplifier output required to balance the two inputs provides a measure of the input current from the detector. The noninverting input is used for an adjustable voltage that allows the setting of V_{out} to a desired level in the absence of signal.

Assume for now that V_+ is near ground ($V_+ \approx 0$), in which case $V_- \approx 0$ and $V_d \approx 0$. Because the op-amp has such high gain and input impedance, it can be driven by very small voltages and currents. From Kirchhoff's current law, we know that $I_d + I_f = I_-$, and, because I_- is so small, we can say that $I_d + I_f \approx 0$. If we assume that both R_f and the detector are purely resistive, then it follows from Ohm's law that

$$\frac{V_{out} - V_d}{R_f} = -\frac{V_b - V_d}{R_d}, \tag{6.1}$$

which becomes

$$V_{out} = -V_b \frac{R_f}{R_d}, \tag{6.2}$$

remembering our assumption that $V_d \approx 0$.

Another way to express the amplifier output is

$$V_{out} = I_f R_f = -I_d R_f. \tag{6.3}$$

This last equation demonstrates one of the key virtues of the TIA: the output voltage depends on the feedback resistance and not on the electrical impedance of the detector. Thus, if the detector current is linear with photon flux, the amplifier output will be linear to the (usually high) degree of linearity that can be achieved for the feedback resistor. The condition on linearity of I_d is met by photodiodes according to equation (5.42), and it is also met for photoconductors (see equation (3.16)) as long as their photoconductive gains are independent of the photon flux (but note, for example, the discussion on dielectric relaxation in Section 3.2.4).

The bias on the detector can be set by adjusting V_+ and/or V_b. The feedback of the amplifier automatically maintains V_- at the set point, thus stabilizing the detector bias. This feature can be essential to maintaining the detector current linearity that was assumed above, as well as to minimizing some of the undesirable detector properties we have discussed for extrinsic photoconductors. In particular, the TIA allows operating photodiodes as a current source into approximately zero bias which both maintains the detector linearity and also suppresses certain types of voltage-dependent noise.

6.3.2 Time dependencies and frequency response

To understand the behavior of the TIA circuit in more detail, we must take into account the nonresistive properties of the circuit elements. To do so, it is convenient to replace the circuit to be analyzed with an electronically equivalent one made up of elementary components such as resistors and capacitors in place of more complex ones such as photon detectors. A Thévenin equivalent circuit is based on voltage sources with series resistances and a Norton equivalent circuit on current sources with parallel resistances. A Norton circuit for the TIA is shown in Figure 6.6. In this case, a current source produces the detector signal, I_d^S, and a second source produces the equivalent noise current, I_d^N. The capacitance and resistance are represented by the discrete elements C_d and R_d for the detector and C_f and R_f for the feedback resistor, which has associated current noise I_f^N. The intrinsic amplifier noise voltage (that is, the amplifier noise voltage prior to modification by external components) is represented by V_A^N. For the time being, set I_d^N, I_f^N, and V_A^N to zero to compute the response to signals. The input voltage to the op-amp is then

$$V_{in} = I_d^S Z_d, \tag{6.4}$$

where Z_d is the detector impedance. We will next determine Z_d.

A standard procedure in alternating current (AC) electrical circuit theory is to treat capacitances and inductances in terms of complex numbers. Thus, the impedance of the capacitor is $1/j\omega C_d$, where j is $\sqrt{-1}$ and ω is the angular frequency ($\omega = 2\pi f$). This impedance can be treated algebraically exactly as if it were a complex resistance. That is, Ohm's law and the rules for computing the net effects of combinations of circuit elements can be applied in exactly the same form as would be used with purely resistive circuits. (Similarly, inductances have an impedance of $j\omega L$, but we will not deal with them here.)

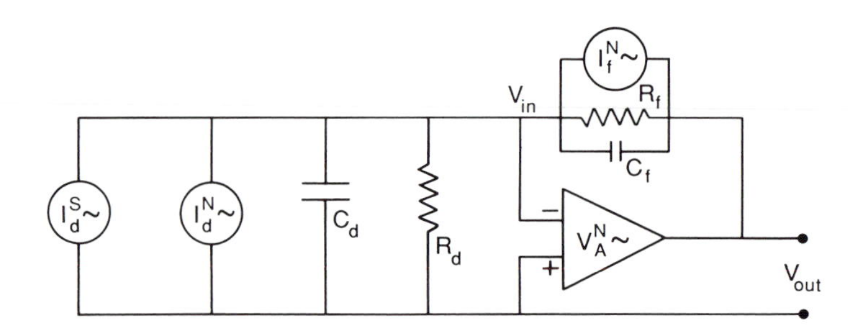

Figure 6.6. Equivalent circuit for the transimpedance amplifier.

Since the detector resistance and capacitance are in parallel,

$$Z_d^{-1} = R_d^{-1} + \left(\frac{1}{j\omega C_d}\right)^{-1} = R_d^{-1} + j\omega C_d, \tag{6.5}$$

or

$$Z_d = \frac{R_d}{1 + j\omega R_d C_d}. \tag{6.6}$$

Since the feedback impedance and detector impedance act as a voltage divider between V_{out} and ground, and since we can assume that $V_{in} \ll V_{out}$, the output voltage is

$$V_{out} = -V_{in}\frac{Z_f}{Z_d}, \tag{6.7}$$

where

$$Z_f = \frac{R_f}{1 + j\omega R_f C_f}. \tag{6.8}$$

Equation (6.7) is the more general form of equation (6.2). Substituting into it from equations (6.4) and (6.8), and multiplying it by $(1 - j\omega\tau_f)/(1 - j\omega\tau_f)$ to put it into a convenient form, we obtain

$$V_{out} = -I_d^S Z_f = -\frac{I_d^S R_f}{1 + j\omega\tau_f} = -\frac{I_d^S R_f}{1 + \omega^2\tau_f^2} + j\frac{\omega\tau_f I_d^S R_f}{1 + \omega^2\tau_f^2}, \tag{6.9}$$

where $\tau_f = R_f C_f$. The amplitude of V_{out} is its absolute value,

$$|V_{out}| = (V_{out}V_{out}^*)^{1/2} = \frac{|I_d^S| R_f}{(1 + \omega^2\tau_f^2)^{1/2}}. \tag{6.10}$$

The signal cutoff frequency f_f is the frequency at which the signal has decreased by a factor of $\sqrt{2}$, or, from equation (6.10),

$$f_f = \frac{\omega_f}{2\pi} = \frac{1}{2\pi R_f C_f} = \frac{1}{2\pi\tau_f}. \tag{6.11}$$

The imaginary part of equation (6.9) gives the phase of V_{out}. This parameter can be ignored for our purposes, although it is quite important in a more complete treatment of the circuit. For example, excessive phase shifts can cause the TIA output to be unstable.

Notice that the output voltage is independent of R_d and C_d. As we found for the simpler analysis, as long as R_f is linear and I_d^S is a linear function of the photon flux, φ, the detector/amplifier system is linear. Moreover, even if $R_d C_d = \tau_d \gg \tau_f$, the frequency response of the system is determined by τ_f, the limit imposed by the feedback resistor. As the detector impedance decreases with increasing frequency, the TIA gain increases to compensate (equation (6.7)). The high frequency performance is improved substantially for photodiodes because the capacitance of the feedback resistor can be made much smaller than that of the detector.

Now we set I_d^S to zero and use I_d^N, V_A^N, and V_f^N to compute the noise, where V_f^N is the voltage noise of the feedback resistor. The equivalent noise of the amplifier, V_A^N, is subject to the amplifier gain. That is, from equation (6.7),

$$\langle (V_{\text{out,A}}^N)^2 \rangle^{1/2} = \langle [V_A^N(f)]^2 \rangle^{1/2} \left(\frac{R_f}{R_d} \right) \left[\frac{1 + j\omega\tau_d}{1 + j\omega\tau_f} \right], \tag{6.12}$$

where we indicate explicitly that V_A^N is likely to have its own frequency dependence (for example, $1/f$ noise) which is modified by the external time constants τ_d and τ_f. The noise voltage amplitude is

$$|\langle (V_{\text{out,A}}^N)^2 \rangle^{1/2}| = \langle [V_A^N(f)]^2 \rangle^{1/2} \left(\frac{R_f}{R_d} \right) \left(\frac{1 + \omega^2 \tau_d^2}{1 + \omega^2 \tau_f^2} \right)^{1/2}. \tag{6.13}$$

Equation (6.13) shows that, although the time constant for signals is reduced by the TIA compared with that of the detector alone, in the interval of time response, $\tau = 1/\omega$, with $\tau_f < \tau < \tau_d$, the amplifier noise is boosted with increasing frequency. f_d is the frequency at which amplifier noise has been boosted by $\sqrt{2}$ (assuming that $\tau_f \ll \tau_d$); that is,

$$f_d = \frac{1}{2\pi R_d C_d} = \frac{1}{2\pi\tau_d}. \tag{6.14}$$

The current noise from the detector is amplified in exactly the same way as the signal current, so, following equation (6.10), the output voltage due to it is

$$|\langle (V_{\text{out,d}}^N)^2 \rangle^{1/2}| = \langle [I_d^N(f)]^2 \rangle^{1/2} \frac{R_f}{(1 + \omega^2 \tau_f^2)^{1/2}}. \tag{6.15}$$

Again, I_d^N contains its own frequency dependencies such as $1/f$ noise.

For an ideal feedback resistor, the current noise I_f^N is Johnson noise; converting from current to voltage noise,

$$\langle (V_f^N)^2 \rangle^{1/2} = \frac{\langle (I_f^N)^2 \rangle^{1/2} R_f}{(1 + \omega^2 \tau_f^2)^{1/2}} = \left(\frac{4kTR_f \,\mathrm{d}f}{1 + \omega^2 \tau_f^2} \right)^{1/2}. \tag{6.16}$$

The total output voltage noise is given by the square root of the quadratic

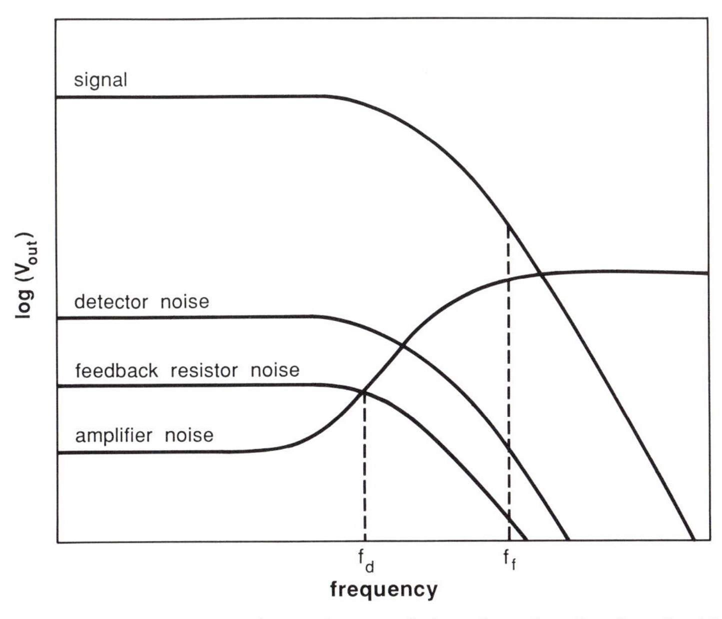

Figure 6.7. Frequency dependence of signal and noise for the TIA.

sum of the contributions from three sources: amplifier, detector, and feedback resistor. Thus

$$\langle (V_{out}^N)^2 \rangle^{1/2} = [\langle (V_{out,A}^N)^2 \rangle + \langle (V_{out,d}^N)^2 \rangle + \langle (V_f^N)^2 \rangle]^{1/2}. \tag{6.17}$$

The dependencies of signal and noise on frequency are shown in Figure 6.7 for the case where V_A^N and I_d^N are independent of frequency, and the feedback resistor contributes only Johnson noise. This figure illustrates the importance of limiting the frequency response of the circuitry to the region where a high signal-to-noise ratio is achieved. If high frequency noise is admitted into the signal chain, it can dominate the signal. Electronic filtering can be used to avoid this problem.

Further discussion of TIAs can be found in Wyatt, Baker, and Frodsham (1974), Hall *et al.* (1975), and Rieke *et al.* (1981). Additional information on amplifiers can be found in Wolfe and Zissis (1978) and Wyatt (1991); the latter reference provides a selection of examples of different applications and is useful in showing the influence of the amplifier on the overall performance of a detection system.

6.4 Integrating amplifiers

A description of the operation of an FET that is equivalent to the one at the beginning of this chapter is that an electric charge, Q, deposited on the capacitance at the gate, C_g, establishes a control voltage, $V = Q/C_g$. The value of the voltage, and hence the charge on the gate, can be read out at will by monitoring the current in the channel of the FET. This way of thinking about FETs is appropriate when they are used in integrating amplifiers.

6.4.1 Simple integrators

A simple integrating amplifier can provide better signal-to-noise ratios than can a TIA when observing signals with long time constants and if it is unnecessary to maintain the detector bias precisely (we will eventually discuss a more complex integrating amplifier that does maintain the detector bias). An additional limitation of integrating amplifiers is that they require a relatively complex set of electronic circuitry (and/or software) to convert their signals to an easily interpreted output format. There are relatively few review articles on integrating amplifiers; two that may be of use are Burt (1988) and Scribner, Kruer, and Killiany (1991).

A simple source follower integrating amplifier is shown in Figure 6.8. Charge leaks from the bias supply through the detector to be deposited on a storage capacitor, C_s. The leakage rate is modulated by the detector impedance, which is a function of the photon flux, φ. Charge can be accumulated noiselessly on the capacitor and read out whenever enough signal has been collected that it stands out well above the amplifier noise. The charge is conducted away after or during readout by closing a reset switch.

FETs are used both to collect the charge and to produce a relatively low impedance output signal that matches well to the following electronics. They can have large enough gate (input) impedances, particularly when operated at low temperatures, that their primary effect on the detector side of the circuit

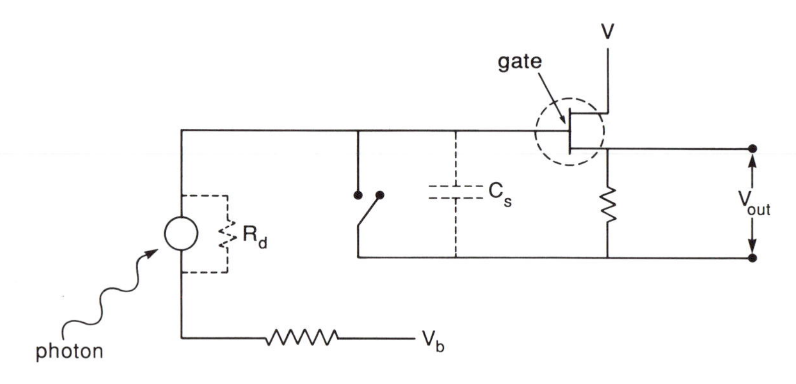

Figure 6.8. Simple source follower integrating amplifier.

is to add the gate capacitance in parallel to the capacitance of the detector and reset switch. For convenience, we define C_s to be the total capacitance at this circuit node. The current in the channel of the FET is a measure of the charge collected on C_s. The uncertainty in the measurement of the charge is typically expressed in electrons and is called the read noise, Q_N.

We would like this collected charge to be proportional to the time-integrated signal current from the detector. To derive the conditions required for this proportionality to hold, refer to the equivalent circuit in Figure 6.9, where we show the input to the FET in terms of the equivalent signal generator I_d^S in the detector, the detector resistance R_d, and the storage capacitor C_s. Assume that the detector produces a constant current, I, beginning at time $t=0$. By integrating the time response of the RC circuit

$$V_g=\int_0^t dV_g=\int_0^t V_0\,e^{-t'/\tau_{RC}}\,d\left(\frac{t'}{\tau_{RC}}\right), \tag{6.18}$$

we find that the voltage on the gate of the FET will be

$$V_g=V_0(1-e^{-t/R_dC_s}). \tag{6.19}$$

For times much less than the RC time constant of the circuit, we can expand the exponential in a series and discard the high order terms to find

$$V_g\approx V_0\left[\left(\frac{t}{R_dC_s}\right)-\frac{1}{2}\left(\frac{t}{R_dC_s}\right)^2+\cdots\right], \quad t\ll R_dC_s. \tag{6.20}$$

The circuit has the desired property of linearity within a tolerance of $(1/2)(t/R_dC_s)^2$ (recall that the last term retained in an alternating infinite series is larger than the net contribution of all the remaining discarded terms). The form of this tolerance shows why a high detector impedance is generally required to achieve successful operation in an integrating amplifier. When R_d is very large, the gate voltage is linear within a tolerance that goes as R_d^{-2}.

The integrating amplifier/detector systems under discussion are subject to

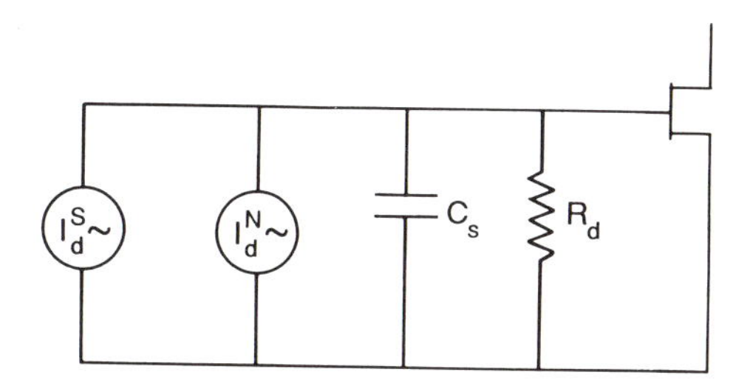

Figure 6.9. Equivalent circuit for integrating amplifier.

a second type of nonlinearity. As signal is accumulated, the bias across the detector is reduced. In the case of a photoconductor, the photoconductive gain and hence the responsivity are reduced proportionally (see equations (3.12) and (3.13)). In the case of a photodiode, the capacitance increases according to equations (5.47) and (5.48), particularly if the detector is being operated near zero bias. Both effects reduce the response of the circuit as signal is accumulated. This source of nonlinearity can be controlled by limiting the change in voltage permitted across the detector before a reset or, since the nonlinearity enters smoothly and monotonically, it can be calibrated and removed from the signal in data reduction. A third potential source of nonlinearity is improper operation of the FET.

From the above discussion, and assuming the linearity conditions are met, the output waveform for a constant rate of charge flow onto C_s is shown in Figure 6.10(a). We assume the FET output is a linear function of its gate voltage and that the gain of the FET is unity. Then the signal can be measured by determining $V_{out} = V_1$ at t_1, resetting, and determining $V_{out} = V_2$ at t_2. The accumulated charge is

$$Q = (V_1 - V_2)C_s, \tag{6.21}$$

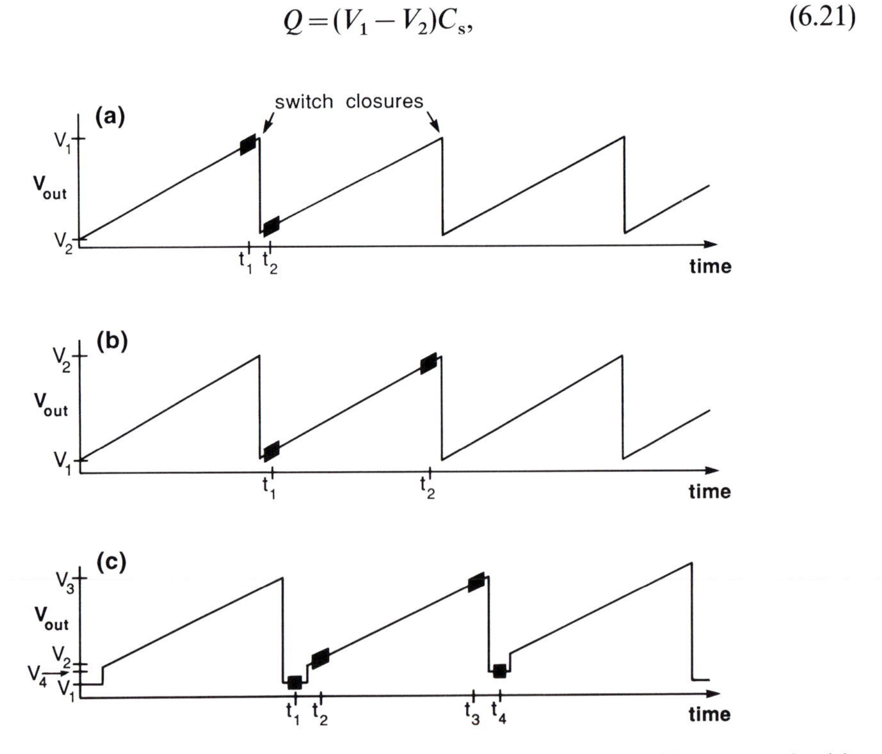

Figure 6.10. Sampling strategies for integrating amplifiers: (a) illustrates double correlated sampling around a reset; (b) is nondestructive reading at the beginning and end of the integration ramp; and (c) is triple correlated sampling.

the usual expression for the charge on a capacitor. This technique is called double correlated sampling.

The charge Q includes any contributions from the detector dark current and leakage through the FET onto its gate, as well as the photocurrent. The limiting noise, assuming other sources of read noise are negligible, is the square root of the total number of charges collected from these currents.

There are at least two other potential sources of noise. One excess noise source is the FET, which will be discussed later. The other is the charge on C_s, which, when sampled in the way described above, is subject to kTC noise (equation (3.43)) that arises in the resistor/capacitor circuit formed when the reset switch is closed. In fact, if the detector, reset switch, and FET gate are replaced by an equivalent circuit with a resistor and capacitor in parallel, the integrating amplifier circuit is an accurate reproduction of the circuit in Figure 3.8, which we used to derive the expression for kTC noise. Upon reset, we therefore have a minimum charge uncertainty (in electrons) on C_s of

$$\langle Q_N^2 \rangle = \frac{kTC_s}{q^2}, \tag{6.22}$$

where q is the charge on the electron.

For example, assume all noise sources except kTC noise are negligible. The smallest practical gate capacitance with a MOSFET/detector combination is of the order of 10^{-13} F; at $T = 4$ K, the resulting noise is $\langle Q_N^2 \rangle^{1/2} = 15$. Such a circuit is said to have 15 electrons read noise.

An alternate readout strategy used to reduce kTC noise is illustrated in Figure 6.10(b). The signal is measured by differencing measurements at the beginning and end of the integration, that is, $V(t_2) - V(t_1)$, with no intervening reset. Assuming the open reset switch and the gate of the FET have high impedance, the virtue of this strategy can be seen from the time constant for changes of charge on C_s,

$$\tau_s = R_d C_s. \tag{6.23}$$

For example, if $C_s = 10^{-11}$ F and $R_d = 10^{15}\,\Omega$, then $\tau_s = 10^4$ s. If $t_2 - t_1 \ll \tau_s$, the kTC noise is 'frozen' on C_s during the integration. More quantitatively, the noise is reduced by the factor for response of an RC filter (the factor multiplying V_0 in equation (6.19)). The disadvantage of this strategy is that it depends both on the FET and on the electronics that follow it to have extremely small low frequency noise. At some level, nearly every electronic device has $1/f$ noise, which can limit the read noise in this situation.

A third strategy, known as triple correlated sampling, includes a third voltage measurement made while the reset switch is closed. This additional measurement tracks the $1/f$ noise of both the FET and the electronics that follow it and allows it to be removed. In other regards, the results are equivalent to the sampling strategy just discussed. The behavior of this read strategy can be seen from Figure 6.10(c), where we measure the two voltage differences $\Delta V_a = V(t_2) - V(t_1)$ and $\Delta V_b = V(t_3) - V(t_4)$. The signal obtained, $\Delta V_b - \Delta V_a$, can be written as

$$\begin{aligned} \Delta V_b - \Delta V_a &= [V(t_3) - V(t_4)] - [V(t_2) - V(t_1)] \\ &= [V(t_3) - V(t_2)] - [V(t_4) - V(t_1)]. \end{aligned} \tag{6.24}$$

It is therefore equivalent to the signal extracted in the sampling strategy in Figure 6.10(b), but with any voltage drift between reset settings subtracted off. Relative to the strategy in Figure 6.10(b), however, the additional measurement contributes to the overall noise by adding the noise of an additional amplifier read.

There is no uniformly 'best' choice among these possibilities. The noise characteristics of the amplifier and the amount of kTC noise determine which of the sampling methods should be used in a given situation.

The first and second readout strategies illustrate the difference between a destructive and nondestructive read. In the first case, the signal is destroyed in the process of reading it out. In the second, it is preserved on the gate of the FET and can be read repeatedly. Nondestructive readouts provide a number of possibilities for noise reduction. For example, voltages can be sampled continuously up the integration ramp and a slope fitted to them, or a series of readings can be made at each end of the ramp and averaged before differencing to determine the signal. Where the dominant noise in each reading is independent of that in additional ones, such strategies can reduce the net read noise of the amplifier significantly.

We now consider the contribution to read noise from the electronic noise of the FET. As can be seen from equation (6.20), the capacitance on the gate of the FET can set a limitation on the circuit performance because the output voltage for a given collected charge is inversely proportional to C_s. Because of electronic noise in the FET (and possibly in the electronics that follow it), there is a minimum voltage difference that can be sensed. As C_s increases, the charge that must be accumulated to produce this minimum voltage difference increases proportionally. As a result, for an FET with given intrinsic voltage noise, the read noise in electrons increases in proportion to the effective gate capacitance. On the other hand, the intrinsic voltage noise of an FET can be reduced by manufacturing it with a larger gate. To first order, the noise goes as the square root of the gate area. Therefore, the read noise of a detector/FET system decreases with increasing gate size but only so long as the gate capacitance does not dominate C_s. A detector/FET system is usually optimized for noise by adjusting the design of the FET so its gate capacitance is similar to that of the detector. Assuming an appropriately designed FET, the read noise will then go roughly as the square root of the detector capacitance.

Even with full optimization, in most practical integrating amplifiers the amplifier noise is a more serious limitation than kTC noise if an appropriate sampling strategy is adopted. It is, consequently, very important to minimize the amount of amplifier noise that is added to the signal. In general, for integrating amplifiers, the response of the FET and following circuitry must extend to much higher frequencies than implied by the integration time between resets. This requirement arises because it is necessary for the amplifier to follow the reset waveform, and it is particularly stringent when such amplifiers are used in arrays in which electronic switching is used to route the output signals from a large number of detectors through a single output line (as will be discussed in Chapter 7). Because of the relatively high frequencies involved, in the following discussion we will ignore the $1/f$ noise of the FET and take its noise output to be white; that is, equal noise power at all frequencies. As a result, the FET noise goes as the square root of the frequency bandwidth of the circuitry (as discussed in Chapter 3). Optimum signal to noise therefore requires that this bandwidth be limited to the smallest value possible that admits adequate signal power.

These arguments can be illustrated qualitatively by considering an example based on a simple integrating amplifier being read out twice, once before and once after reset. We will largely ignore normalization constants (and signs) because we are interested only in the frequency behavior of the signal components. If the time coordinate is greatly magnified, the signal can be represented by a step function, such as sgn(t). The signal voltage amplitude as a function of frequency is proportional to the absolute value of the Fourier transform of sgn(t); referring to Table 2.3 and setting the normalization constant to unity, we have

$$V_S(f) = \left[\left(\frac{\mathrm{j}}{\pi f}\right)\left(\frac{-\mathrm{j}}{\pi f}\right)\right]^{1/2} = \frac{1}{\pi f}. \tag{6.25}$$

Assuming white noise, the frequency dependence of the noise is

$$\langle [V_N(f)]^2 \rangle = \mathbb{C}, \tag{6.26}$$

where $\mathbb{C}$ is a constant.

Without frequency filtering, the net output for signals is just the integral of equation (6.25) over frequency (this integral may be modified slightly by the manner in which the signal is actually measured). Therefore, the signal grows only logarithmically with increasing frequency bandwidth. The net noise output (again with no further frequency filtering) is the square root of the integral of equation (6.26) over frequency and grows as the square root of the frequency bandwidth. Without some reduction in noise toward high frequencies, the signal-to-noise ratio in the measurement will tend toward zero as the bandwidth increases. To maximize the signal-to-noise ratio, the electronic frequency response must be adjusted to reject as much noise as possible while still passing the low frequencies that contain the signal.

To avoid having amplifier noise limit the detector performance, we need to collect enough charges on the integrating capacitor so that the intrinsic statistical noise exceeds the amplifier noise, that is, that the square root of the number of collected charges is greater than the read noise. In general, the FET will have a maximum output voltage, V_{max}, above which nonlinearity or saturation will occur. The maximum number of charges that can be collected without driving the output beyond V_{max} is known as the well depth. Continuing to assume unity gain, the well depth is

$$N_{max} = \frac{C_s V_{max}}{q}. \tag{6.27}$$

Although this relationship suggests that the well depth could be increased by increasing C_s, we have seen a variety of negative effects that would ensue from large values of C_s.

It is possible for the statistical noise to dominate the amplifier noise only if N_{max} is larger than the square of the read noise in electrons. If the amplifier satisfies this requirement, the integration time is long enough to collect sufficient charge, and dark currents and other such phenomena do not provide significant charge compared to the signal, then the detector/amplifier system can be photon noise limited. Assuming negligible dark current, the photon-noise-limited dynamic range of the system can be described in terms of the well depth and read noise as

$$\mathbb{R} = \frac{N_{max} - Q_N^2}{Q_N^2}. \tag{6.28}$$

It is the factor in signal strength over which the system can operate near photon-noise-limited sensitivity. For example, a system with $N_{max} = 10^6$ and $Q_N = 100$ can detect signals varying in strength by up to a factor of 99 and still remain nearly photon noise limited.

Another specification of dynamic range refers to the ratio of N_{max} to the minimum signal that can be detected at $S/N = 1$, that is, to the read noise.

For the parameters just given, the dynamic range measured in this way is 10^4; the circuit will be read noise limited over most of this range.

6.4.2 Capacitive transimpedance amplifier (CTIA)

There is a class of circuit that integrates but still maintains the detector bias as well as provides intrinsically linear output. An example is the capacitive transimpedance amplifier, or CTIA, shown in Figure 6.11. A current can be generated from the feedback capacitor to the input of the amplifier by varying the output voltage of the amplifier appropriately,

$$I_f = C_f \frac{dV}{dt}. \tag{6.29}$$

As a result, the amplifier can balance itself by varying its output to produce an I_f that equals the output current of the detector, I_d. The output voltage of the amplifier is then proportional to the total charge generated by the detector, while the input voltage of the amplifier, that is, the detector bias, is maintained. The system is reset by closing a switch around the feedback capacitor, thus discharging it and (momentarily) establishing a conventional TIA such as we analyzed earlier. We already know that the TIA will drive

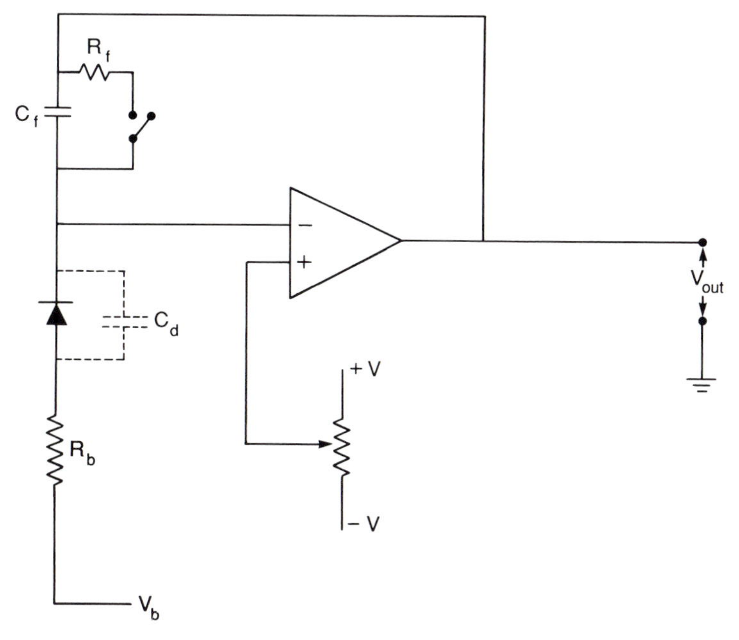

Figure 6.11. Capacitive transimpedance amplifier.

itself to a stable set point, so the circuit will re-establish itself for a repeat integration.

The gain of the CTIA can be computed from equation (6.7) and the expression $Z_f = 1/j\omega C_f$ to be

$$V_{out} = -V_{in}\left(\frac{C_d}{C_f}\right). \tag{6.30}$$

When the conditions for linearity as in equation (6.20) are satisfied, $V_{in} = Q/C_d$, giving

$$V_{out} = -\frac{Q}{C_f}. \tag{6.31}$$

If the capacitance of the feedback can be made independent of voltage, the output of the CTIA will be linear with collected charge. Equation (6.31) also suggests that C_f must be kept relatively small to provide a large output signal and to minimize the influence of the electronic noise of the amplifier on the achievable ratio of signal to noise.

The well depth in the CTIA is set by the voltage differential across the feedback capacitor, giving

$$N_{max} = \frac{C_f V_{max}}{q}. \tag{6.32}$$

From the method of operation of the CTIA, there can be reset noise associated with the capacitance C_f. Without affecting our conclusion, we can simplify the following calculation by assuming that the output of the amplifier with the reset switch closed has been set to ground by adjusting the voltage to the noninverting input, and that $V_b = 0$, as is often the case with a photodiode. When the switch is opened, there will be a charge uncertainty of

$$\langle Q_f^2 \rangle^{1/2} = (kTC_f)^{1/2} \tag{6.33}$$

on C_f. However, the bottom plate of C_f and the upper plate of C_d must be at the same voltage, giving us

$$V_{in} = \frac{Q_f}{C_f} = \frac{Q_d}{C_d}. \tag{6.34}$$

If $\langle Q_d^2 \rangle^{1/2}$ represents the charge uncertainty on the detector capacitance corresponding to $\langle Q_f^2 \rangle^{1/2}$, we have

$$\langle Q_d^2 \rangle^{1/2} = \left(\frac{C_d}{C_f}\right) \langle Q_f^2 \rangle^{1/2} = \left[kT \left[C_d \left(\frac{C_d}{C_f}\right) \right] \right]^{1/2}, \tag{6.35}$$

where we have substituted for $\langle Q_f^2 \rangle^{1/2}$ from equation (6.33). That is, the reset noise is equivalent to the noise from simple integrators (see equation (6.22)), except that the effective capacitance is the detector capacitance multiplied by the ratio of detector to feedback capacitances. If a sufficiently low noise amplifier can be constructed to allow large values of C_f, the reset noise can be significantly reduced from that obtained with simple integrators. Of course, sampling strategies to circumvent reset noise can also be employed with CTIAs.

The CTIA is clearly the readout circuit born with a silver spoon in its mouth. Here, however, as in many other books with mediocre plot lines, one should not be surprised to discover in the next chapter that the CTIA does not entirely live up to our expectations when turned out into the real world.

6.5 Performance measurement

So far we have discussed performance attributes of detectors and readouts from a theoretical viewpoint. Many of these parameters can be measured in a relatively straightforward way. The techniques will be illustrated in terms of the detector types already discussed, but with slight variations they can also be applied to the electronic detectors in following chapters. A more extensive description of performance measurement procedures can be found in Vincent (1990).

Parameters of an amplifier/detector system that we may wish to measure include: (1) responsivity (as a function of frequency), (2) linearity, (3) dynamic range, (4) spectral response, (5) noise (also as a function of frequency), and (6) gate capacitance (for integrating amplifiers). Other performance parameters such as quantum efficiency and NEP can be derived from those listed.

The responsivity of a detector/amplifier system – the volts or amperes of output per watt of input signal – can be measured by having the system view a blackbody source of known temperature, calculating the input signal power, and measuring the output electrical signal. To be confident that the signal is derived only from the blackbody source, a second set of measurements must be taken with the output of the source blocked but no other measurement conditions changed, and the signal in this case must be subtracted from that when viewing the source.

One possible implementation of this procedure is shown in Figure 6.12, as it might be applied to a nonintegrating amplifier. A calibrated blackbody source is placed behind a baffle plate and allowed to emit through an aperture of accurately determined area (the source must be larger than the aperture). The output of the source is chopped at a frequency f by rotating a toothed

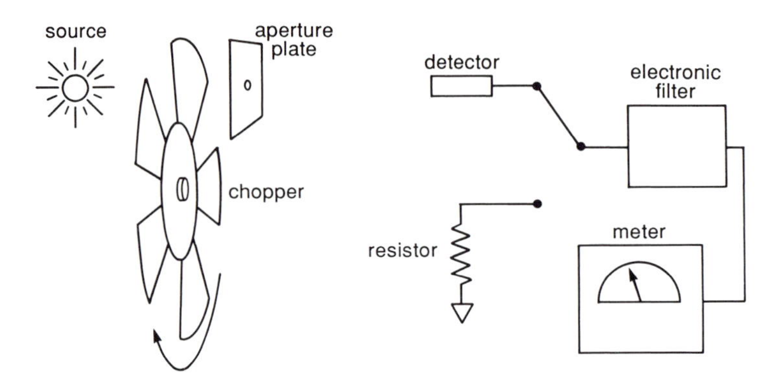

Figure 6.12. Test arrangement for detector calibration.

wheel between the baffle plate and the source (if the wheel were placed in front of the plate, spurious signals might be produced by reflected light or thermal emission from the wheel). The net chopped output signal is the difference between the source emission and that of the chopper wheel through the aperture (the emission of the wheel can be significant at infrared wavelengths). The detector response to this signal should be taken to an electronic bandpass filter carefully tuned to the fundamental chopping frequency, f, to remove any components at frequencies that are overtones of the fundamental. The output of this filter can be amplified, if necessary, and measured with an rms voltmeter. With corrections for the electronic gains of any amplifiers in the test apparatus, it is possible to compute the rms detector response, $V_{rms}(f)$, at the modulation frequency f. The response as a function of frequency can be determined by repeating the above measurement at different rotation rates of the notched wheel or by substituting wheels with different numbers of teeth.

Because the detector views the aperture over the source from a known distance, the power incident on the detector can be calculated as described in Chapter 1. However, in making our measurement the electronic filter rejected all frequencies in the output signal other than the one at the fundamental frequency f. We need to correct the calculation of the power received from the source for the fact that the signal as modulated by the chopper wheel contains components at overtone frequencies other than f. A Fourier series allows us to express the frequency content of a periodic function in a manner analogous to Fourier transformation for nonperiodic ones. If we take the chopped signal to be a square wave of amplitude P_D

calculated from equation (1.13), it can be represented by the series

$$P(t)=\frac{P_{\mathrm{D}}}{2}+\frac{2P_{\mathrm{D}}}{\pi}\left[\cos(\omega t)-\frac{\cos(3\omega t)}{3}+\cdots\right], \tag{6.36}$$

where $\omega=2\pi f$. The $\cos(\omega t)$ term is the one at the fundamental frequency. For a cosine wave, the rms value is the amplitude divided by $\sqrt{2}$ or, from equation (6.36), the rms power received by the detector is $\sqrt{2}P_{\mathrm{D}}/\pi=0.4502\,P_{\mathrm{D}}$. The responsivity of the detector is then

$$S(f)=\frac{V_{\mathrm{rms}}(f)}{0.4502P_{\mathrm{D}}}. \tag{6.37}$$

The factor we have derived for conversion of a square wave signal to an rms value at the fundamental frequency is called the waveform factor (sometimes it is also termed the modulation factor, but this term could cause confusion with the usage in Chapter 2). Strictly speaking, our derivation for a square wave is only valid for the limiting case of an infinitesimal aperture in the baffle plate in Figure 6.12. However, the error introduced by assuming square wave modulation is no more than 1% for aperture diameters up to 17% of the gap between teeth in the chopper wheel. Where larger apertures or other modulation schemes are used, consult Vincent (1990) for a discussion of the appropriate corrections.

To determine the linearity of the detector, one must make accurate relative variations in the input signal power; a high degree of accuracy in relative signal strength can be achieved by varying the size of the aperture over the light source (although caution must be used not to make this aperture so small that diffraction effects become important). The range of variations can be extended by changing the distance between the source and the detector, changing the source temperature, or interposing neutral density filters to reduce the source flux. This same set of measurements can determine the dynamic range. The spectral response is measured by interposing spectral filters between the source and detector. Greater resolution can be achieved by using a monochromator in place of filters.

To measure the noise, the detector is usually arranged so as not to view the source. To determine the noise in a 1 Hz bandpass, the signal *could* be passed through an electronic filter with a bandpass of exactly 1 Hz centered on f and the output signal measured with an rms voltmeter. A difficulty in this method is that noise occurs at all frequencies, unlike the signal, which is modulated at a specific frequency. As a result, a noise measurement made in this fashion could depend critically on the properties of the electronic bandpass filter. A convenient way of removing this criticality and the resulting uncertainties is to build a 'standard noise source', essentially a high quality resistor that produces Johnson noise, which can be suitably amplified. The level of noise can be calculated from equation (3.42). The output of the noise source can be attached to the bandpass filter and measured with an rms voltmeter; the output of the detector can be measured in an identical fashion. The noise of the detector is then the computed Johnson noise for the standard noise source times the rms voltage reading for the detector divided by the rms voltage for the standard noise source. The filter properties affect each noise measurement equally and do not need to be known accurately.

Noise as a function of frequency can be measured by repeating the above procedure at a series of frequencies. However, it can be more convenient to replace the electronic filter and rms voltmeter with equipment that records the noise signal at all frequencies. An example would be to digitize the signal and record it in a computer with no filtering or other modification of its frequency characteristics. The variation of noise with frequency is computed by Fourier transformation of the recorded signal.

The NEP is the noise (for example, in V $Hz^{-1/2}$) divided by the responsivity (in this case, in V W^{-1}). The detective quantum efficiency can be determined in a series of steps. First, the noise is measured while the detector views a constant level of signal power (for example, the source in Figure 6.12 without the chopper running). Next, the response to this signal is measured and combined with the noise to yield the output signal-to-noise ratio. The level of signal power received by the detector can be calculated as in Chapter 1 and used to derive the input signal-to-noise ratio as discussed in Chapter 2. The DQE is then derived as in equation (2.11).

Properties of a detector/integrating amplifier combination can be measured using techniques similar to those just described. The net response is obtained as the difference of signals obtained with the calibrated source alternately in view and hidden from view synchronously with the read cycle of the amplifier. These data can be obtained by synchronizing the action of the chopper wheel with the reset/readout cycle of the amplifier. This measurement gives, for example, volts per watt of photon energy received by the detector. By measuring the gains of the integrating FET and any other amplifier stages, this measurement can be converted to volts per watt at the integrating capacitance, C_s. Extending the measurements to obtain linearity, dynamic range, and spectral response is fairly straightforward. The voltage noise is measured by removing the source, accumulating a series of amplifier reads, and computing the root-mean-square scatter in the output voltage. It again can be referred to the input with knowledge of the gain of the FET.

To convert the responsivity to units of current and the voltage noise to read noise requires knowledge of the integrating capacitance, C_s. These conversions are necessary to understand the operation of the circuit and the detector separately. This information can be obtained in a variety of ways depending on the design of the integrating circuit; in many cases, a controlled charge can be injected onto C_s and the resulting voltage change measured. Even if the circuit is not built in this way, there is a useful general technique to obtain C_s as long as the readout does not add noise above the square root of the number of collected charges. In this technique, one compares the output noise, $\langle V_N^2 \rangle^{1/2}$, with the signal, V. The signal is proportional to the number of collected charges and the noise to the square root of this number, that is,

$$V = \frac{Nq}{C_s}, \tag{6.38}$$

and

$$\langle V_N^2 \rangle^{1/2} = \frac{N^{1/2} q}{C_s}, \tag{6.39}$$

where N is the number of collected charges. Consequently,

$$C_s = \frac{qV}{\langle V_N^2 \rangle}. \tag{6.40}$$

In the more realistic case where there is a noise contribution from the amplifier, the same argument can be made by comparing signal and noise over a range of signal values so the amplifier noise can be separated from the statistical uncertainty in the number of charges. Caution is advised in this situation since some amplifiers have an amplitude-dependent noise component that could be confused with the statistical noise.

6.6 Examples

6.6.1 Readout performance

Assume a photodiode is operated near zero bias and at a temperature of ~ 170 K, where its zero bias resistance $R_0 = 2 \times 10^{12}\ \Omega$ and $C = 36$ pF. Compare the performance of this detector combined with the following readout circuits: (a) a TIA operating at 1Hz bandwidth with $R_f = 10^{11}\ \Omega$ and $C_f = 1$ pF; (b) a simple source follower integrating amplifier operating with the signal determined by sampling, resetting, and sampling again with 0.5 s between resets; and (c) a simple integrating amplifier operating as in (b) but with the signal derived from samples at the beginning and end of the integration ramp (without resetting). In all cases, assume that $1/f$ noise is negligible and that none of the systems is shot noise limited. The diode properties (other than capacitance and resistance) will cancel in the comparisons. Just to be specific, though, we will assume $\eta = 0.6$ and that we are operating at 0.9 μm.

(a) For the TIA, the limiting noise is Johnson noise, and from equation (6.16) with $\omega = 2\pi f = 6.28$Hz and $\tau_f = R_f C_f = 0.1$ s, we get

$$\frac{\langle (V_f^N)^2 \rangle^{1/2}}{df^{1/2}} = 2.59 \times 10^{-5}\ \mathrm{V\ Hz^{-1/2}}.$$

Similarly, from equation (3.42) with $R = R_0$,

$$\frac{\langle (I_d^N)^2 \rangle^{1/2}}{df^{1/2}} = 6.85 \times 10^{-17}\ \mathrm{A\ Hz^{-1/2}}.$$

Taking this value as the current noise from the detector, from equation (6.15) we get

$$\frac{\langle (V_{out,d}^N)^2 \rangle^{1/2}}{df^{1/2}} = 5.80 \times 10^{-6}\ \mathrm{V\ Hz^{-1/2}}.$$

Combining the two noise contributions quadratically, the total noise is

$$\frac{\langle (V^N)^2 \rangle^{1/2}}{df^{1/2}} = 2.65 \times 10^{-5}\ \mathrm{V\ Hz^{-1/2}}.$$

From equations (5.1) and (6.10), the signal voltage is

$$|V_{out}^S| = 8.14 \times 10^{-9} \varphi\ \mathrm{V\ s},$$

where φ is the photon arrival rate. The power incident on the detector is $P = hc\varphi/\lambda$, so the voltage responsivity of the circuit is

$$S = \frac{|V_{out}^S|}{P} = 4.10 \times 10^{16} \lambda(\mathrm{m})\ \mathrm{V\ W^{-1}},$$

or for $\lambda = 0.9 \times 10^{-6}$ m,

$$S = 3.69 \times 10^{10}\ \mathrm{V\ W^{-1}}.$$

The NEP is

$$\frac{\langle (V_f^N)^2 \rangle^{1/2}}{S\, df^{1/2}} = \frac{2.65 \times 10^{-5}\ \mathrm{V\ Hz^{-1/2}}}{3.69 \times 10^{10}\ \mathrm{V\ W^{-1}}} = 7.18 \times 10^{-16}\ \mathrm{W\ Hz^{-1/2}}.$$

This result implies that the detector/amplifier would detect a flux of

$\varphi = 3255\,\mathrm{s}^{-1}$ at a signal to noise of unity ($df = 1$ Hz). However, interpreting NEP in this fashion gives an overly optimistic estimate of the system performance, since it does not allow for the comparison of the signal viewing the source with that not viewing the source, a step that is virtually always necessary for detection at low levels. To include this effect, we assume a specific measurement strategy: that the signal is chopped in a square wave manner and that the detector output is measured as the rms voltage. The response is then reduced by the factor 0.4502, with no change in the noise. Hence, a signal level of $\varphi = 7230\,\mathrm{s}^{-1}$ is required for a signal to noise of unity.

(b) Assuming that $\varphi = 7230\,\mathrm{s}^{-1}$, we need to compute the signal to noise that would be achieved by the integrating amplifier. First calculate the signal current in electrons per second:

$$I^{S} = \eta\varphi = 4338 \text{ electrons s}^{-1}.$$

The output noise during one readout is given by equation (6.22) to be

$$\langle Q_N^2 \rangle^{1/2} = \frac{(kTC)^{1/2}}{q} = 1816 \text{ electrons.}$$

In analogy with the measurement strategy adopted for the TIA, a measurement is made by spending 0.5 s 'on source' and 0.5 s on background. In the 0.5 s on source, we accumulate (0.5 s)(4338 electrons s^{-1}) = 2169 electrons. The noise from the two reads is $\sqrt{2} \times$ (1816 electrons) = 2568 electrons. Thus, $S/N = 0.84$, so this detector/amplifier arrangement is 1.0/0.84 = 1.2 times *less* sensitive than the TIA system in part (a).

(c) The arguments involving signal calculation are the same as in case (b). In this readout scheme, the noise per reset is diminished by a factor $(1 - e^{-t/RC})$. The time constant of this detector is $\tau = R_0 C = 72$ s. The kTC noise in 0.5 s is then

$$(1 - e^{-t/RC})(1816 \text{ electrons}) \approx 13 \text{ electrons.}$$

In two reads, the total noise is then $\sqrt{2} \times$ (13 electrons) = 18 electrons. We then get $S/N = 2169/18 \sim 120$, so this readout scheme gives a S/N about 120 times greater than the one using the TIA.

6.6.2 Performance measurement

Suppose the detector of the example in Chapter 3 is calibrated as illustrated in Figure 6.12. Let the source be a blackbody at 1300 K that shines through a round aperture 0.5 mm in radius. The optical filter transmits light in the range $0.99\,\mu\mathrm{m} \leqslant \lambda < 1.01\,\mu\mathrm{m}$ with an average efficiency of 60%. The aperture over the source is at a distance of 1 m from the detector, and the net gain of the detector amplifier and electronic filter is 100. An oscilloscope measures a sinusoidal signal from the amplifier/filter of 1.1×10^{-10} A, peak-to-peak. When the view of the source is blocked off, the ammeter measures a signal of 1.2×10^{-10} A (rms); when the input is switched to the noise calibrating resistor, which has a value of 100 000 Ω, and whose noise output is amplified by 1000, the ammeter indicates a signal of 1.5×10^{-9} A (rms). Determine: (a) the power incident on the detector; (b) the responsivity; (c) the noise; (d) the NEP; (e) D^* for the detector at $\lambda = 1.00\,\mu$m; and (f) the DQE.

(a) From equation (1.5), the spectral radiance is 1.85×10^9 W m^{-3} ster^{-1}. Given the blackbody area of 1.96×10^{-7} m^2, bandwidth of 2×10^{-8} m, solid angle of the detector viewed from the source of 7.85×10^{-7} ster, and filter efficiency of 0.6, it can be shown that the power received at the detector is 3.40×10^{-12} W.

(b) The signal into the oscilloscope can be converted to an output from the detector by dividing by the gain of the detector amplifier, giving 1.1×10^{-12} A peak-to-peak. The peak-to-peak measure gives twice the sinusoidal amplitude; therefore to convert to rms we divide by $2\sqrt{2}$, giving 3.89×10^{-13} A (rms). The input signal is converted to rms by applying the waveform factor, giving $0.4502\ (3.40 \times 10^{-12}\ \mathrm{W}) = 1.53 \times 10^{-12}$ W (rms). The responsivity is then

$$S = \frac{3.89 \times 10^{-13}\ \mathrm{A(rms)}}{1.53 \times 10^{-12}\ \mathrm{W(rms)}} = 0.254\ \mathrm{A\,W^{-1}}.$$

(c) From equation (3.42), the Johnson noise from the resistor (assumed to be at room temperature ~300 K) is 4.07×10^{-13} A Hz$^{-1/2}$. The noise from the detector can be obtained by comparing its signal when the source is blocked off with that from the resistor, correcting for gains:

$$\frac{\left[\dfrac{1.2 \times 10^{-10}\ \mathrm{A}}{100}\right]}{\left[\dfrac{1.5 \times 10^{-9}\ \mathrm{A}}{1000}\right]} \times (4.07 \times 10^{-13})\ \mathrm{A\,Hz^{-1/2}} = 3.3 \times 10^{-13}\ \mathrm{A\,Hz^{-1/2}}.$$

(d) The NEP is given by the noise divided by the responsivity,

$$\mathrm{NEP} = \frac{3.3 \times 10^{-13}\ \mathrm{A\,Hz^{-1/2}}}{0.254\ \mathrm{A\,W^{-1}}} = 1.3 \times 10^{-12}\ \mathrm{W\,Hz^{-1/2}}.$$

(e) D^* is the square root of the area (in centimeters squared) divided by the NEP, or

$$D^* = \frac{0.1\,\text{cm}}{[1.3 \times 10^{-12}\,\text{W}\,\text{Hz}^{-1/2}]} = 7.7 \times 10^{10}\,\text{cm}\,\text{Hz}^{1/2}\,\text{W}^{-1}.$$

(f) From equation (1.1), the energy of a 1 μm photon is 1.99×10^{-19} J, and using the power at the detector from (a), we find that $\varphi = 1.71 \times 10^7$ photons s^{-1}. The noise in a 1s integration is the square root of the number of photons received, multiplied by the energy per photon, or 8.2×10^{-16} J. Converting from a 1s integration to the equivalent frequency bandpass via equation (3.39) and expressing as NEP, we get NEP $= 1.16 \times 10^{-15}$ W Hz$^{-1/2}$. The achieved NEP from (d) is a factor of 1121 higher; the DQE is then $2/(1121^2) = 1.6 \times 10^{-6}$. The 2 in the numerator arises because of the convention that the $\sqrt{2}$ increase in G–R noise for a photoconductor compared with a photodiode is not included in the DQE.

6.7 Problems

6.1 Repeat Example 6.6.1 for a net integration of 100 s. Assume the integrating amplifiers perform two integrations of 50 s each.

6.2 Evaluate the nonlinearity of a simple integrator with the photodiode designed in the example at the end of Chapter 5. Compute the response in volts per charge for diode biases of 0, -1, and -2 V.

6.3 Plot the signal and noise as a function of frequency from 1 to 100 Hz for the TIA and detector in Example 6.6.1. Assume the amplifier electronic noise is $V_A^N = 1 \times 10^{-7}$ V Hz$^{-1/2}$ independent of frequency. Take the detector noise to be $I_D^N = (1 \times 10^{-15}$ A Hz$^{-1/2})(1/f)$, where f is in hertz.

6.4 Evaluate a test of the detector in Problem 4.1. Assume that the conditions of measurement are the same as in Example 6.6.2. The response is measured through a filter that transmits with 60% efficiency between 19.8 and 20.2 μm. With the blackbody chopped with a room temperature chopper, the detector/amplifier give a net signal of 0.40 nA. The detector noise signal (detector blocked off) is 4.0×10^{-14} A Hz$^{-1/2}$; the noise viewing the source is exactly five times larger. Compute NEP, D^*, and DQE.

6.5 Given the following data set for a simple integrating amplifier, derive the node capacitance and the amplifier noise:

Signal (mV)	Rms noise (μV)
0.32	32
0.96	35
3.2	44
9.6	63
32	106
96	178
320	321

6.6 The signal from an integrating amplifier is extracted by a destructive read, with a sample followed by resetting followed by the second sample. The amplifier noise is white. The output is taken to a simple RC low pass filter, with time constant $\tau = RC$ adjusted to maximize the ratio of signal to noise. If Δt is the time interval between samples, derive the optimum value of Δt in terms of τ.

7
Arrays

Photoconductors and photodiodes are combined with electronic readouts to make detector arrays. The progress in integrated circuit design and fabrication techniques has resulted in continued rapid growth in the size and performance of these solid state arrays. In the infrared, these devices are based on a (conceptually) simple combination of components that have already been discussed; an array of integrating amplifiers is connected to an array of detectors. These devices can provide many thousands of detector elements, each of which performs near the fundamental photon noise limit for most applications between wavelengths of 1 and 30 μm. For visible and near infrared detection, monolithic structures can be built in silicon, forming arrays with millions of high performance pixels. The most highly developed of these visible detectors is the charge coupled device, or CCD, which combines an array of intrinsic photoconductors, a sequentially addressable array of integrating capacitors, and an FET output amplifier, all built together on a single wafer of silicon. Operated at low temperatures, CCDs can reach fundamental detection limits for virtually all applications between the soft X-ray and ~ 1 μm, except when rapid time response is required

7.1 Infrared arrays

In theory, a rudimentary detector array could be produced by building a supply of detectors and amplifiers and connecting them together. For more than a few pixels, however, this procedure is tedious, although it has frequently been pursued. A much more streamlined construction method is possible using integrated circuit technology to miniaturize the amplifiers so that each one is no bigger than the detector it serves. The amplifiers are produced in a grid (or other suitable pattern), each with an exposed contact for the input signal. The detectors are produced in a mirror-image grid and have exposed output contacts. Bumps of indium solder are deposited on both sets of contacts, and the detector and amplifier pieces are 'flip chipped' by carefully aligning them and squeezing them together. When the indium bumps deform under this pressure, their naturally occurring oxide coatings tear, exposing the bare metal. The indium metal on the two contacts welds, forming the necessary electrical connections between the detectors and amplifiers as well as providing the mechanical strength to hold them together. In some cases, epoxy is flowed into the space between the readout and the detectors to increase the bonding strength. The resulting device is called a bump bonded direct hybrid array; it is illustrated in Figure 7.1.[1]

Our breezy discussion of indium bump bonded arrays glosses over many difficulties. Fundamentally, the success probability of each individual processing step is less than unity, sometimes much less. The probability of obtaining a satisfactory array at the end of all the steps is the product of all the intermediate probabilities, so achieving high yields can be difficult. A low yield means many arrays are thrown out along the way; consequently the price of a finished device is high.

The salient advantage of the technique, however, is that the readouts and detectors can be optimized separately in terms of choice of materials and processing. Readouts are usually made in silicon, for which there is an extremely well developed technology for fabrication of high performance ultra-miniature electronic circuits. The detectors can be made of InSb, HgCdTe, or any other material that gives optimum performance for the intended application.

Even for extrinsic silicon detectors, the processing steps for the electronics and the infrared detectors can be sufficiently different that it is advantageous to produce them separately and bump bond them together – for example, high

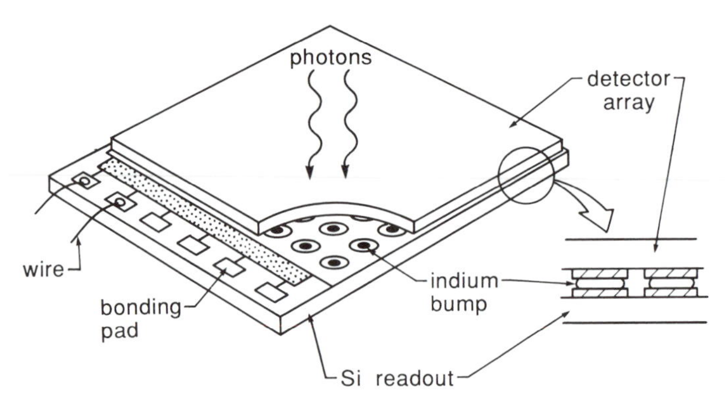

Figure 7.1. Infrared direct hybrid array.

temperatures used in producing readouts may damage the detectors. In addition, when the detectors are produced together with the readouts, the results may be unsatisfactory because both functions must share the area on the wafer of semiconductor material. For example, the use of a single piece of silicon for both detectors and readouts combined with the relatively simple processing to make Schottky diodes can allow manufacture of arrays with large numbers of pixels. However, since the silicon area must be divided between the diodes and their readouts, the percentage of the array area that is sensitive to photons – the 'fill factor' – is low, which can result both in a further reduction in realizable quantum efficiency from the low values intrinsic to the Schottky diodes and also in difficulties in extracting quantitative measurements because the response of the array is not uniform on the scale of its pixels. As a result, high performance Schottky diode arrays are usually bump bonded so they can achieve a fill factor approaching 100%.

Limitations arise in hybrid array construction because of the necessity for very accurate dimensional control during the construction of both the readout and the detector arrays. For detectors of material other than silicon, this difficulty is exacerbated because the devices are constructed and stored at room temperature but must be cooled for operation. Dimensions must be held within tolerance limits over a large temperature range in spite of the differing thermal contraction and expansion properties of the detector material and its silicon readout. With commonly used detector materials, it is possible to attain device sizes up to about 1 cm square while still retaining adequate dimensional control to survive cooling. To produce a 256×256 array thus requires cell sizes on the order of 40 μm on a side. Efforts to produce arrays with more pixels center on further reducing the size of the unit cell and on thinning the detector or readout layers so that they can stretch to accommodate the contraction of the other material. An additional difficulty with constructing large arrays is controlling the bonding forces so that the indium bumps are sufficiently compressed without breaking the thin wafers of material that carry the detectors and readouts or having the wafers slip relative to each other causing adjacent bumps to touch.

There are, in addition, a number of restrictions placed on the readout circuits that are used with infrared arrays. The necessity of having small unit cells requires that the circuits have a minimal number of components. A second concern is electroluminescence of the circuits. As mentioned in Chapter 1, solid state devices (for example, LEDs) can *emit* light if a current is passed through them. Even though silicon is an indirect semiconductor and this process is suppressed, low-level self-emission by the readout can be a deadly problem in a high performance detector array.

It has been found that the simple integrator circuits based on MOSFETs can collect charge on their gate capacitors equally well when they are off as when they are on. Turning these circuits off causes electric charges to be introduced at various points within the circuit through unavoidable point-to-point capacitances; however, when power is restored, the circuit recovers its previous 'on' state so accurately that its read noise is not degraded. By leaving the MOSFET amplifiers off while the circuit is accumulating a signal, the power requirements for these arrays can be kept very small. Low power consumption is an important consideration in the design of cooling apparatus, and it is also helpful in keeping the detector array at the proper operating temperature. Moreover, the problems with electroluminescence can be minimized if power is provided to the readout only for the brief instant when its output is actually being measured. This operating mode cannot be used with a CTIA because this circuit must be on continuously if it is to remain in balance, so reducing its power requirements and controlling its self-emissions can be more difficult than with other types of readout. The CTIA does retain its intrinsic advantages discussed in Chapter 6, and arrays at the current state of the art use a variety of readout types, including CTIAs, depending on the application.

There is another important restriction placed on the detectors in a hybrid array. These devices and their contacts are constructed on one side of a wafer of suitable semiconductor, and this side is subsequently placed in contact with the readout. Therefore, the photons must enter from the other side, an arrangement called backside illumination. The detector wafer must be thin enough to allow the photo charge carriers to migrate from their generation sites to the inputs of the readout amplifiers. For example, in the case of photodiodes, not just the diode alone but the entire wafer carrying the diode must be thin enough to have good collection efficiency (as expressed in equation (5.24)). The required thickness may be 10 μm or less and must be held to the correct value over the entire array; for example, there may be a tolerance of 1 μm in thickness uniformity across a 1 cm array. Two approaches have been taken to meet this requirement. In the first, the detector array is produced on a wafer that is sufficiently thick to be mechanically strong. After placing the detector array in contact with the readout to provide additional strength, the resulting sandwich is mechanically thinned to the final detector thickness. In the second, a thin layer of detector material is grown onto a substrate of some other material that is both mechanically rugged and transparent at the wavelengths of interest. The detectors are grown with their contacts on the exposed surface, which is bump bonded to the readout. The detectors can then be backside illuminated through the transparent substrate.

The readout is given the final job of reducing the number of array output lines from one per pixel to a manageable number (for example, one, two, or four for the *entire array* of thousands of pixels). The signals appear sequentially on these output lines, where they can be processed serially by the following electronics. It is said that the signals have been 'multiplexed' through a single set of electronics, and the entire readout is frequently referred to as a multiplexer, or MUX, even though it has a number of additional functions.

A particularly simple implementation of the multiplex switching action addresses a column of unit cells by turning on the power to their amplifiers; the cells in other columns that are attached to the same output line but which have their power off will not contribute signal. This arrangement is illustrated in Figure 7.2. The operation can be understood in terms of the upper left cell. Signal is collected from the photodiode on the gate of T_1. Readout of the signal is enabled by using R_1 and C_1 to close the switching transistors T_2, T_3, and T_4. They apply power to T_1 and connect it to the output bus so a reading can be taken through the output amplifier. Pulsing the reset line connects V_R to the photodiode through T_5 and T_3 to re-establish the detector bias. A second reading can be taken, if desired, with the reset switch, T_5, closed. This switch is then opened and another reading taken of the output

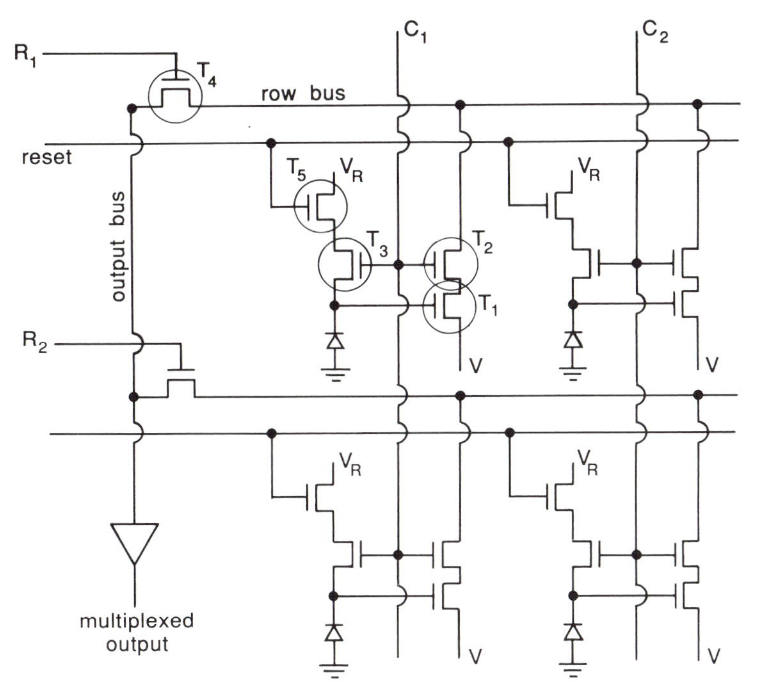

Figure 7.2. Four cells of the readout for an infrared array.

of T_1, after which the cell is turned off by opening T_2, T_3, and T_4. Note that no other cell in the array has been disturbed by this sequence of events.

An important aspect of some multiplexing schemes is that appropriate control electronics allow the MUX to address any unit cell or combination of unit cells on the array. This feature is called random access, and it allows sub-units of the array to be used whenever it is beneficial (for example, when time response is desired that is faster than can be achieved by switching through the entire array). As pointed out in the preceding paragraph, the array in Figure 7.2 operates in this manner.

Currently, indium bump bonded direct hybrid arrays are used extensively for measurements at wavelengths between 1 and 30 μm and are under development out to 40 μm. They are seldom used at shorter wavelengths because very high performance monolithic arrays based on intrinsic absorption in silicon are available (see below). Because the readout electronics for these latter devices can be built on the same piece of silicon as the detectors, the yield problems typical for bump bonded arrays are avoided. At wavelengths longer than 40 μm, conventional hybrid arrays become less useful because of the unsuitable characteristics of photoconductors used in the far infrared. Between 1 and $\sim$10 μm, arrays can be based on well developed photodiode technology in HgCdTe, InSb, PtSi, or other materials, and between 1 and 6 μm they can achieve quantum efficiencies of 80–90% (if antireflection coated) and read noises of 10–70 electrons. Between roughly 4 and 40 μm, arrays can be based on extrinsic silicon photoconductors or silicon BIB detectors, with quantum efficiencies of 30–80% and read noises of $\sim$50 electrons. Reviews of infrared array technology can be found in Norton (1991) and Scribner, Kruer, and Killiany (1991). The limiting noise in infrared arrays is reviewed by Nelson, Johnson, and Lomheim (1991).

There is no well-established technology for building arrays that operate between 40 and 120 μm, despite the availability of high performance germanium photoconductors in this spectral region. There are a variety of reasons for this situation. Germanium does not lend itself to photoconductor arrays as well as silicon does. Absorption lengths are long, necessitating thick detectors and low photoconductive gains with transparent contact geometries. The shallow impurity levels required for long wavelength detection also require that the detectors be operated at very cold temperatures to control thermally excited currents; hence, the detectors must be isolated from the heat produced by their readouts. Diffusion lengths are long, leading to crosstalk between pixels. Most importantly, there has been no application for far infrared arrays that has commanded sufficient resources to overcome these difficulties.

7.2 Charge coupled devices (CCDs)

For intrinsic silicon detectors, it becomes advantageous to process the readout and detector arrays on a single piece of silicon. The readout can be designed so the collected charge carriers are conveyed to a readout amplifier outside the photosensitive area, so there need be no competition for real estate between detectors and amplifiers, and therefore high fill factors can be achieved. The integration of detector and readout onto a single monolithic piece of silicon reduces the number of processing steps (compared with hybrid arrays), and therefore increases yields and reduces costs. This approach leads to very high performance detector/amplifier combinations such as the charge injection device (CID) and the charge coupled device (CCD).

7.2.1 Operation of a single pixel

Refer to Figure 7.3 for the following discussion. A thick oxide layer has been grown onto a piece of silicon. A metal electrode has been evaporated onto the oxide, which as an insulator blocks the passage of the electrons from the silicon onto the electrode (see Figure 7.3(a)). This structure is called a metal–oxide–semiconductor, or MOS, capacitor; we have already encountered it as the key element in a MOSFET (Chapter 6). The electrical properties of the silicon have been controlled in this example by doping it p-type; because extrinsic absorption is relatively weak compared with intrinsic absorption, this doping has little effect on the photoresponse.

Suppose the bulk semiconductor is grounded, and at time $t = 0$ a positive voltage V_g is put on the electrode (see Figure 7.3(b)). Because of the p-type doping, there will be virtually no conduction electrons to drift toward the electrode. The voltage will cause any holes in the immediate vicinity of the electrode to drift away, creating a depletion region. This process is virtually identical to the creation of a depletion region in a BIB detector, and it can be described mathematically as in equations (4.7) and (4.8). Unlike the BIB, however, any free charge carriers created in the silicon will accumulate in the depletion region because they cannot penetrate the oxide layer to the electrode.

The voltage on the electrode provides the bias for an intrinsic photoconductor in the silicon. Any photons absorbed by the silicon will produce a photocurrent that will flow between the contacts. The free electrons that are created within the depletion region or that diffuse into it will drift toward the electrode and collect opposite it at the Si–SiO_2 interface. The detector is operated with its photoconductive gain very close to unity, so nearly every photoelectron is collected. The electrons will accumulate against the oxide

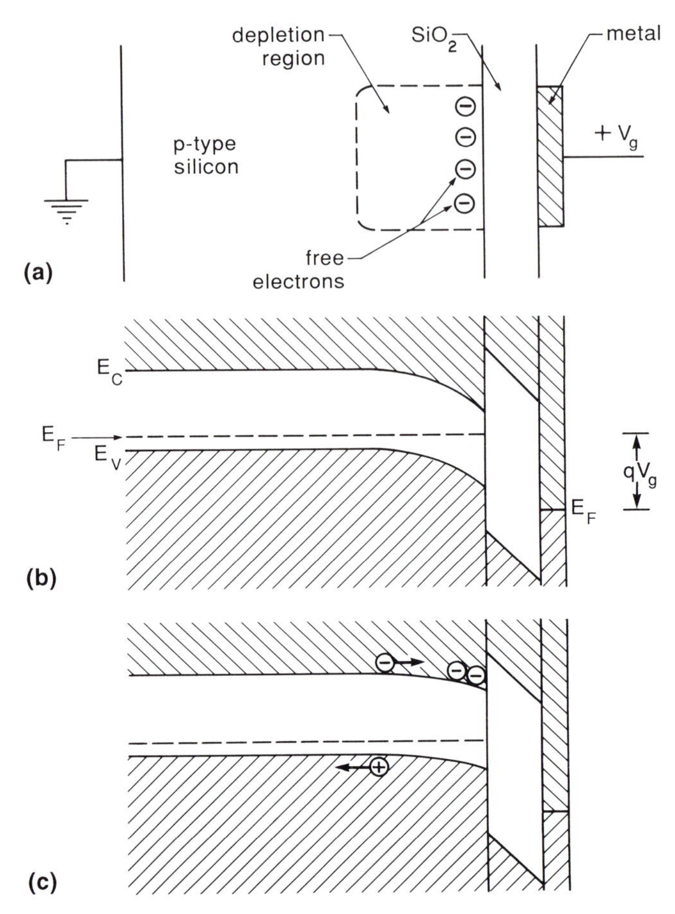

Figure 7.3. Charge collection at a single pixel of a CID or CCD: (a) illustrates the collection of free charge carriers at the silicon–oxide interface; (b) shows the band diagram immediately after application of a voltage V_g to the gate to form a well; and (c) shows the filling of the well as it collects charge carriers.

layer because there are virtually no opportunities for them to recombine in the depletion region. Assuming that all the electrons generated in the photoconductor diffuse into the depletion region, the electrons collected against the insulating layer in a long exposure will accurately reflect the number of photons absorbed by the silicon photoconductor during this time.

Free electrons can also be created by thermal excitation; at room temperature, the thermal dark current can dominate the photocurrent and may limit the maximum permissible integration time. In low light level applications, the dark current must be suppressed by cooling the array to 130–160 K, where dark currents in the best devices become virtually unmeasurable. In the following discussion, we make our usual simplifying assumption that the device is operated at a low enough temperature that thermal generation can be ignored.

Although we have assumed p-type bulk semiconductor and positive V_g, the same arguments would hold with n-type semiconductor and a negative V_g; in such a case, holes from any photocurrent would collect at the storage well. Similar structures can be built on other semiconductors besides silicon, in which case they are termed metal–insulator–semiconductor (MIS) devices. Infrared arrays can utilize MIS structures grown on a small-bandgap semiconductor. However, these devices tend to have undesirably small breakdown voltages (leading to small well capacity) and poor charge transfer (leading to high noise and compromised imaging). These difficulties have led to the general adoption of hybrid arrays for the infrared with the readout fabricated on a piece of silicon.

As Figure 7.3 is drawn, the electrode of the MOS capacitor would block photons from illuminating the depletion region directly. If the electrode is transparent, which can be achieved by making it of very thin silicon and doping it heavily so it will be electrically conducting, then the depletion region can be illuminated through it (the oxide layer is very transparent through the visible and ultraviolet). This geometry is referred to as front illumination. Another important benefit of doped silicon electrodes is that electrical shorts between gates introduced by manufacturing flaws are usually not so highly conducting as to be fatal to the operation of the detector. Early CCDs that used metal electrodes suffered from very low yields because of such shorts.

Heavily doped thin silicon electrodes, however, are not transparent in the blue; as shown in Figure 3.5, direct transitions can occur with absorption of photons of wavelength shorter than 0.41 μm, and the absorption coefficient goes up correspondingly. Even at wavelengths somewhat longer than this value, conventional silicon electrodes degrade the detector quantum efficiency. If silicon nitride is used as an insulator (in place of silicon oxide), it is found that significantly thinner silicon electrodes can be made and the transmittance improved. Electrodes can also be made of materials that are more transparent than silicon, such as doped indium tin oxide. Alternatively, the problems associated with transparent electrodes can be circumvented with CCDs that are backside illuminated, that is, illuminated through the side opposite the electrodes.

Backside illumination leads to another set of issues that must be circumvented for good performance in the blue and ultraviolet. As viewed by the photons, the depletion region in a backside illuminated CCD lies behind a layer of silicon. Because of the very short absorption lengths for blue or ultraviolet photons, they generate charge carriers primarily near the back surface of the detector. The charge carriers must diffuse from these creation sites across any intervening silicon and into the depletion region. The physics of this process is very similar to that already discussed with reference to photodiodes, and the quantum efficiency will be as described in equation (5.24) (it can be worse than this equation predicts if significant numbers of charge carriers get caught in surface state traps at the backside of the silicon). A variety of techniques can be used to ensure that virtually all the charge carriers diffuse into the depletion region.

The most direct method to encourage full collection of charge carriers from blue and ultraviolet photons is to thin the detector. If much of the intervening silicon is removed, the photons will be absorbed close enough to the depletion region that the charge carriers will diffuse efficiently into it. However, if the detector is thinned below an absorption length, its quantum efficiency will suffer because most photons will produce no charge carriers. The absorption lengths for silicon are a strong function of the wavelength (see Figure 3.5), so an optimum thickness will depend on the wavelength at which maximum efficiency is desired. For example, at $\lambda = 1\ \mu$m, the absorption length is $\sim 80\ \mu$m, while at $\lambda = 0.4\ \mu$m, the absorption length is only $\sim 0.3\ \mu$m. This large range of absorption lengths would pose an insoluble dilemma, except that the diffusion length for electrons in silicon is relatively long, up to $250\ \mu$m in intrinsic material at 150 K, and still $10-50\ \mu$m at the same temperature in the moderately doped material found in CCDs. It has been found that good blue quantum efficiency can be achieved by thinning to $15-20\ \mu$m. From comparison with the absorption length at $1\ \mu$m, however, it is clear that thinning to this degree will reduce the signals in the near infrared. In addition, it is difficult to achieve uniform thickness, and thus uniform response, at wavelengths where very thin detectors are needed.

A further important step in improving blue sensitivity is to treat the back surface of the CCD. For example, flooding the CCD with intense ultraviolet light is found to enhance the response. The ultraviolet photons cause photoemission of electrons in the bulk silicon, and some of these free electrons migrate to the backside surface, where they fall into traps in the oxide layer there. The trapped electrons persist after the ultraviolet radiation is removed and create an electric field that repels photoexcited electrons and prevents them from being trapped at the backside so they will diffuse toward the storage wells. Similar results can be achieved by flooding the chip with gases such as chlorine or nitric oxide (NO). Another approach is to add a junction with a thin metal layer (or a thin layer of appropriately doped silicon) that repels the photoelectrons from the back surface. Platinum has been used in this application because it does not diffuse into the silicon; however, these platinum 'flash gates' have proved to be unstable under vacuum. A junction can also be created by ion implanting boron in a thin layer at the backside of the CCD. Still another approach is to create a 'biased flash gate' by adding a thick insulator layer and a transparent electrode, for example of indium tin oxide, and putting a potential on the electrode that repels the photoexcited electrons. When combined with appropriate thinning, a variety of these techniques can produce good blue response with a high degree of uniformity and with quantum efficiency limited primarily by reflection losses. At wavelengths short of about 0.3 μm, however, ultraviolet flooding is required for this performance level, both because of the very short absorption lengths in the silicon and because electrode materials such as indium tin oxide also become absorbing.

Photons with wavelength shorter than $\sim$0.3 μm typically generate more than one electron–hole pair when they are absorbed in the CCD. The 'quantum yield' is defined as the number of electron–hole pairs per interacting photon. For energetic photons (that is, those of wavelength shorter than 0.1 μm), it is found that 3.65 eV is required per electron-hole pair; the excess energy above the bandgap is dissipated as heat in the silicon lattice (Janesick and Elliott, 1991). For X-rays, the number of charge carriers is therefore a measure of the energy of the photon. The uncertainty in this energy measurement consists of the usual uncertainties imposed by the operation of the CCD for lower energy photons, plus a term called Fano-noise due to the statistical nature of the loss of energy to thermal excitation of the lattice.

7.2.2 Readout of the MOS capacitor by charge injection

The processes described in the preceding section will proceed through the state in Figure 7.3(c); eventually the field from the collected electrons balances that from V_g and there is no longer any net field to attract additional electrons (that is, the energy bands and E_F are horizontal through the well). The number of electrons that the MOS capacitor can hold is the well capacity. It is given by

$$Q_W = C_0(V_g - V_T), \tag{7.1}$$

where V_g is the voltage on the electrode, V_T is the threshold voltage for formation of a storage well, and

$$C_0 = \frac{A\kappa_0\varepsilon_0}{X_0}, \tag{7.2}$$

where A is the electrode area, κ_0 is the dielectric constant of silicon dioxide (=4.5), ε_0 is the permittivity of free space, and X_0 is the thickness of the SiO_2

layer. For example, if $V_g - V_T = 3$ V, $X_0 = 0.1\ \mu$m, and the electrode is 15 μm on a side, we find that $Q_W \sim 1.7 \times 10^6$ electrons.

Because the filled well can no longer collect charge, it must be discharged to continue detection. The simplest way to discharge a MOS capacitor is to adjust its voltage so that any collected charge is injected into the silicon substrate (see Figure 7.4). Of course, we do not just want to get rid of the charge, we want to measure it! To do so, we can measure the current in the substrate, I_{ss}. If an empty MOS capacitor is pulsed, the current will show a typical capacitively coupled differentiated waveform (Figure 7.5); if there is a stored charge, it will add to this capacitively coupled current and make the waveform asymmetric in proportion to the additional charge injected. If the output waveform is integrated, the switching transients from voltage changes on the capacitor will cancel, but the collected charge will appear on the output of the integrator. Since a current equal to I_{ss} must also flow in the reset line to remove the stored charge, the accumulated signal can also be measured by monitoring the reset current. A detector that is read out in one of these fashions is called a charge injection device, or CID.

A CID array readout can operate by injecting the charge from one electrode and collecting it under a neighboring one. If these electrodes are connected as in Figure 7.6, with one set on column drivers and the second on row drivers, individual pixels can be addressed. To illustrate, the well potential and the 'sea' of charge carriers are drawn schematically under the electrodes in the figure. Row 2 has been selected for readout by closing its switch to the amplifier line. Charge has been transferred to the row electrodes along column 2 by pulsing the C_2 line. If the amplifier output is sensed and the charge-holding

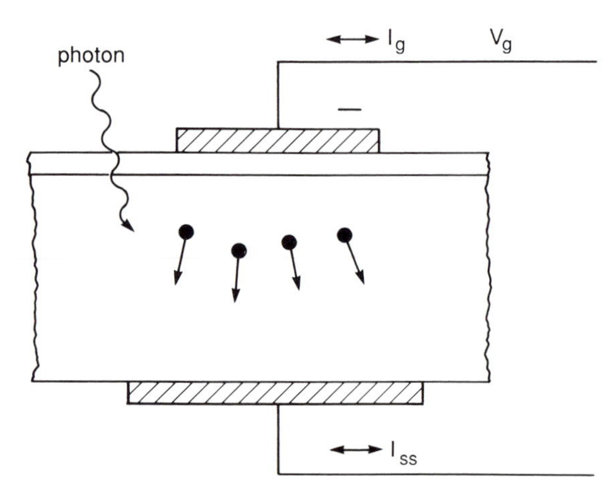

Figure 7.4. Charge injection.

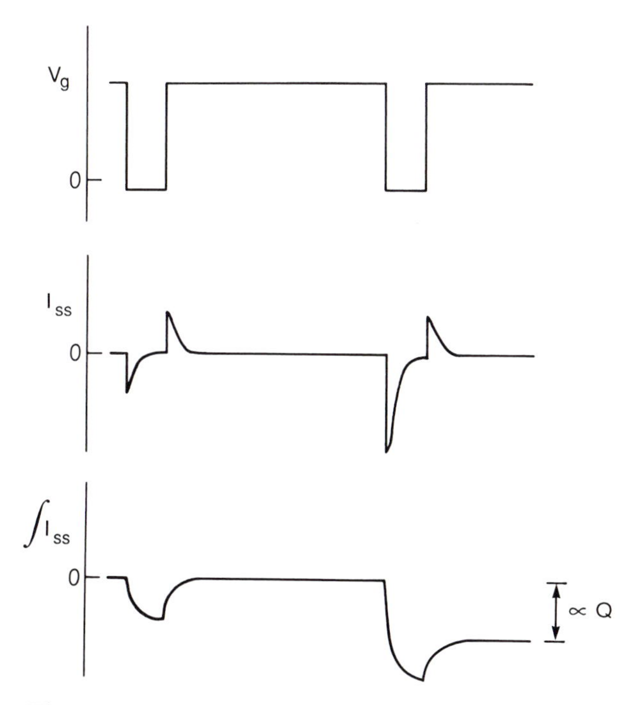

Figure 7.5. Charge injection electrical waveforms.

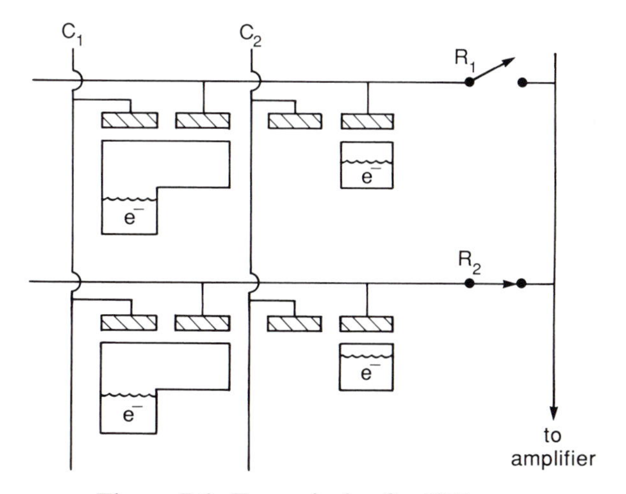

Figure 7.6. Four pixels of a CID array.

voltage then restored to C_2 (as in C_1), the charge in column 2 will go back under the column electrodes. The charge collected by the pixel in the lower right can be read out repeatedly by toggling C_2 to transfer it back and forth between the column electrode and the row electrode. As a result of this repetitive nondestructive readout process, a series of independent measurements of the charge can be made and averaged. Under many circumstances, the signal-to-noise ratio will be improved by the square root of the number of reads.

After the array has been read out, the collected charge is injected into the bulk silicon by setting both the row and column electrodes to repel it. Charge collection is repeated as desired by exposure to light.

7.2.3 Charge coupled readouts

The CID just described suffers from large read noise because an entire row of MOS capacitors is connected at one time to the output amplifier, leading to a large effective gate capacitance (see discussion in Chapter 6). In theory, this problem could be overcome by using repeated nondestructive reads to improve the final signal to noise, but the resulting readout would be extremely slow. The CCD provides an alternate detector architecture in which the surrounding electrodes are electrically isolated from each other except while charge is transferred from one to another, leading to substantially lower read noise.

In the CCD, the collected charges are passed from one electrode to another along columns to one edge of the array and then passed along the edge to an output amplifier. As in the discussion of a single pixel, we will assume that the photo charge carriers are electrons and that the silicon has been doped p-type. Refer to Figure 7.7 for the following discussion, and assume that we

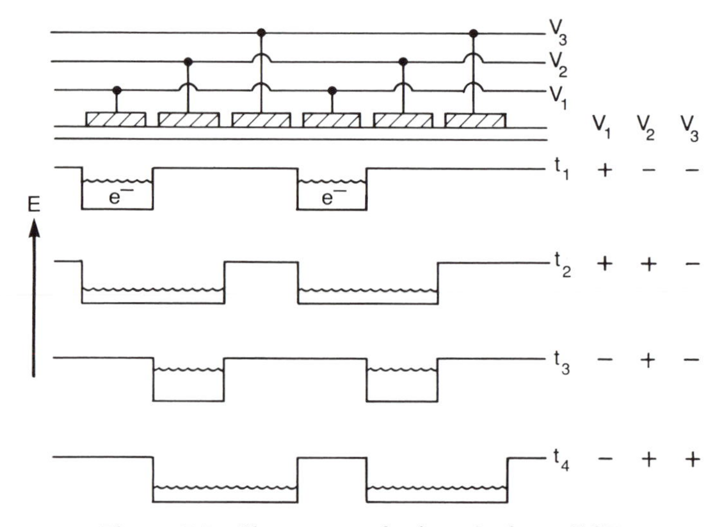

Figure 7.7. Charge transfer in a 3-phase CCD.

want to move the collection of electrons under the left electrode toward the right. The motion of the electrons is initiated by pulsing a neighboring electrode positive so the electrons can collect in the newly created depletion region under it. Shortly later, the electrons are repelled from the original electrode by pulsing it negative, thus driving the transfer to completion. The negative voltage on the original electrode maintains the electrical isolation between the collection of electrons and other electrodes in the CCD. Once the electrons have all been driven into the new depletion region, the entire process can be repeated to shift them to a third electrode.

This process is usually described in terms of electrons flowing between the storage wells under the electrodes. Figure 7.7 shows a row of MOS capacitors and beneath them four configurations of their storage wells in a time sequence. Assume that we start with V_1 positive at time t_1 and hold it at that voltage for an extended period of time to collect signal electrons in the resulting wells. At t_2, we begin the transfer of this signal by placing a positive bias on V_2 and enlarging the storage well. At t_3, we set V_1 negative, contracting the well and driving the signal electrons under the electrode adjacent to the one that initially collected them. At t_4, we set V_3 positive to create a condition similar to that at t_2, thus continuing to pass the charge down the column of MOS capacitors. The pattern of biases on the electrodes acts both to isolate electrically the successive packets of signal charge that are being transferred and to disconnect the successive storage wells from the charge packet as it is passed along.

Figure 7.7 illustrates a 3-phase CCD with three sets of electrodes and voltage lines. The description above, however, applies equally well to a 4-phase CCD. Much more flexibility in CCD design is achieved by making use of the dependence of well depth on the doping of the silicon and the thickness of the SiO_2 insulator layer. This behavior follows from equations (4.8) and (4.10), with slight modifications to allow for the fact that the dielectric constant in the CCD SiO_2 insulator (4.5) is lower than that of the Si insulator (11.8) for the silicon BIB detector. For our purposes, it is not worth repeating the derivation with this modification. We note that the field strength where the well is at the surface of the insulator will be determined by an expression similar to equation (4.10) with equation (4.8) substituted for w. Examination of these two equations shows that this field strength is a function of the gate voltage (V_b in the notation used for the BIB detectors), the oxide layer thickness (analogous to t_b), and the doping (N_A). For example, if the electrodes in a 4-phase CCD are placed at different depths in the oxide and connected in pairs such that the well generated by one member of the pair for a given voltage is deeper than the well for the other member, the CCD method of

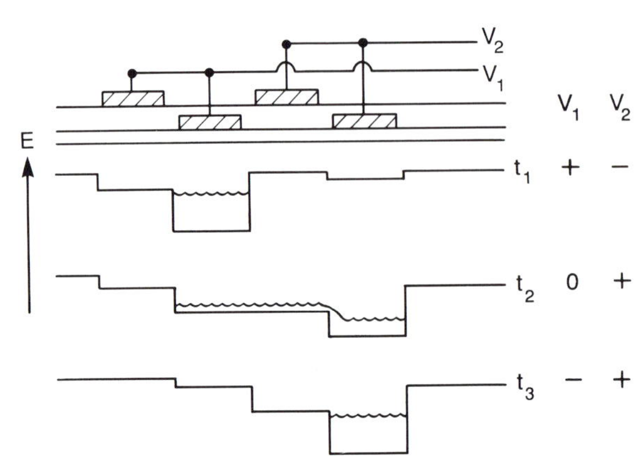

Figure 7.8. Charge transfer in a 2-phase CCD.

operation can be applied to a 2-phase device, as shown in Figure 7.8. Other modes of operation are possible if the electrodes of a CCD are driven with a more complex waveform than a simple, abrupt change of voltage level, if more complex systems of electrodes are used than the ones we have described, or if additional doping is introduced under some electrodes, for example by ion implantation.

Although for simplicity we have drawn the CCD electrodes as simple plates deposited on a plane oxide layer, in practice such a construction only works well if the gaps between electrode edges are kept extremely small. It is easier to construct CCDs that have electrodes on two or three different levels separated by thin layers of insulating material; this construction allows some degree of overlap at the electrode edges and facilitates efficient transfer of charge by eliminating unwanted potential minima between the electrodes.

We would like the CCD to have high charge transfer efficiency (CTE), that is, to move virtually all the charges in the transfer process. Poor CTE results in a blurring of the image due to charge trailing behind, and it degrades the noise because charge packet losses occur in processes that are subject to the usual statistical fluctuations (see below).

A number of mechanisms influence the CTE. Referring to Figure 7.7, if there is a large amount of charge in the well at t_1, electrostatic repulsion will cause the charge carriers to spread into the wider well at t_2. This self-induced drift mechanism is similar to the drift under external fields discussed in Chapter 3. We can derive an approximate time constant for the process as follows. Assume we have N_0 electrons in one well and none in the other; the wells are separated by the electrode spacing L_e. Then the voltage difference between these wells is $V_{21} = N_0 q / C_0$, where C_0 is from equation (7.2). Thus, $\mathscr{E}_{21} = N_0 q / L_e C_0$. From equation (3.6), the average drift velocity $\langle v_{21} \rangle = -\mu \mathscr{E}_{21}$. We define the self-induced drift time constant to be $\tau_{SI} = -L_e / \langle v_{21} \rangle$. Substituting for $\langle v_{21} \rangle$ and $\mathscr{E}_{21}$ gives

$$\tau_{SI} \sim \frac{L_e}{\mu \mathscr{E}_{21}} = \frac{L_e^2 C_0}{\mu N_0 q}. \tag{7.3}$$

More rigorously (Carnes, Kosonocky, and Ramberg, 1972),

$$\tau_{SI} = \frac{2 L_e^2 C_0}{\pi \mu N_0 q}. \tag{7.4}$$

For example, taking the electrode parameters given after equation (7.2) and $N_0 = 3 \times 10^5$, we find $\tau_{SI} \sim 0.002\,\mu s$.

For small amounts of charge, thermal diffusion tends to drive the electrons across the storage well. Sufficient time must elapse to allow virtually all the charges to diffuse into the next well, thus giving good CTE. Drawing on the discussion of diffusion in Chapter 5, but redefining symbols such that they are applicable to the CCD transfer geometry, the exponential time constant for thermal transfer of charges is (from equation (5.18))

$$\tau_{th} \sim \frac{L_e^2}{D}, \tag{7.5}$$

where τ_{th} is analogous to the recombination time, the electrode spacing, L_e, is analogous to the diffusion length, and D is the diffusion coefficient. More rigorously, it can be shown (Carnes, Kosonocky, and Ramberg, 1972) that

$$\tau_{th} = \frac{4 L_e^2}{\pi^2 D}. \tag{7.6}$$

Calculating D from Table 3.1 and the Einstein relation (equation (5.13)), we find that $\tau_{th} \sim 0.026\,\mu s$ for the electrode parameters following equation (7.2) and $T = 300$ K.

For full wells, $\tau_{th} \gg \tau_{SI}$, so initially the electrostatic-driven transfer will dominate. The time constant of this process goes as N_0^{-1}, so eventually it will slow down, and the transfer will be completed by thermal diffusion. We can define a critical number of charge carriers in the well, N_0^{crit}; if the number of charges in the well is below this number, thermal diffusion is the dominant transfer process. We calculate N_0^{crit} by setting $\tau_{th} = \tau_{SI}$:

$$N_0^{crit} = \frac{\pi D C_0}{2 \mu q}. \tag{7.7}$$

For operation at very low light levels, a very high CTE is required; the timescale for charge transfer is then dominated by thermal diffusion. The example at the end of this chapter uses these results to illustrate the timing involved in obtaining good CTE.

The CTE normally refers to the unit imaging cell. For a CCD that operates with m phases ($m=2$, 3, or 4 for the examples we have discussed) and a charge transfer mechanism that is characterized by a simple exponential, such as thermal diffusion, the CTE takes the form:

$$\mathrm{CTE}=(1-e^{-t/\tau})^m, \tag{7.8}$$

where τ is the time constant (for example, equation (7.6)).

Charge transfer can also be driven by a third mechanism: the fringing fields between the electrodes. Contrary to our illustrations, the well produced under one electrode is not completely independent of the voltages on neighboring electrodes. Particularly for CCD geometries in which the depletion regions are relatively far below the electrodes, the fields generated by neighboring electrodes overlap; these fringing fields enhance transfer efficiency both by producing appropriate slopes in the floors of the potential wells and by rounding off their edges. The effect of fringing fields depends on the CCD construction details and on the characteristics of the voltage signals used to transfer charge. Fringing fields usually need to be calculated through numerical rather than analytic techniques. A more detailed discussion, including other charge transfer mechanisms, can be found in Séquin and Tompsett (1975).

The CTE can be affected by imperfections in the construction of the CCD. For example, an imperfection in the field produced by an electrode can lead to a minimum in its potential well which can trap residual charge, or a potential minimum between electrodes may obstruct charge transfer between them. Traps that degrade the CTE can be produced either by design errors or by flaws in the processing such as lifting of the edges of the electrodes, diffusion of implanted dopants, lattice defects in the silicon, or unwanted impurities. In such cases, complete transfer may not be possible in any time interval, no matter how long. This sort of problem can be reduced by adjusting electrode voltages, time constants, etc., as well as by slowing down the overall read cycle. Nonetheless, the readout times calculated from electrostatic repulsion, thermal diffusion, and fringing fields usually represent only lower limits.

Fringing fields are potentially troublesome. If the columns are not adequately shielded from each other, the fringing fields can allow electrodes in one column to affect charge transfer in a neighboring column; electrons may even be transferred across columns. To prevent such an occurrence,

CCDs are manufactured with strongly doped (and hence conductive) barriers between their columns, which are held at ground potential to shield adjacent columns. The electric fields do not extend into these 'channel stops', ensuring that the columns are electrically isolated from each other so that charge can be transferred efficiently along the intended direction.

Since poor charge transfer efficiency results in a portion of the signal being trailed behind when an image is transferred out of the CCD, the CTE can be measured by imposing a sharply confined image onto the array and measuring carefully the amount of charge that is deferred to later pixels upon readout. One technique is based on illuminating the CCD with monoenergetic X-rays to inject large and repeatable amounts of charge into single pixels. From knowledge of the X-ray energy and the quantum yield and measurements of the collected charge, the CTE can be calculated accurately. CTE can also be measured by observing the resulting image spreading; this latter technique, however, is not sensitive to CTE problems that spread the lost charge over many pixels.

Several readout architectures for CCDs are illustrated in Figure 7.9. The simplest (shown in Figure 7.9(a)), called line address architecture, shifts the contents of the columns successively into an output register, or row, that transfers the charge packets to the amplifier. In this scheme, the array is illuminated while transfers occur. To avoid smearing the image, the CCD must either be read out quickly (in comparison with the time it takes to build up the picture) or be closed off with a shutter during readout. This requirement is usually no problem for very low light levels, for example, those encountered

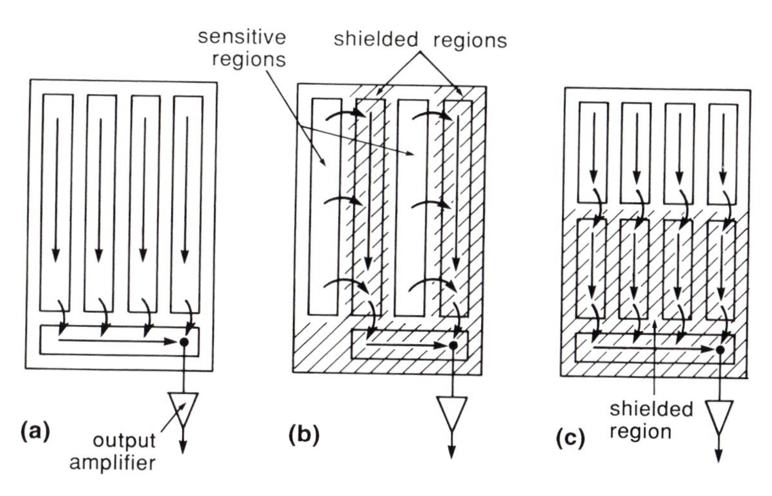

Figure 7.9. CCD readout architectures: (a) line address; (b) interline transfer; and (c) frame transfer.

in astronomy. For the broadcast industry, where neither of these solutions is satisfactory, other architectures are used in which the charge is first shifted into a portion of the CCD that is protected from light and then read out. For example, Figure 7.9(b) shows interline transfer architecture, in which the charges are first shifted laterally into columns that are protected from light by a layer of evaporated metal and then shifted along these protected columns to the output register. In frame/field transfer architecture (Figure 7.9(c)), the charge is shifted quickly to an entire adjacent CCD section that is protected from light. Readout of this section can proceed at a slower rate as required.

The readout scheme shown in Figure 7.9(a) allows the CCD to be operated in time-delay integration (TDI) mode. A fixed scene is swept over the array by moving the telescope or other optics in a manner that confines the motion of the image to occur along the columns of the array. The clocking along the columns is adjusted to move the charge at exactly the same rate as the scene, therefore building up the image while compensating for the motion of the object. Two advantages of this mode of operation are that the data are read out at a constant rate rather than in the bursts that come with reading the entire array, and corrections for fixed pattern noise (see Section 7.3.1) are made easier by averaging the response over columns.

The CCD transfer process can also be used to combine signals from adjacent pixels without introducing extra noise. For example, as shown in Figure 7.10,

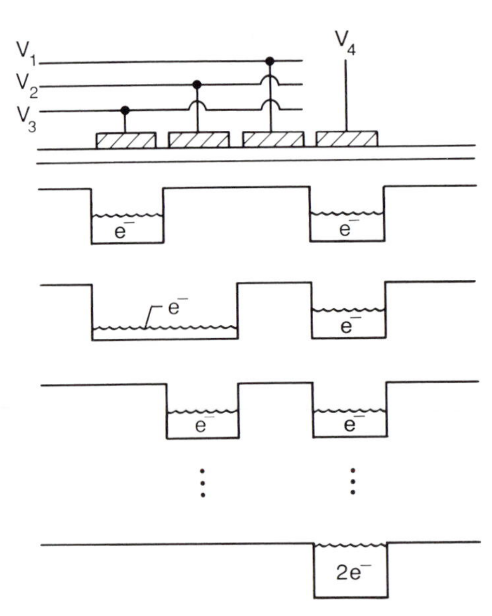

Figure 7.10. Charge combination in a CCD. The charge packets are combined under the electrode at V_4.

the columns can be advanced i pixels worth without clocking the output row, thus accumulating the contents of i pixels into the wells of the output row before this row is read out by the output amplifier. The combined pixel shape would then be rectangular with dimensions 1 pixel along rows and i pixels along columns. Similarly, the output row can be advanced j pixels worth into the output amplifier before a reading is taken, thus combining charge along rows. When combined with the combination along columns, the equivalent pixel size becomes i pixels along columns and j pixels along rows. If the CCD noise is dominated by the output amplifier read noise and these pixels were combined by reading them out *individually* and then averaging the results, the noise would be increased by a factor of $(ij)^{1/2}$ compared to the noise that is obtained if charge from the same pixels is combined on the CCD before reading out the signal.

The charge collection and transfer process in CCDs is subject to a number of noise mechanisms. As with all detectors, CCDs are subject to shot noise; because they do not allow recombination, this noise is just the square root of the number of collected charges. There are two additional noise mechanisms associated with the charge transfer. The first arises from poor CTE. If the transfer inefficiency $\varepsilon = 1 - \mathrm{CTE}$ is the fraction of the charges not transferred, a total of εN_0 charges is left behind. If the CTE were perfect ($=1$), then no charges would be left and there would be no noise associated with the transfer process. However, if the transfer inefficiency is greater than zero and is Poisson-distributed, there is an uncertainty of $(\varepsilon N_0)^{1/2}$ in the number of charges left behind and hence an equal uncertainty in the number transferred out of the unit cell. Assuming an identical cell precedes the one under discussion, there will be a similar uncertainty in the number transferred into the cell. Thus, for n transfers, the net uncertainty is

$$N_{n,\mathrm{TL}} = (2\varepsilon n N_0)^{1/2}. \tag{7.9}$$

The second mechanism arises from the trapping of charge carriers. Since the Si–SiO_2 interface along which charge is transferred represents a discontinuity in the Si crystal, it contains incomplete bonds which act as traps. On average, these traps will reach some equilibrium occupancy level and will not affect the total charge transferred. This occupancy level, however, is subject to statistical fluctuations, causing charges to be added to and subtracted from a charge packet randomly. For $l = mn$ transfers, where n is the number of cells and m the number of transfers per cell, the net uncertainty is (Carnes and Kosonocky, 1972)

$$N_{n,\mathrm{T}} = (2kTlN_{ss}A)^{1/2}, \tag{7.10}$$

where N_{ss} is the density of traps (usually given in units of $(\mathrm{cm}^2\,\mathrm{eV})^{-1}$), and A is the interface area.

Interface trapping noise can be reduced in two ways. One approach is to maintain a low level of charge in the wells virtually continuously so that the traps are always filled. Doing so requires that a small, uniform charge level be introduced immediately after each readout; this charge is called a fat zero. Although this procedure can significantly reduce interface trapping noise, it is not completely effective because the well tends to spread as more charge is accumulated, leading to the exposure of additional traps. In addition, a fat zero increases the net noise by the factor $(1+N_{fat}/N)^{1/2}$, where N_{fat} is the average number of charges placed in a well by the fat zero and N is the average number collected from other sources (and we assume the read noise is negligible). With a low noise CCD operated at low signal levels, the noise added by a fat zero can seriously degrade the performance.

A more satisfactory way to reduce interface trapping noise is to modify the structure of the CCD so that the charges are kept away from the oxide layer. Such CCDs are said to have a buried channel. A thin (typically $\sim 1\,\mu$m thick) layer of silicon with the opposite dopant type of that in the bulk silicon of the detector substrate is grown under the oxide layer. Such a construction is shown in Figure 7.11, where we continue the example of a CCD with a p-type

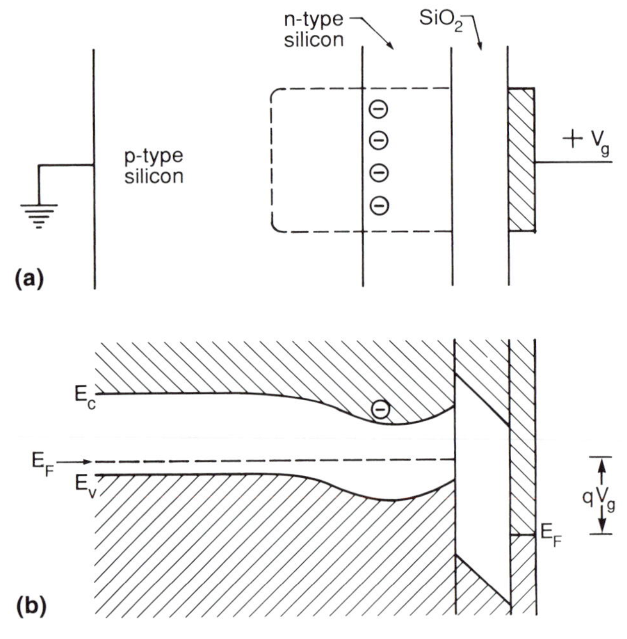

Figure 7.11. Physical arrangement (a), and band diagram (b) for a buried channel CCD.

substrate and electrons as the active charge carriers, but now with a thin n-type layer between the SiO_2 and the p-type substrate. The two layers, n-type and p-type, form a junction. At the edge of the CCD, a heavily doped n-type region makes contact from the surface through a hole in the SiO_2 to the thin n-type layer. A positive voltage is applied to this contact until the junction is sufficiently strongly back-biased that the n-type region in it has been completely depleted. The depletion region of the CCD then consists of two zones: one just under the SiO_2 layer has a net excess hole density and the second, in the p-type region, has an excess negative ion density. When the CCD electrode is pulsed positive, the space charge in these zones acts to create a potential well within the n-type material. With appropriate selection of layer depths and doping, and the voltages used to operate the device, this well forms away from the Si–SiO_2 interface. Since the trap density is much smaller in bulk silicon than at the crystal surface, interface trapping noise is very small. In very demanding applications, the density of traps in the bulk silicon may still limit the noise as in equation (7.10) and a fat zero may be required to maintain ultra-high transfer efficiency.

A further advantage of a buried channel is that the distance between the electrode and well has been increased from $\sim 0.1\,\mu$m for surface channel to $\sim 0.5\,\mu$m, resulting in stronger fringing fields and better CTE. A disadvantage is that the well capacity is reduced because the distance between the electrode and the well is greater (see equation (7.2)).

To understand these arguments more quantitatively, we refer to Figure 7.12. Figure 7.12(a) shows a cross-section through the CCD; Figure 7.12(b) shows the electric field in various zones of the CCD; and Figure 7.12(c) shows the negative of the corresponding electrostatic potential. In the case of p-type semiconductor, the potential energy of the minority carriers (electrons) is proportional to the negative of the electrostatic potential (if the potential is in units of electron-volts, the constant of proportionality is unity). Therefore, Figure 7.12(c) can equivalently be taken as a diagram of the electron potential energy. We note that the potential reaches a minimum at $x = \ell$, which is removed from the Si–SiO_2 interface that lies at $x = \ell_2$. We will use Poisson's equation to derive and quantify this result.

We use a coordinate system in which x is 0 at the left edge of the depletion region and increases as we move through the CCD. We first discuss Figure 7.12(b). For negative x, the material is assumed to be sufficiently conducting that $\mathscr{E}_x(x<0) = 0$. The portion of the depletion region in the p-type silicon has a density of N_A negative ions. Using Poisson's equation in this region,

$$\frac{d\mathscr{E}_x}{dx} = -\frac{qN_A}{\kappa_0 \varepsilon_0}, \qquad 0 \leqslant x \leqslant \ell_1, \tag{7.11}$$

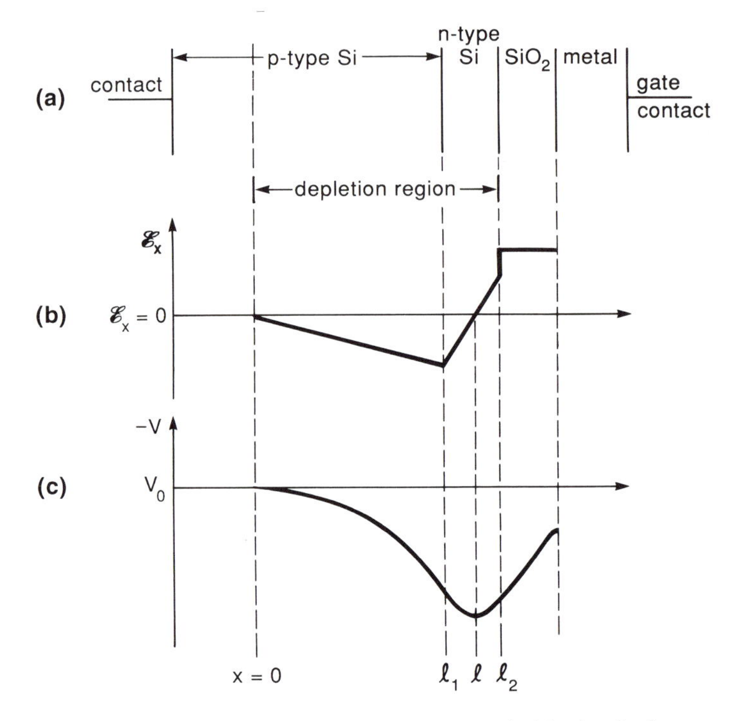

Figure 7.12. Principle of operation of a buried channel: (a) physical arrangement; (b) electric field; and (c) negative voltage, proportional to electron energy level.

and applying the boundary conditions $\mathscr{E}_x(x=0)=0$ and $V(x=0)=V_0$, and the definition $\mathscr{E}_x=-\mathrm{d}V/\mathrm{d}x$, we can show that

$$\mathscr{E}_x=-\frac{qN_\mathrm{A}}{\kappa_0\varepsilon_0}x, \qquad 0\leqslant x\leqslant \ell_1, \tag{7.12}$$

and

$$V-V_0=\frac{qN_\mathrm{A}}{2\kappa_0\varepsilon_0}x^2, \qquad 0\leqslant x\leqslant \ell_1. \tag{7.13}$$

The portion of the depletion region in the n-type silicon has a density of N_D positive ions. Starting from Poisson's equation for this region,

$$\frac{\mathrm{d}\mathscr{E}_x}{\mathrm{d}x}=\frac{qN_\mathrm{D}}{\kappa_0\varepsilon_0}, \qquad \ell_1\leqslant x\leqslant \ell_2, \tag{7.14}$$

and imposing the boundary conditions that both $\mathscr{E}_x$ and $V-V_0$ should be continuous across the boundary at $x=\ell_1$, we find that

$$\mathscr{E}_x=\frac{qN_\mathrm{D}}{\kappa_0\varepsilon_0}x-\frac{q\ell_1}{\kappa_0\varepsilon_0}(N_\mathrm{D}+N_\mathrm{A}), \qquad \ell_1\leqslant x\leqslant \ell_2, \tag{7.15}$$

and

$$V - V_0 = -\frac{qN_D}{2\kappa_0\varepsilon_0}x^2 + \frac{q(N_D + N_A)\ell_1}{\kappa_0\varepsilon_0}\left(x - \frac{\ell_1}{2}\right), \qquad \ell_1 \leqslant x \leqslant \ell_2. \quad (7.16)$$

We have now described the situation in the depletion region, but for completeness we continue the journey along increasing x into the remainder of the CCD. The magnitude of the discontinuity in $\mathscr{E}_x$ at ℓ_2 is equal to the inverse of the ratio of dielectric constants in the materials (assuming there is no surface charge trapped at the oxide/silicon interface). The field is constant within the insulator (again assuming there is no charge trapped within it). Within the metal electrode, the field should be zero. The potential varies linearly across the insulator and should be continuous at ℓ_2. The net bias across the detector is given by the difference between the voltage at the electrode and V_0 (V_0 would normally be at ground). These field and voltage behaviors are drawn in Figure 7.12 but will not be derived explicitly here.

The condition for formation of a buried channel is that the minimum in the electron potential energy should occur within the silicon. The location of this minimum is obtained by setting $dV/dx = 0$, where V is from equation (7.16). After a little manipulation, this condition leads to

$$\frac{\ell - \ell_1}{\ell_1} = \frac{N_A}{N_D}, \quad (7.17)$$

where the value of x at the minimum is denoted by ℓ. A buried channel results if $\ell < \ell_2$.

To specify the operating parameters that allow a buried channel, we start with an expression for the voltage across the n-type layer. If the voltage at the p-type/n-type interface is V_{p1}, then by analogy with equation (7.13),

$$V_n - V_{p1} = \frac{qN_D}{2\kappa_0\varepsilon_0}(x - \ell_1)^2. \quad (7.18)$$

$\ell < \ell_2$ will be possible up to some maximum of $V_n = V_{max}$. The SiO_2 layer in the CCD is usually much thinner than indicated in the schematic diagram in Figure 7.12; as a result, we can take $V_{max} \approx V_{gmax}$, the maximum gate voltage that allows a buried channel. Setting $\ell = \ell_2$ in (7.17) and $x = \ell_2$ in equation (7.18), we solve this equation to find

$$V_{gmax} = \left(\frac{qN_D}{2\kappa_0\varepsilon_0}\right)(\ell_2 - \ell_1)^2\left(\frac{N_D}{N_A} + 1\right). \quad (7.19)$$

For example, if $N_D = 10^{15}\,\text{cm}^{-3}$, $N_A = 10^{14}\,\text{cm}^{-3}$, and $(\ell_2 - \ell_1) = 1\,\mu\text{m}$, $V_{gmax} \approx 8.4$ V.

The construction of a buried channel CCD is governed by equation (7.19). Its operation also depends on maintaining the device at a sufficiently high temperature that the junction creating the buried channel does not freeze out. From a practical point of view, buried channels are only useful for CCDs that operate above 70–100 K.

It should be noted that overfilling the well of a buried channel CCD will cause a portion of the collected charge to contact the oxide layer, causing a reversion to surface channel behavior. In most applications, the well depth should be defined as the maximum number of charge carriers that can be collected before charge contacts the oxide; the performance of the CCD is degraded beyond this point, for example by additional noise and by trailing of the image because of charges retained in surface state traps. A higher level of overfilling can cause the charge to spill out of the wells altogether and onto adjacent pixels. Extreme levels of overillumination can fill large numbers of surface traps and lead to excessive 'dark' current that can persist for long periods of time (days) so long as the CCD is held at low temperature.

Another operating mode for CCDs can be explained by referring to Figure 7.12. Imagine that we maintain the voltage of the n-type layer but drive the gate voltage increasingly negative. We will eventually reach a condition where the voltage at ℓ_2, the Si–SiO_2 interface, is equal to that of the substrate, V_0 ($=0$ V). Since the channel stops are at V_0, at this point holes flow from them into the Si–SiO_2 interface. As the gate potential is driven farther negative, more holes flow to this interface and maintain it at V_0 so that the full gate potential falls across the SiO_2 layer and the potential well is maintained in the shape established by the potential on the n-type layer and the doping. This condition is described by saying the surface potential is pinned at the frontside of the CCD, and/or that the n-channel has inverted at the Si–SiO_2 interface.

Without frontside pinning, large dark currents can be produced when electrons are excited thermally into a trap at the Si–SiO_2 interface and then excited again into the conduction band. Under frontside pinned conditions, the large number of charge carriers at this interface fill the surface traps and suppress this dark current mechanism. Adjustment of the electrode design and use of implants to adjust the local surface potentials allows construction of multi-pinned-phase CCDs, which can collect charge under one set of electrodes while all phases are in an inverted condition. During readout, the voltages on the phases are manipulated to pass the charge to the output amplifier as in other CCDs. Such devices minimize requirements for cooling the detector to control dark current and can be operated in modes that eliminate blooming and provide a number of additional advantages (Janesick and Elliott, 1991).

At the end of the charge transfer described above, the signal must be transferred to an output amplifier, which can be grown on the same piece of silicon as the CCD. The input of the amplifier is electrically connected to a suitable structure to sense the charge from the CCD. For example, an electrode attached to the gate of the output amplifier FET (and called a floating gate) may be deposited inside the oxide layer beneath a transfer electrode; as the charge passes over this electrode, it is sensed capacitively by the floating gate. This process provides for nondestructive readout of the charge. Since charge is transferred over the floating gate and into a storage well in the CCD, it can be carried to additional floating gates and read again. In an alternate arrangement, the amplifier gate is attached to a heavily doped contact diffused into the silicon (and called a floating diffusion). In this case, the charge is transferred onto the diffusion and conducted to the gate, and the signal is reset in the process of reading it out; that is, this design leads to destructive readout.

For illustration, we consider an integrating amplifier circuit similar to those discussed previously in Chapter 6 and replace the detector and bias supply with the floating gate output of the CCD. Such an arrangement might look like that in Figure 7.13. Charge packets are transferred successively onto the MOSFET gate and sampled, and the system is reset. If we are not careful with our sampling strategy, however, this circuit will be limited by kTC noise. If $T = 150\,\mathrm{K}$ and $C = 0.1\,\mathrm{pF}$, $Q_N = 90$ electrons.

We need to adopt a readout strategy in which the information is obtained

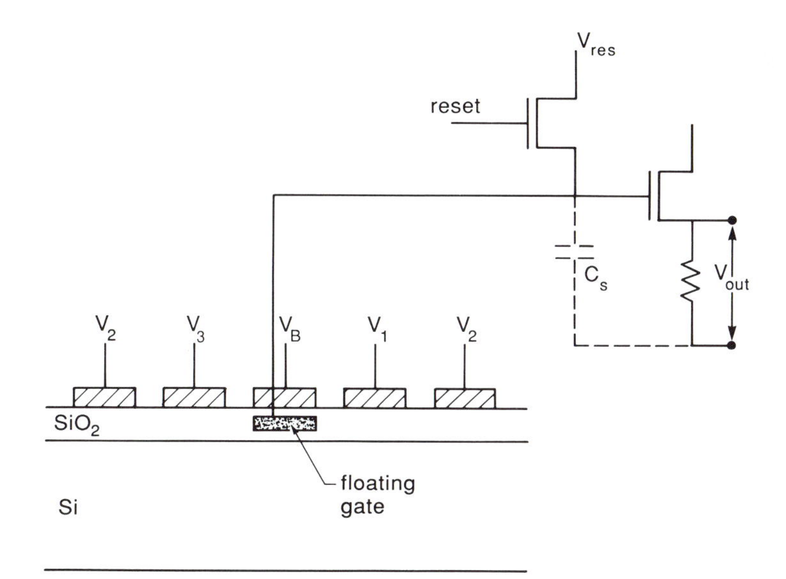

Figure 7.13. Simple CCD floating gate output amplifier.

without closing the reset switch, thus freezing the kTC noise on the gate so it does not appear in our signal. To do so, we first close the reset switch to remove any residual charge from the gate; we then open the switch and obtain the first measurement of the amplifier output. This output is the sum of the charge deposited on the gate by the switch and kTC noise. We then transfer charge from the CCD structure onto the gate and obtain a second measurement to complete the double correlated sample. Because the CCD structure transfers charge while maintaining very high impedance isolation, there will be no additional kTC component in the second measurement; taking the difference of the first and second measurements will give the collected charge from the signal alone (subject only to amplifier noise). Finally, the reset switch is closed to remove this charge packet, and the cycle is repeated for the next one.

An additional reduction in noise can be made by using the distributed floating gate amplifier illustrated in Figure 7.14. Here, the charge packets are transferred successively by one CCD structure onto the gates of a series of amplifiers. The outputs of these amplifiers are added into a single charge packet by feeding them into a second CCD structure which is clocked synchronously with the input. After the desired number of amplifications has occurred, the charge is transferred to a final output stage, and the contents of both CCD registers are reset. Using this technique, CCD arrays can achieve read noises of only a few electrons. An alternate approach (called a skipper readout) is to transfer the charge back and forth between two amplifiers,

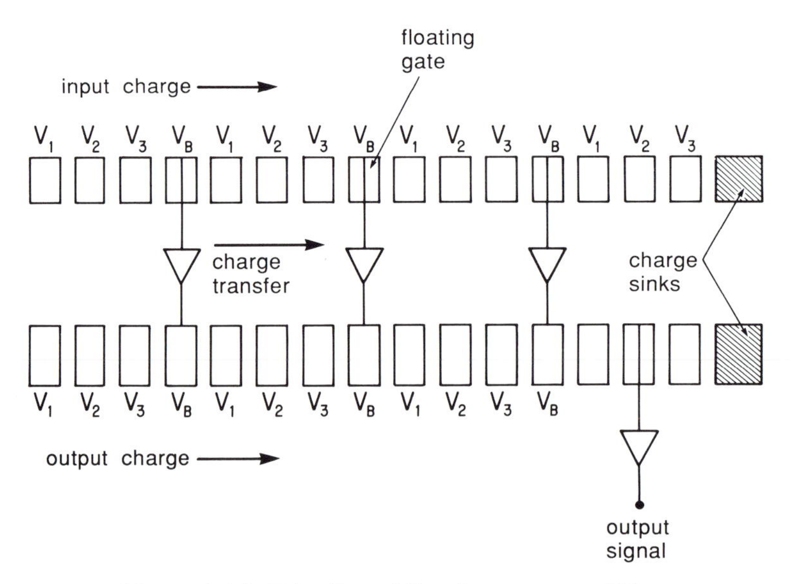

Figure 7.14. Distributed floating gate amplifier.

reading it each time. With enough time to allow ~100 reads, the accumulated charge can be determined to a root-mean-square noise of less than one electron (Janesick *et al.* in Jacoby, 1990).

Where low noise is required with short readout times, CCDs can be made with multiple output amplifiers that read out subsections of the array.

Currently available CCDs have dimensions ranging roughly from 500 × 500 to 4000 × 4000 pixels and achieve noise levels of one to five electrons (rms). Peak quantum efficiencies (achievable from ~0.3 to ~0.9 μm) can be 80–100% if the device is anti-reflection coated, and with ultraviolet treatment to enhance short wavelength response, quantum efficiencies of ~60% can be achieved at wavelengths short of 0.3 μm. These performance levels require operation at an optimum temperature of ~150 K. CCDs can also be used as X-ray detectors, in which case a single photon produces many free electrons. A rough indication of the input photon energy can be derived from the number of electrons produced. Reviews of CCD operation and performance can be found in Séquin and Tompsett (1975), Barbe and Campana (1977), Hobson (1978), Beynon and Lamb (1980), and Janesick and Elliott (1991), while historically important articles are available in Melen and Buss (1977). Jacoby (1990) provides a collection of papers on the most recent developments regarding these detectors. Buil (1991) gives a very complete and readable description of the use of CCDs, including the necessary electronics and software and data reduction techniques.

7.3 Array properties

We now expand the discussion to describe two aspects of performance that affect all types of solid state arrays.

7.3.1 Fixed pattern noise

It is impossible to make the individual pixels of an array absolutely identical. The differing pixel properties distort the electronic representation of the pattern of illumination on the array and therefore constitute a form of noise. Unlike the kinds of noise discussed up to now, this noise is nonrandom and remains stable from one picture to the next; it is therefore called fixed pattern noise. A series of calibration steps can, in theory, remove fixed pattern noise virtually completely. This procedure is frequently known as flat fielding; to achieve the full performance potential of modern semiconductor arrays, very accurate flat fielding is essential. We will describe the basic steps required for this type of calibration.

There are three kinds of variation in pixel properties that can contribute to fixed pattern noise: offset, dark current, and response.

Offset variations[2] have amplitudes independent of both signal level and integration time. They may arise because the readouts have different 'zero' voltage levels for different pixels; for example, processing nonuniformities or errors in layout can lead to differing FET characteristics across the readout of an infrared array. In addition, the electronic clocking signals that control the multiplexing and charge transfer in the arrays will deposit capacitively coupled charge in differing amounts within the array. The output electronics may have an offset such that zero output does not correspond to zero signal.

Dark current variations have amplitudes that depend on integration time but not on signal level. Detector dark current (for example, thermally excited current) is an obvious example, but similar behavior will be caused by photon leaks that reach the array when it is nominally blocked off from signals.

Response variations have amplitudes that depend on both integration time and signal level, normally as the total integrated signal in the exposure. The individual pixels will have different responsivities, and the wavelength dependence of their quantum efficiencies may also differ. These effects can arise because of gradients in detector thickness or doping across the array. Because the arrays are made of thin, parallel layers of material, it is also likely that interference of light occurs within the array; when observing in narrow spectral bands, the array may even exhibit interference fringes across its face.

Three kinds of calibration frames are required to correct data frames for these three effects. An offset frame of very short integration time should be obtained with the detector blocked off from light. A dark current frame of long exposure is taken with the detector blocked off from light signals. A response frame is obtained by exposing the array to a uniform illumination source. All of these frames need to be obtained at high signal to noise so that they do not degrade the signal to noise of the final reduced data.

One data reduction strategy starts by subtracting the offset frame from the dark, response, and data frames. The dark current frame is then scaled to the exposure times of the response and data frames and subtracted from them. Finally, the data frame is divided by the response frame. If the data frame has been illuminated by a uniform background, the result should be a uniform output level corresponding to the ratio of the exposure on the data frame to that on the response frame. Any images of a source should appear on top of this uniform output. Other strategies can give equivalent results. For example, if the data are taken in pairs of frames with equal exposure on uniform background and on the object of interest, the background frame can be subtracted from the object frame. If the background has not varied, this procedure removes not only background terms but also those due to offsets and dark currents. The resultant image can be divided by the response frame (corrected for offsets and dark current) to complete the reduction.

In carrying out this calibration, one must allow for the probability of time variability in the array parameters, and also take into account the inherent nonlinearities in the array. The response frames should be obtained as close as possible in time to the observations. Although the offset and dark current frames are often stable enough that they need not be obtained in extremely close proximity to the data frames, the possibility of variations does exist, particularly for the dark current if the array temperature is not stable. If the array is nonlinear and the dark current large, linearity corrections are required in estimating the dark current for a different integration time than the one for which it was measured. Also, the array may have traps that prevent an initial portion of the charge from any exposure from appearing at the output. In this case dark current and response frames at a number of integration times must be obtained to compute the necessary corrections.

Extremely accurate corrections become a necessity when the sources being observed are weak compared with the surrounding background signal. Because of the potential color dependence of detector responses, to obtain an adequate response frame the array needs to be illuminated with light of a similar color to the background. When nonlinearities are a concern, the illumination level should also be similar. The effects of any charge trapping must also be allowed for by making the response and data frames have similar exposure times. One approach that can lead to extremely accurate flat fields is to alternate observations of the source plus background radiation with observations of background in an adjacent direction and then to generate the response frame from the ones of the background. Another implementation of this strategy is to move the sources on the array between exposures so that a response frame can be deduced from the suite of exposures, ignoring for each frame the regions occupied by sources. These strategies have permitted flat fielding to an accuracy exceeding one part in 10^5.

7.3.2 Crosstalk and MTF

Even a perfectly concentrated spot of illumination on the center of a pixel in an array will produce some response on the neighboring pixels. The response of a neighboring pixel expressed as a percentage of the response of the illuminated pixel is called the crosstalk.

A number of mechanisms can contribute to crosstalk. Optical crosstalk occurs when photon paths in the detector material pass through more than one pixel. It is obviously more of a problem in arrays that (a) have thick detectors with low absorption efficiency, (b) use reflective back surfaces to enhance quantum efficiency by increasing the photon path in the detector, and (c) are illuminated with fast optical beams, where part of the light enters the array at a large angle from the normal. Fortunately, the high refractive indices of most detector materials act to bend the incoming light toward the direction normal to the face of the array and therefore to reduce optical crosstalk. The extent of this effect can be calculated from Snell's law:

$$\sin(\theta) = n \sin(\theta_d), \tag{7.20}$$

where θ is the incident angle relative to the normal to the surface, θ_d is the angle relative to this normal inside the detector, n is the refractive index of the material, and it is assumed that the light is incident from a medium with refractive index close to unity.

Crosstalk also occurs electrically. In all detector array types, lateral diffusion of charges can carry them to an electrode other than the one under which they were created, or fringing fields and capacitances between electrodes can cause one electrode to poach charges from another. A general way to express resolution is through the modulation transfer function (MTF) – see Section 2.4.4. Holloway (1986) shows that the MTF due to lateral diffusion of charge carriers from one pixel to another in an array with high absorption efficiency and in the limit of small pixels is

$$\mathrm{MTF} = \frac{\mathrm{sech}\left(\dfrac{d(1+k^2)^{1/2}}{L}\right)}{\mathrm{sech}\left(\dfrac{d}{L}\right)}, \tag{7.21}$$

where d is the thickness of the detectors, L is the diffusion length, and $k = 2\pi/\ell$, where $1/\ell$ is the spatial frequency, and ℓ is in units of the diffusion length.

Resolution is also affected by the finite size of the detector elements in an array. Assuming no gaps between pixels, it can be shown that the MTF due to this cause is:

$$\mathrm{MTF} = \left|\frac{\sin(\pi l f)}{\pi l f}\right|, \tag{7.22}$$

where l is the pixel width and f is the spatial frequency. MTFs for other mechanisms are discussed, for example, by Hobson (1978).

When electronic detectors are illuminated by a very bright source, there can be a spreading of the images beyond that described by the MTF. This behavior is referred to as blooming. For example, in CCDs, it can appear as a spreading of the image along the columns of the device as charge spills from a filled well into adjacent ones.

7.4 Example

We consider a 500×500 pixel, 3-phase, buried channel CCD operated at 170 K with 25 μm electrodes. The output amplifier has an effective capacitance of 2×10^{-13} F, and it is being read out as follows:

(1) The charge is transferred to the gate of the output amplifier and a reading is taken.

(2) The amplifier is reset.

(3) A second reading is taken.

(4) The signal is computed as the difference of these two readings.

To maximize the time spent integrating on signal, the array is read out in a total time of 0.2 s with 10^5 collected charges in each detector well. Assume the mobility scales as $T^{-3/2}$ between 300 K and 170 K, starting from the value in Table 3.1.

List the contributions to the noise that will be observed in the signal, and estimate each one (in electrons) for the pixel farthest (that is, most transfers) from the output amplifier.

(a) There will be a noise of $N^{1/2}$ in the number of collected charges, or 316 electrons.

(b) Because the signal is determined by resetting between samples, there will be kTC noise:

$$Q_N = (kTC)^{1/2} = 2.16 \times 10^{-17}\,\mathrm{C}, \quad \text{or } 135 \text{ electrons.}$$

(c) There will be noise because of incomplete charge transfer. The transfer time is dominated by diffusion. To compute the time constant, we need the diffusion coefficient $D = \mu kT/q$. From Table 3.1, with mobility scaling as $T^{-3/2}$, we get

$$\mu_n = 1350\,\mathrm{cm^2\,V^{-1}\,s^{-1}} \left[\frac{300\,\mathrm{K}}{170\,\mathrm{K}}\right]^{3/2} = 3165\,\mathrm{cm^2\,V^{-1}\,s^{-1}}.$$

We then obtain $D = 4.64 \times 10^{-3}\,\mathrm{m^2\,s^{-1}}$. From equation (7.6), with $L_e = 25 \times 10^{-6}$m, we get

$$\tau_{th} = 5.46 \times 10^{-8}\,\mathrm{s}.$$

Estimating the capacitance from equation (7.2), with $X_0 = 0.1\,\mu$m,

$$C = \frac{(25 \times 10^{-6}\,\mathrm{m})^2(4.5)(8.85 \times 10^{-12}\,\mathrm{F\,m^{-1}})}{1.0 \times 10^{-7}\,\mathrm{m}} = 2.49 \times 10^{-13}\,\mathrm{F},$$

and, from equation (7.7),

$$N_0^{\text{crit}} = \frac{\pi(4.64 \times 10^{-3}\,\text{m}^2\,\text{s}^{-2})(2.49 \times 10^{-13}\,\text{F})}{2(0.3165\,\text{m}^2\,\text{V}^{-1}\,\text{s}^{-1})(1.60 \times 10^{-19}\,\text{C})} = 35\,800.$$

Passing the charge from the farthest pixel through the output amplifier requires 500×3 transfers down the column and 500×3 transfers along the output row. The column transfers occur sufficiently slowly that the transfer efficiency should be high. The output row must make $500 \times 500 \times 3$ transfers in 0.2 s, or 1 in $2.67 \times 10^{-7}\,\text{s} = 4.89\tau_{\text{th}}$. We therefore lose a fraction $\varepsilon = e^{-4.89} = 7.52 \times 10^{-3}$ of N_0^{crit} for the thermally driven transfers (see Section 7.2.3), or a net loss of $7.52 \times 10^{-3} \times 35\,800 = 270$ electrons per transfer. Since we started with 10^5 electrons, the net charge transfer efficiency is

$$\text{CTE} = 1 - \left(\frac{270}{10^5}\right) = 1 - 2.7 \times 10^{-3} = 0.9973.$$

Substituting into equation (7.9),

$$N_{\text{n,TL}} = [(2)(2.7 \times 10^{-3})(1500)(10^5)]^{1/2} = 900 \text{ electrons}.$$

The total read noise is then

$$N_{\text{tot}} = (316^2 + 135^2 + 900^2)^{1/2} = 963 \text{ electrons},$$

so the CCD is operating at about three times above the background-limited noise level of 316 electrons.

7.5 Problems

7.1 Comment on the feasibility of producing a buried channel CCD with doping to allow low noise operation as an extrinsic photoconductor in the mid-infrared spectral region (for example, near 10 μm).

7.2 Consider the CCD in the example, but illuminate it at a level that produces 1000 electrons per pixel in an integration. Describe (numerically where appropriate) the changes required in the operation of the chip to stay within 30% of background-limited operation.

7.3 Use the following three 5×5 pixel frames to calibrate the data frame for fixed pattern noise. The boxes give counts for each pixel under the conditions indicated. Derive the counts per second in units of the counts per second in the response frame exposure with all corrections for fixed pattern noise.

Data frame, viewing source, 100 s integration:

1097	1110	1095	1116	1127
1075	1091	1150	1087	1150
1154	1135	1149	1122	1169
1113	1144	1148	1136	1186
1108	1123	1141	1190	1171

Frame viewing uniform illumination, 50 s integration:

5005	5102	4907	5029	4970
4853	4924	5155	4813	5055
5134	5035	5073	4888	5010
4957	5121	5014	4951	5124
4876	4939	4981	5182	4997

Frame shut off from illumination, 100 s integration:

111	111	111	132	151
122	122	122	144	163
133	133	133	156	175
144	144	144	168	187
155	155	155	179	199

Frame shut off from illumination, 0.1 s integration:

10	0	−10	0	10
10	0	−10	0	10
10	0	−10	0	10
10	0	−10	0	10
10	0	−10	0	10

7.4 Estimate the optical crosstalk for an array of silicon detectors 100 μm square and 0.9 mm thick, illuminated at $f/2$. Assume the absorption length is 0.9 mm and the refractive index of silicon is 3.4. Ignore reflection from the back face of the detector.

7.5 Consider a CCD that has a series of m pixels that have been illuminated uniformly. After n transfers with imperfect CTE, the output signal will show a deficit of ΔV_1 at the first pixel because of charge that has been lost to following pixels. The following pixels will show increasing signals, with the maximum, V_m,, at pixel m. Show that the CTE is given approximately by

$$\left(1-\frac{\Delta V_1}{V_m}\right)^{1/n}.$$

For n large and CTE ~ 1, derive the approximation

$$\mathrm{CTE} \sim 1-\left(\frac{1}{n}\frac{\Delta V_1}{V_m}\right).$$

7.6 Consider a CCD similar to the one in Figure 7.12. Let the thickness of the n-type region be 1 μm and the doping $N_D = 10^{16}\,\mathrm{cm}^{-3}$; and let the thickness of the p-type region be 40 μm and its doping $10^{14}\,\mathrm{cm}^{-3}$. Assume the diffusion length in the p-type region is 20 μm. Estimate the quantum efficiency in the blue spectral region (ignoring reflection losses and surface state charge trapping).

7.7 Prove equation (7.22). Consider a periodic image with exactly two pixels per period; show that the MTF at the image spatial frequency is 0.64. This degree of image resolution is called Nyquist sampling and gives the minimum number of pixels to recover the full information in the image.

7.8 Suppose an array has detectors with high absorption efficiency of thickness 10 μm and with a diffusion length of 20 μm. What is the minimum pixel size so that the MTF of a Nyquist-sampled signal is $\geqslant 0.5$?

Notes

1 In the following discussion, we shall see that a variety of restrictive conditions must be met by both the readout and the detector array to permit this type of construction. Where these conditions cannot be met, an alternative type of array can be built using 'Z-plane technology'. In this approach, the readouts are built in modules that are mounted perpendicular to the sensitive

face of the detectors (rather than parallel as in the direct hybrid approach). The readout modules can extend a significant distance behind the detectors. However, these modules are made thin enough that the detector/readout modules can be mounted in a mosaic to provide a large sensitive area. These arrays allow bulkier readouts than bump bonded hybrids, they help isolate the detectors from luminescence and heat produced by the readout, and with mosaicing they can be built in larger formats than bump bonded hybrids. However, the added construction complexity makes them unpopular unless these attributes have a very high priority.

2 Offset variations are frequently called bias variations, but we have avoided this nomenclature to prevent confusion with the detector bias voltage.

8

Photoemissive detectors

Photoemission refers to a physical process in which a photon, after being absorbed by a sample of material, ejects an electron from the material. If the electron can be captured, this process can be used to detect light. Photoemissive detectors use electric or magnetic fields or both to accelerate the ejected electron into an amplifier, enabling it to be detected as a current or even as an individual particle. These detectors are capable of very high time resolution (up to 10^{-9} s) even with sensitive areas several centimeters in diameter. They can also provide excellent spatial resolution either with electronic readouts or by displaying amplified versions of the input light pattern on their output screens. They have moderately good quantum efficiency (10–30%), particularly at ultraviolet and visible wavelengths. They are unmatched in sensitivity at room temperature or with modest cooling, leading to many important applications. They can be readily manufactured with 10^6 or more pixels. If the photon arrival rate is low enough that they can distinguish individual photons, the detectors are extremely linear.

8.1 General description

A photoemissive detector is basically a vacuum tube analog of a photodiode; in fact, the simplest form of such a detector is a vacuum photodiode, illustrated in Figure 8.1. The device is biased by placing a negative potential across its cathode, analogous to the p-type region of the diode, and anode, analogous to the n-type region. The vacuum maintained in the tube between cathode and anode is depleted of everything, not just free charge carriers; it corresponds to the depletion region of the diode. A photon releases an electron from the photocathode; it is accelerated by the electric field maintained by the voltage supply and collected at the anode. The resulting current is a measure of the level of illumination of the cathode. This kind of detector performs extremely

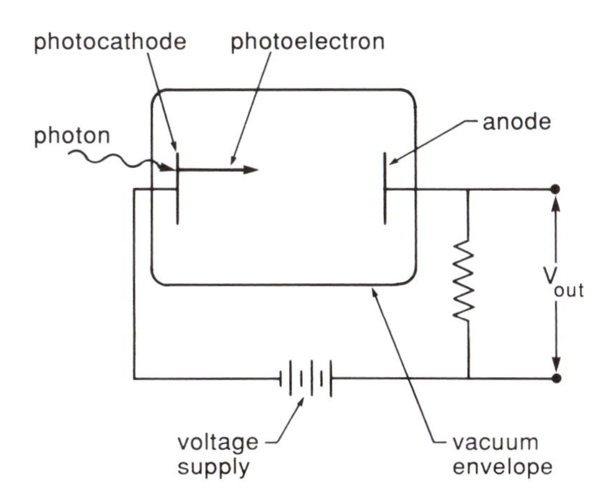

Figure 8.1. Operation of a vacuum photodiode.

well even at room temperature because of the very high impedance of the physical vacuum that forms its depletion region. An excellent general discussion of these detectors is provided by Csorba (1985).

A conduction electron is produced in the semiconductor material of the cathode when it absorbs a photon with energy greater than the intrinsic bandgap of the cathode material. This electron diffuses through the material until it reaches the surface, where it may escape into the vacuum. The negative potential of the photocathode helps accelerate the emitted electrons into the vacuum, and the electric field between cathode and anode drives the electron across the vacuum to the anode, where it is collected and provides a signal. The efficiency of the photocathode is determined by three mechanisms: (1) the absorption efficiency for photons; (2) the transport losses of electrons as they migrate from the absorption sites to the photocathode surface; and (3) surface barrier losses resulting from the inhibition of electron emission from the photocathode.

The key to achieving good performance with these detectors is to maximize the efficiency of the cathode. To elaborate on its operation and the selection of suitable materials for its construction, we first describe the simplest form of photoemission – that from a metal. Consider the metal–vacuum interface; the band diagram is shown in Figure 8.2. Because the conduction and valence states are adjacent, the available electron states will be filled to the Fermi level. An electron on the surface of the metal will require some energy to escape into the vacuum; we define the energy level in the vacuum to be $E = 0$. The work function, W, is defined as the energy difference between the Fermi level and the minimum escape energy. If a photon with $E > W$ is absorbed, it can raise an electron high into the conduction band. As this electron diffuses

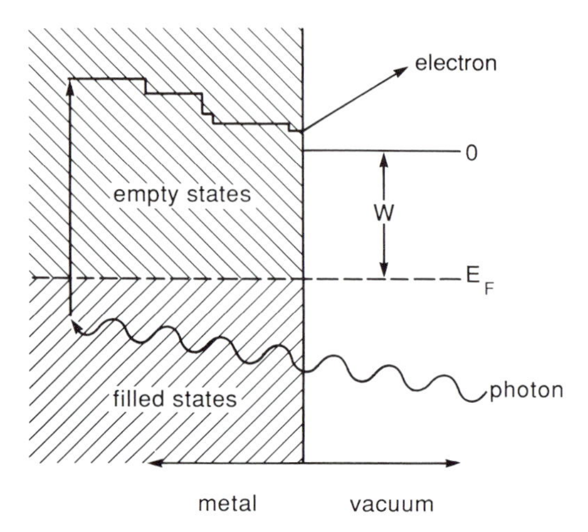

Figure 8.2. Energy band diagram of the photoelectric effect in a metal.

through the material and begins to thermalize, it will lose energy to the crystal lattice. As long as it is not in thermal equilibrium, it is referred to as a 'hot electron'. If it reaches the metal surface with energy $E \geqslant 0$, it has a reasonably high probability of escaping into the vacuum, as described by the theory of the photoelectric effect. As discussed here, however, the process is not very useful for a photon detector because the high reflectivity of metals would lead to very low quantum efficiency.

Semiconductors are used as photoemitters in detectors because they have much lower reflectivity than metals. The more complex properties of these materials require expanded notation to describe photoemission. Refer to the energy level diagram in Figure 8.3, which applies to a p-type material. As before, W is the energy difference between the Fermi level and the minimum energy the electron needs to escape the surface. The electron affinity, χ, is the difference between E_c (the minimum energy level of the conduction band) and the minimum escape energy. Unless the material has a very small value of $E_F - E_v$ or is at a high temperature, the Fermi level contains few electrons, so a photon of energy W can rarely produce photoemission. Usually an energy $\geqslant(E_g + \chi)$ is required. Finally, even for electrons that reach the surface with sufficient energy, the escape probability is always less than unity. Escape probabilities are difficult to predict theoretically, and are normally determined through experiment. The simplest photoemissive materials useful for photon detectors are semiconductors that have $(E_g + \chi) \leqslant 2\,\mathrm{eV}$ and electron escape probabilities of about 0.3. Unlike metals, these materials have reflectivities of only 15–50%. Thus, in the most favorable cases, the quantum efficiencies are 50–85% times 0.3, or 15–25%.

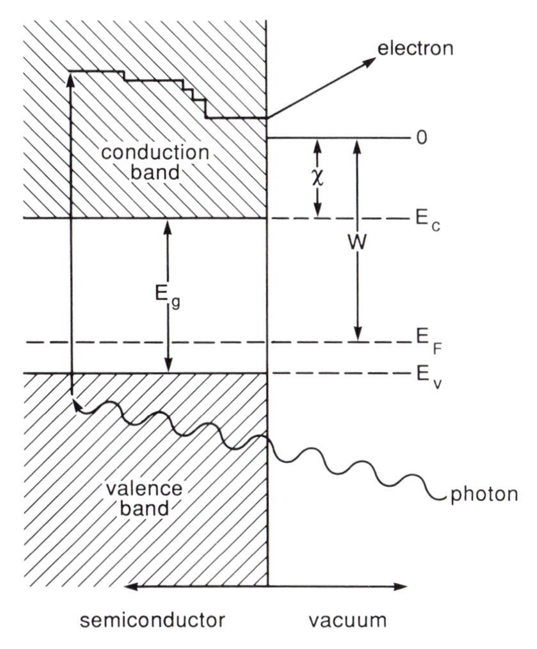

Figure 8.3. Energy band diagram of the photoelectric effect in a semiconductor.

Photon absorption occurs at some depth in the material, as has been emphasized in Figures 8.2 and 8.3. To escape from the material in the reflection geometry illustrated in these figures, the free electron must diffuse back through this depth to the surface without losing too much energy to the crystal lattice. Transparent photoemitters are often used rather than reflective ones. For them, the thickness of the photoemitter must be sufficient for good absorption of the incident photons but not too thick to inhibit the diffusion of electrons out the other side. In either case, to make an effective detector, the absorption length for photons needs to be of the same order or less than the diffusion length for electrons.

As we have seen, the absorption coefficient, $a(\lambda)$, in semiconductors is typically between 10^4 and $\geqslant 10^5\ \text{cm}^{-1}$ for direct transitions and between about 10^2 and $\geqslant 10^3\ \text{cm}^{-1}$, or about a factor of 100 smaller, for indirect transitions. For hot electrons, the diffusion length, L_H, is difficult to predict theoretically. Measurements indicate that a typical value is $10^{-2}\ \mu\text{m}$ for direct transitions and $\sim 1\ \mu\text{m}$, or a factor of 100 greater, for indirect transitions. Thus, to first order, the absorption lengths and diffusion lengths compensate, and the probabilities that photoexcitation will produce electrons that can escape from the surfaces of direct and indirect semiconductors are comparable. Practical photocathodes can be produced in which virtually all the photoelectrons contribute to the signal (within the limits set by the reflectivity and the electron escape probability discussed above).

Hot electrons will eventually thermalize, falling down to energy levels near the bottom of the conduction band. In this case, the diffusion length is given by equation (5.18) and is of the order of 1–10 μm for direct semiconductors that are suitable for high quality detectors. The thermalized electrons can therefore reach the surface from far deeper in the photocathode than can the hot ones discussed so far; unfortunately, for most materials the thermalized electrons do not have sufficient energy to escape from the surface once they get there. This problem is particularly severe for detectors operating in the red and near infrared because the photon energy is already low. The emission of low energy electrons can be enhanced substantially, however, by using multi-layer photocathodes; the most extreme forms have negative electron affinities, enabling thermalized conduction electrons to escape.

As an example, consider what happens if we use GaAs as the photoemitter and coat it with a thin layer of Cs_2O. For the sake of illustration, assume that we begin with two distinct materials (GaAs and Cs_2O) and bring them into physical and electrical contact as shown in Figure 8.4; in reality the Cs_2O is evaporated onto the GaAs. Assume that the GaAs is doped p-type and the Cs_2O n-type (neutral or n-type GaAs will not produce satisfactory results – see Problem 8.1). The energy levels of the two materials prior to contact are shown in Figure 8.4(a). When the two materials come into electrical contact with each other, the charge carriers will flow until the Fermi levels are equal, producing the situation illustrated in Figure 8.4(b). In this energy band structure, *any* electron excited into the conduction band of the GaAs has sufficient energy to escape from the material if it can diffuse through to the thin layer of Cs_2O before it recombines in the GaAs. Therefore, a photon needs to have only $E \geqslant E_g$, not $E \geqslant (E_g + \chi)$, to produce a free electron, thus providing good response in the red and near infrared. An additional advantage of this configuration is that the escape probability is raised to about 0.45, providing peak photocathode quantum efficiencies of 35–40%. Figure 8.5 shows the spectral response of a variety of photocathodes, including some that have negative electron affinity. Negative electron affinity photocathodes are discussed by Spicer (1977), Zwicker (1977), and Escher (1981); a detailed description of other photocathode types is given by Sommer (1968).

Mechanisms other than photoexcitation can also release electrons from the surfaces of photoemissive materials, producing dark current that interferes with the observation of low levels of light. At room temperature, dark current is usually dominated by the emission of thermally excited electrons. For the sake of simplicity, we assume that all electrons reaching the surface with

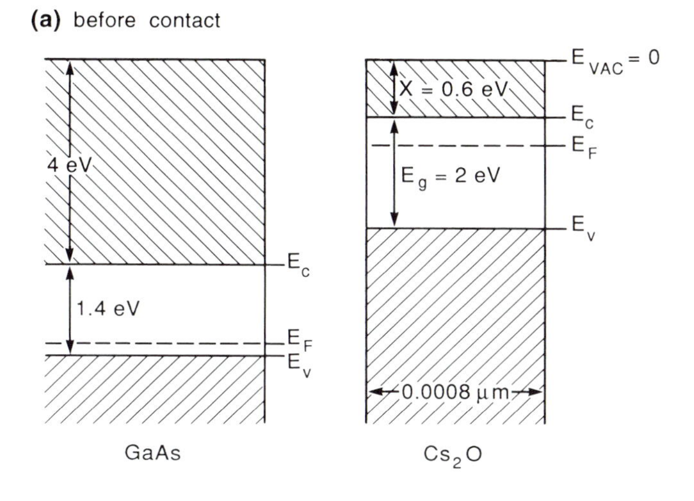

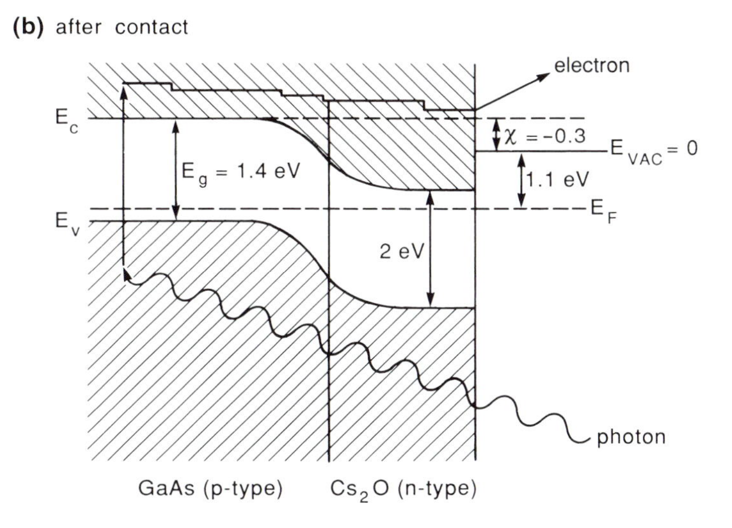

Figure 8.4. Operation of a negative electron affinity photocathode.

sufficient energy to escape are emitted and that emitted electrons are immediately replaced by other conduction electrons (assumptions appropriate for a metal). A calculation similar to that described in Section 3.2.6 predicts a dark current of

$$I_{\mathrm{d}}=q\left(\frac{4\pi m_{\mathrm{e}}}{h^{3}}\right)(kT)^{2}A\,e^{-W/kT}, \tag{8.1}$$

where A is the area of the photoemissive surface. This expression is called the Richardson–Dushman equation. It normally overestimates I_d for semiconductors, but it remains a useful approximation.

Equation (8.1) suggests two means of controlling dark current. The first is to cool the detector. Temperatures of 193–253 K (−80 to −20°C) usually reduce the thermal component of the dark current below other components; the required temperature depends on the photoemissive material. The second means, applicable only to nonimaging devices, is to restrict the detector area, which can be done by making it small in the first place or, for some detectors, by using external magnetic fields to deflect photoelectrons from all but a small area of the photoemissive surface away from the collecting and amplifying apparatus.

In addition to the thermal component of dark current, there are nonthermal components that are not removed by cooling. For example, electrons are released when high energy particles (cosmic rays or local radioactivity) pass through the photoemissive material; these electrons produce very large output pulses. When residual gas molecules in the detector vacuum strike the photocathode surface, they also produce large pulses called ion events. The device may also have a detectable level of electrical leakage, which allows currents to pass directly from the voltage input to the signal output terminal. The latter two items must be controlled during the manufacture of the detector

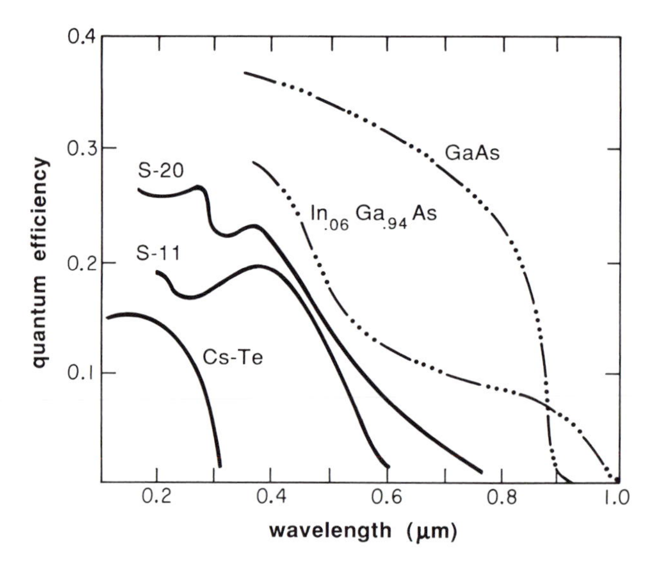

Figure 8.5. Spectral response of some representative photocathodes, after Zwicker (1977). The solid lines are for conventional materials, and the dot-dash lines are for negative electron affinity ones.

by: (1) establishing and maintaining a high quality vacuum within the device to eliminate residual gas; (2) designing the device with extremely high electrical resistance between the input and output; and (3) maintaining a high level of cleanliness so the high resistance between input and output established in manufacture is not compromised by contamination. For any given device, these nonthermal dark current mechanisms can significantly decrease the performance from the theoretical value, which is based primarily on the photocathode quantum efficiency.

Traditional (that is, positive electron affinity) photoemissive surfaces are prepared by placing the substrates to carry the surfaces under ultra-high vacuum and cleaning them thoroughly by heating and by bombarding them with high energy ions. The desired materials are then deposited by evaporating them at high temperature from bulk samples while still under vacuum. Negative electron affinity photocathodes are usually grown epitaxially on their substrates. In either case, the completed cathode is promptly sealed in a high vacuum within the envelope of the tube. The high vacuum must be maintained to avoid serious degradation of the performance. To the degree that the production processes are not fully controlled or the vacuum in the tube is degraded, performance parameters can vary significantly from one device to another that is nominally identical.

When low light levels are observed, the number of photoelectrons escaping from the photocathode needs to be amplified. Electron optics can be used to guide the electrons precisely while they are in the vacuum, bringing them to the inputs of high performance amplifiers. These high performance amplifiers and optics come in a variety of forms and will be described in Section 8.3. It is the accuracy and relative ease with which electrons can be guided and amplified in the vacuum that accounts for much of the power of photoemissive detectors.

8.2 Quantitative results

We can make quantitative performance predictions for photocathodes, particularly for those with negative electron affinity, based on the properties of photodiodes. Results from Chapter 5 accurately describe the performance of photoemissive detectors, although we must keep in mind that for the photodiode we assumed that the probability that the electrons would escape into the depletion region was unity. We have seen that this probability will be only about 25–50% for the photoemissive detector.

For example, suppose φ photons s^{-1} fall on the photocathode, producing $n_e = \eta\varphi$ electrons s^{-1}, where η is the quantum efficiency. The signal current is then

$$I_s = q\eta\varphi. \tag{8.2}$$

Representing the power in the incident flux as in equation (3.11), the responsivity is

$$S = \frac{\eta\lambda q}{hc}, \tag{8.3}$$

which is identical to equation (5.2). Similarly, the noise in the photocurrent can be obtained from equation (5.3) to be

$$\langle I_N^2 \rangle = 2q^2\varphi\eta\, df. \tag{8.4}$$

So, as in equation (5.4), the NEP becomes

$$\mathrm{NEP} = \frac{hc}{\lambda}\left(\frac{2\varphi}{\eta}\right)^{1/2}. \tag{8.5}$$

The performance may instead be limited by the dark current. If the dark current is Poisson in behavior, that is, caused by single electron events similar to those responsible for the photocurrent, we have (instead of equation (8.4))

$$\langle I_N^2 \rangle = 2qI_d\, df, \tag{8.6}$$

where I_d is the dark current, and the NEP becomes

$$\mathrm{NEP} = \frac{hc}{\eta\lambda}\left(\frac{2I_d}{q}\right)^{1/2}. \tag{8.7}$$

If there is a substantial dark current component from ion events, cosmic ray events, or other causes that do not obey the same Poisson statistics as the photon events, the noise will be degraded from that given in equation (8.6), and the NEP will be greater than that given by equation (8.7).

The wavelength of operation and the photon flux are usually determined by the desired measurements and are not under our control. We can then see from the expressions in equations (8.5) and (8.7) that the primary measures of detector performance are the quantum efficiency and the dark current. The total NEP is the quadratic combination of the contributions from photocurrent and dark current given by equations (8.5) and (8.7).

The analogy with photodiodes can be extended to details of photoemissive detector operation. In particular, the quantum efficiency of a photocathode can be determined from the material properties and the thickness of the photoemissive layer using a calculation exactly parallel to the one for the quantum efficiency of the diode, see Section 5.2.2.

8.3 Practical detectors

8.3.1 Photomultiplier tube (PMT)

A commonly used photoemissive detector is the photomultiplier, illustrated schematically in Figure 8.6. The voltages required for operation of the photomultiplier are provided by a single supply, the output of which is divided by a chain of resistors as shown in Figure 8.7. Photons eject electrons from a photocathode that is held at a large negative voltage (of order 1500 V relative to the anode). The electrons are accelerated and focused by electric fields from additional electrodes ('electron optics') until they strike dynode 1, the surface of which is made of a material that emits a number of electrons when struck by one high energy electron. These secondary electrons are accelerated into the surface of dynode 2 by the potential maintained by the resistor divider

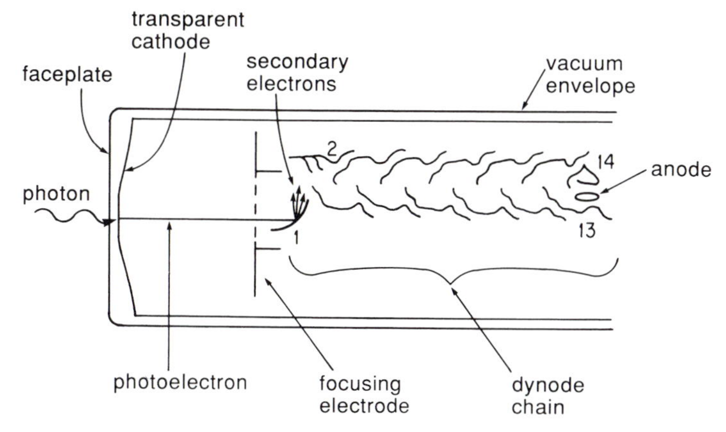

Figure 8.6. Design of a photomultiplier, after Csorba (1985).

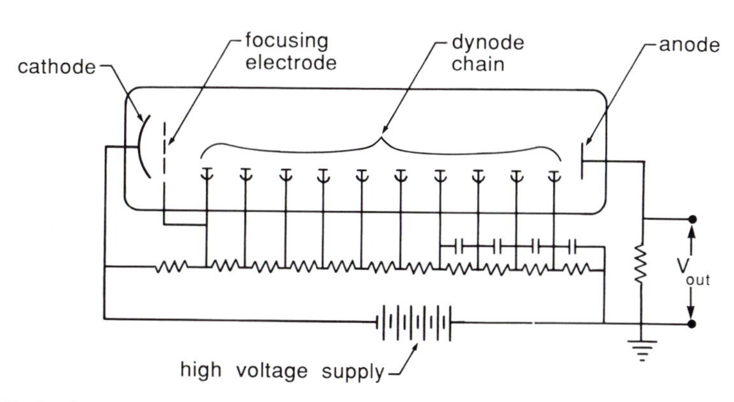

Figure 8.7. Voltage divider for a photomultiplier to give fast pulse response under situations where there is a high output current.

chain. At this second dynode they produce more electrons. Dynode-to-dynode voltages of 100–200 V give good yields of secondary electrons without unnecessarily increasing the unwanted side effects of high voltages (for example, corona and breakdown). This process of acceleration and electron emission is repeated down the chain of dynodes, and the amplified pulse or current is collected at the anode. By designing the tube geometry carefully, the output pulse width can be made as short as a few nanoseconds, allowing single photons to be counted with extremely high time resolution.

To be specific about the action of the dynode amplifier, assume that φ photons s^{-1} strike the photocathode in some time interval t; the number of electrons emitted is

$$n_1 = \varphi \eta t = \eta n_0, \tag{8.8}$$

where n_0 is the total number of photons striking the cathode in time t. Assuming that dynode 1 has gain d_1, the signal emerging from this dynode will be

$$n_2 = d_1 \eta n_0 = d_1 n_1, \tag{8.9}$$

and the net signal after amplification by m identical dynodes each with gain d is

$$n_{out} = d^m n_1. \tag{8.10}$$

This kind of amplifier is called an electron multiplier.

The image dissector scanner is a variation on the photomultiplier. In this device, the electron optics are designed to focus the output electrons from the photocathode onto a plate with only a small aperture for them to pass through. As a result, the dynode chain receives photoelectrons from only a small area of the cathode. Deflection coils in the electron optics allow the sensitive region to be shifted in position across the photocathode. If the sensitive spot is scanned over the cathode, the image dissector scanner operates as an areal detector array, but one that is sensitive to only one pixel at a time.

The process of emission by the dynodes is very similar to the photoemission that occurs at the cathode, except that the escaping electrons are produced by the interaction of an energetic electron with the material rather than by the interaction of a photon. Nonetheless, the same considerations apply regarding the escape of excited electrons from the material; as a result, similar materials (including some with negative electron affinity) are used for both the cathode and the dynodes.

Although in the simplistic discussion above, the photocathode properties determine the sensitivity of a photomultiplier, a variety of other effects also play a role. First, electrons released by the cathode must be focused on the first dynode; losses in this process appear as a reduction in the effective quantum efficiency. In the following discussion, we combine this loss with the photocathode quantum efficiency into an effective quantum efficiency, η'. From equation (8.8), the intrinsic signal to noise at the entrance to the first dynode is then

$$\left(\frac{S}{N}\right)_0 = (\eta' n_0)^{1/2}. \tag{8.11}$$

The losses in the electron optics are increased when electrons are deflected by magnetic fields; for example, unshielded photomultipliers can show variations in response for different orientations relative to the earth's magnetic field. Artificial sources of magnetic fields, such as electric motors, can also modulate the collection efficiency of the electron optics. Fortunately, thin sheets of high permeability magnetic material ('mu metal') can provide adequate magnetic shielding.

In addition, the signal-to-noise ratio is reduced if the dynodes have inadequate gain. Assuming that the electron multiplication in the dynodes follows Poisson statistics, the noise in the signal current emerging from the first dynode (equation (8.9)) will have two components: $d_1(n_1)^{1/2}$ (the amplified noise in n_1) and $(n_2)^{1/2}$. Combining these two noise sources quadratically, the total noise is

$$(d_1^2 n_1 + n_2)^{1/2} = (d_1^2 \eta' n_0 + d_1 \eta' n_0)^{1/2}. \tag{8.12}$$

The signal to noise emerging from dynode 1 is then (from equations (8.9), (8.11), and (8.12))

$$\left(\frac{S}{N}\right)_1 = \frac{d_1 \eta' n_0}{(d_1^2 \eta' n_0 + d_1 \eta' n_0)^{1/2}} = \frac{(\eta' n_0)^{1/2}}{\left(1 + \frac{1}{d_1}\right)^{1/2}}$$
$$= \left[\eta' n_0 \left(\frac{d_1}{d_1 + 1}\right)\right]^{1/2}. \tag{8.13}$$

Taking the expression in (8.13) as the input to a similar calculation, it can be shown that the signal to noise emerging from a chain of dynodes is

$$\frac{S}{N} = \frac{(\eta' n_0)^{1/2}}{\left[1 + \frac{1}{d_1}\left(1 + \frac{1}{d} + \frac{1}{d^2} + \cdots\right)\right]^{1/2}}$$
$$\approx \left[\frac{\eta' n_0}{1 + \frac{1}{d_1}\left(\frac{1}{1 - 1/d}\right)}\right]^{1/2} = \left[\eta' n_0 \left(\frac{d_1'}{d_1' + 1}\right)\right]^{1/2}, \tag{8.14}$$

where we have taken the gains of the subsequent dynodes to be equal to each other and to d. Here

$$d_1' = d_1\left(\frac{d-1}{d}\right). \tag{8.15}$$

From equation (8.14), it is clear that good performance depends on the first dynode having a large gain, d_1; large gain for the remaining dynodes is desirable but less critical. Comparing equation (8.14) with equation (8.11), it can be seen that the effective quantum efficiency, η', delivered at the output current is degraded compared to the value delivered to the input of the dynode chain. The factor by which it is reduced is

$$f_i = \frac{d_1'}{d_1'+1}. \tag{8.16}$$

For example, if $d_1' = 3$, the detective quantum efficiency is reduced by about 25%.

Fortunately, dynodes made of materials with negative electron affinity can yield $d_1 = 10$–20. Because they must carry larger currents than the initial dynode, the later dynodes are often made of a material with more robust electrical properties but lower gain. Equations (8.15) and (8.16) show that such designs can largely avoid degradation in signal to noise from the action of the dynode chain. To realize the potential of these designs, it is important to maintain a large accelerating voltage in the first few stages of the photomultiplier so that these stages operate at maximum gain; if the overall gain of the tube must be reduced (if, for example, the background light level is relatively high), either the voltages should be reduced at later stages or the signal should be removed before it reaches these last stages.

Although the analysis above is based on sensing the current in the photomultiplier, these detectors are frequently operated in a pulse counting mode. In this case, fast electronic circuits at the photomultiplier output identify all output pulses above some chosen threshold value. The threshold is selected to weed out small pulses produced by noise in the electronics or other events not associated with photon detection. With high quality photomultipliers, a range of threshold settings exists in which virtually all photon events are accepted but with rejection of the great majority of noise events from the dynode chain (of course, noise events arising from electrons being accelerated into the first dynode are indistinguishable by this means from photon events).

A disadvantage of this approach appears at relatively high illumination, when the pulse rate can become sufficiently large that pulses begin to overlap. A variety of problems then occur; for instance, pulses above the threshold are missed because they arrive too close in time to other pulses, and pulses below the threshold can be counted if they arrive in coincidence. The result is that the photomultiplier is no longer linear; the current mode of operation is therefore generally preferred for high levels of illumination.

At low illumination, however, pulse counting is usually preferred, primarily because the mechanisms that produce large pulses (for example, ion events or cosmic rays striking the photocathode) do not contribute disproportionately to the measurement. In fact, by setting a second threshold above which pulses are not counted, it is possible to eliminate many of these events from the signal completely. This high degree of noise immunity is not shared by current mode operation, in which the entire charge in a pulse contributes current. Even under ideal conditions, pulse counting has an advantage in terms of achievable signal to noise because to first order it is immune to the variations in pulse shape and size that occur due to statistics in the multiplication process. In other words, each event above the threshold (that is, each detected photon) is counted equally.

Nonetheless, the action of the dynode chain also degrades the signal to noise in pulse counting, in this case when the statistics of the multiplication process drop the pulse height below the detection threshold. Again, it is important for the first dynode to be operated with high gain; assuming the dynode multiplication follows Poisson statistics, the probability of no response whatever is

$$P(0)=e^{-d_1}. \tag{8.17}$$

For example, $d_1=3$ would yield $P(0)=0.05$. For any realistic setting of the pulse threshold, there would be significant additional losses beyond $P(0)$.

So far, we have concentrated on quantum efficiency and related parameters that influence signal to noise in photomultipliers, but stability and linearity are also important in most applications. The voltages required by the photomultiplier are typically provided by a circuit similar to the one shown in Figure 8.7. The dynode multiplication d varies roughly linearly with applied voltage; hence, with the total tube gain of d^m, we derive a condition on the power supply stability necessary to obtain a given stability in the gain of the dynode chain:

$$\frac{G+\Delta G}{G}=\left(\frac{d+\Delta d}{d}\right)^m=\left(\frac{V+\Delta V}{V}\right)^m, \tag{8.18}$$

where G is the total gain, V is the power supply voltage, and for simplicity we have taken all the dynode stages to have equal gain. Simplifying through the approximation that $\ln(1+x)\approx x$ for small x, we find that

$$\frac{\Delta V}{V}=\frac{1}{m}\left(\frac{\Delta G}{G}\right), \tag{8.19}$$

that is, to maintain gain stability of 1% requires a power supply with stability better than 0.1% for a typical tube having $m\approx 10$. Where the photomultiplier is used in current mode, the gain stability affects the accuracy of measurements directly. In pulse counting mode, it is possible through judicial choice of the accepted pulse threshold value to provide significantly reduced sensitivity to gain variations.

If it is also important that the output current of the tube be linear with input signal, care must be taken in designing the resistor chain that supplies voltages and currents to the dynodes so the signal current flowing in the tube does not significantly alter the gains of the amplifier stages. A common rule of thumb is that the current through the resistor divider chain should be at least ten times the maximum output current of the tube. A more quantitative approach is to replace the dynode-to-dynode signal currents with equivalent resistances and to analyze the resulting circuit to be sure that the voltages at the dynodes are not affected beyond the tolerance set by the desired degree of linearity in gain.

Particularly at low light levels (for which the signal tends to be resolved into a series of pulses), the analysis suggested above needs to be based on the peak current in a pulse, which may be far larger than the time-averaged current. Because it is impractical to supply the divider chain with the large currents required for the pulses, it is common to install capacitors in parallel with the last few resistors in the divider chain. The capacitors are selected to store sufficient charge to maintain gain stability for the duration of a pulse.

At large signal levels, space charge effects inside the tube can also affect the linearity. For this reason, as well as to avoid damage from excess currents, it is often desirable to remove the signal from an intermediate stage (and not supply voltage to the following ones) if the signal is excessively large.

From the above discussion, even at low light levels pulse counting has only a slight advantage over current mode operation in ideally behaved cases. However, it is often preferred because of the noise immunity, stability, and high degree of linearity it affords. A much more detailed analysis of these issues is provided by Young (1974).

8.3.2 Microchannels

A different form of electron multiplier can be made from a thin tube of lead-oxide glass, typically of 8–45 μm inner diameter and with a length-to-

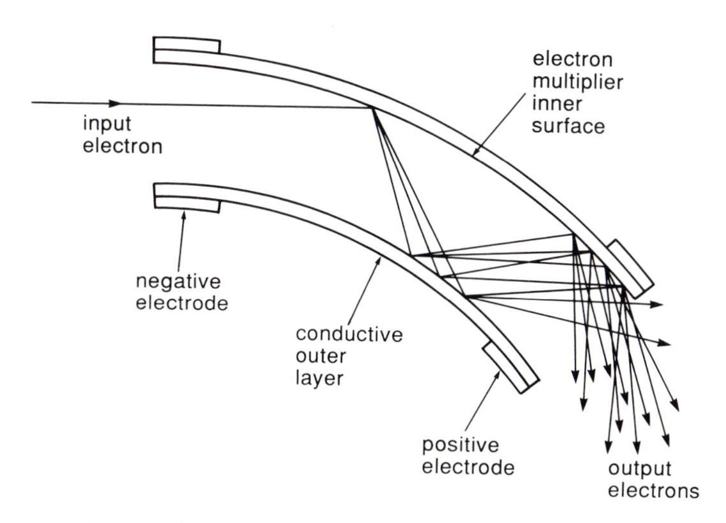

Figure 8.8. Operation of a single microchannel.

diameter ratio of about 40. The inner surface of the tube is fired in a hydrogen atmosphere, leading to a breakdown of the glass surface and formation of a layer of PbO, which acts as an electron multiplier. A large voltage is maintained across the ends of the tube (see Figure 8.8); the tube must be manufactured in such a way that a suitable voltage gradient is set up along its length. Electrons from a photocathode are accelerated and directed into the tube by electron optics. The tubes are mounted at an angle to the electron beam and/or they are curved so the incoming electrons collide with their walls. These collisions create secondary electrons that are accelerated by the voltage along the tube until they strike the walls again and produce more electrons. The net result is similar to the multiple amplifications in a dynode chain, and can produce gains nearly as large.

As in the preceding discussion on dynode resistor chains for photomultipliers, one requirement on the interior surface of the glass tubes is that it be able to conduct sufficient current to replace the charge lost in the electron cascade without varying the voltage significantly. This goal can only be achieved for relatively small total currents ($< 1\ \mu$A). Using currently available materials, the gain for one amplification in a microchannel cannot be made much larger than two or three; from equation (8.14), the intrinsic signal to noise of the photocathode is thus degraded significantly. Microchannels also tend to have relatively high dark currents, $\approx 10^{-11}$ A (compared to $\approx 10^{-16}$ A for a dynode chain). As a result of these limitations, microchannels are usually not competitive with dynode chains for single-channel detectors such as photomultipliers. Microchannels can, however, be readily manufactured in close-packed arrays, called microchannel plates, which can provide electron multiplication whilst retaining spatial or imaging information. As a result, they are an essential element in important types of image intensifier.

8.3.3 Image intensifiers

A photocathode, electron focusing optics, and a phosphor output screen or a suitable recording medium can be combined in a number of ways to intensify light whilst retaining imaging information. Phosphor output image intensifiers are reviewed briefly by Wampler (1974) and in depth by Csorba (1985).

Various implementations of image intensifier will be discussed below in the order of increasing complexity. They are all arranged so that photoelectrons produced at a photocathode are accelerated through a vacuum into a phosphor, where they are absorbed and produce a number of output photons, providing gain over the single photon that was initially absorbed by the cathode. The most sophisticated of these designs produce gains of 100 or more. As with other photoelectric detectors, very high performance can be achieved at room temperature because of the high resistance of the vacuum in the device. Nonetheless, to achieve extremely low noise, these devices are cooled to reduce the thermal emission of electrons from their photocathodes.

The first two ingredients in the image tube recipe should be familiar. The third – a phosphor – is a material that de-excites by releasing energy in the form of light; this concept was introduced in Chapter 1. In the present case, the excitation is provided by energetic electrons. A common family of phosphors is based on ZnS or ZnSCdS, which is activated with silver and chlorine. The silver acts as a p-type impurity and the chlorine as an n-type one; in equilibrium, the chlorine donates its electron to the silver so both impurities are charged. The luminescence process is illustrated in Figure 8.9. The energetic electron creates free electrons and holes in the sulfide. These charge carriers wander until they are captured by a charged impurity atom. The de-excitation

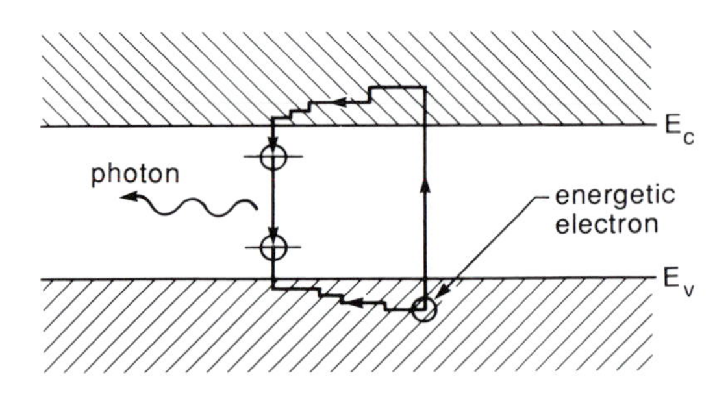

Figure 8.9. Process of luminescence of a phosphor.

(with emission of light) occurs when the electron transfers from the chlorine to the silver to re-establish the equilibrium condition. Typically, a well designed phosphor has an efficiency of about 25%; that is, about 25% of the energy lost by the electrons appears as light, with the remainder being deposited as heat.

To prevent the emitted light from escaping back into the image tube where it would be detected by the photocathode, the phosphor is coated on the cathode-facing side with aluminum. This coating can be thin enough that it does not impede the passage of the energetic electrons. The aluminum also acts as an electrode, maintaining the voltage potential of the phosphor.

In the simplest arrangement used to make an image intensifier, the phosphor is the anode (positive electrode) and is placed close to the photocathode (see Figure 8.10(a)). The electrons are accelerated over this short gap, which must be kept small enough to prevent them from spreading appreciably in the lateral direction. This device is called a proximity focused image intensifier.

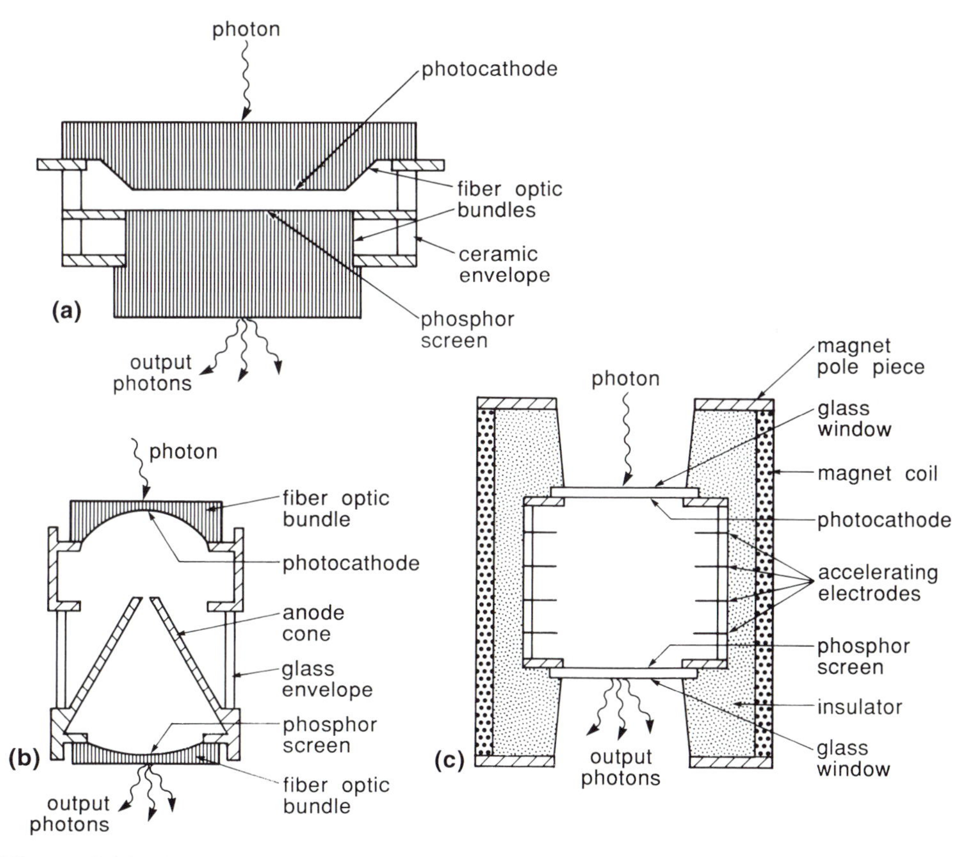

Figure 8.10. Cross-sectional diagrams of a variety of image intensifer types: (a) proximity focused; (b) electrostatically focused; and (c) magnetically focused. After Csorba (1985).

Its performance is limited by the size of the electric potential that can be held without breakdown across the small gap between cathode and anode. In addition, the small gap between photocathode and phosphor makes it particularly difficult to avoid contaminating the input signal with amplified light from the output.

Another type of device uses a microchannel plate as an electron multiplier between the cathode and anode (phosphor). The microchannels overcome the breakdown problem of the proximity focused device and provide much higher gain through their extra stage of multiplication. A disadvantage is that the effective quantum efficiency is reduced through a number of effects: (1) 25–45% of the photoelectrons are absorbed in the walls of the glass tubes between the microchannels and never enter the multipliers; (2) the multiplication per stage is sufficiently low that there is a significant probability that the photoelectron will produce no secondary electron when it strikes the wall of the microchannel; and (3) the low gain increases the noise as described by equation (8.14). The microchannel plate may also produce gain nonuniformities. Nonetheless, because they provide large light amplification in a small volume, these devices are popular and are called second generation, or 'Gen II', image intensifiers. A similar design with an advanced photocathode is sometimes termed a third generation intensifier.

A third class of image intensifier uses the field between the cathode and anode to accelerate the electrons as in the proximity focused tube. However, the separation between cathode and anode is increased compared to that in the proximity focused device, and a suitable set of electrodes is placed surrounding the space between the cathode and anode (see Figure 8.10(b)). These electrodes act as an electron lens by creating an electrostatic field to deflect the electrons and to keep them from spreading excessively in the lateral direction, refocusing them onto the phosphor at the end of their flight. The velocity dispersion of the photoemitted electrons perpendicular to the tube axis is not corrected by this type of electron optics, so it limits the sharpness of the image formed on the anode. In addition, the electrostatic electron optics suffer from significant pin cushion distortion. Because the optics in these tubes suffer substantial field curvature, the photocathode and the phosphor must lie on curved surfaces to obtain sharp images. Flat output faces are obtained by matching the curved face plates to dense bundles of fiber optics. Despite these limitations, these devices are used widely because they are simple and efficient. They are called first generation, or 'Gen I', image intensifiers.

A fourth system is similar to the Gen I intensifier except it uses magnetic instead of electrostatic focusing (see Figure 8.10(c)). As before, the electrons are accelerated from cathode to anode by the electrostatic potential maintained between these electrodes. The space between the electrodes, however, is surrounded by either an array of permanent magnets or an electromagnet that produces an axial magnetic field. The defocusing due to the dispersion in electron emission velocity is reduced by an order of magnitude over that achieved by electrostatic focusing; image diameters can be less than 10 μm. Moreover, pin cushion distortion is reduced. The chief remaining distortion is caused by nonuniformity in the magnetic field, which can result in the outer parts of the image being rotated with respect to the inner parts. This 'S' distortion can be held to $\leqslant$0.1 mm over a 40 mm diameter tube with carefully designed systems. The major disadvantage of this kind of intensifier is that its magnet makes it heavy and bulky. Assuming that the tube is to be cooled to reduce thermal emission, an electromagnetic coil requires substantial cooling power.

The fifth, and for us final, version of image intensifier is the electronographic tube. In this device, the phosphor of the four previous detectors is replaced with a photographic emulsion. The emulsion is exposed directly by the electrons accelerated from the photocathode as in the electrostatic or magnetic tubes.

To illustrate the operation of electron optics used in many varieties of photoemissive detector, we consider a magnetically focused image intensifier (see Figure 8.10(c)). If the coils around the evacuated region are large enough (and conform to other requirements), they will produce a highly uniform magnetic field, B, running down the axis of the tube from cathode to anode (or vice versa). This field will have no effect on the electron velocity component that runs along this axis, $v_{\parallel}$, but any velocity components perpendicular to the tube axis, $v_{\perp}$, will be subject to a force perpendicular to $v_{\perp}$ and of magnitude $qv_{\perp}|B|$. Similarly, if the electrostatic accelerating field is purely along the tube axis, it will influence $v_{\parallel}$ only and leave $v_{\perp}$ unchanged. The electron will therefore undergo a spiral motion, with the radius of the spiral determined by the balance between magnetic and centripetal forces:

$$r = \frac{m_e v_{\perp}}{q|B|}. \tag{8.20}$$

In one complete cycle of this spiral, the electron will return to a position exactly equivalent to the one where it started, but translated along the tube axis by

$$d_c = \int_0^{t_c} v_{\parallel}(t)\,dt, \tag{8.21}$$

where t_c is the time required for one complete cycle. This time is

$$t_c = \frac{2\pi r}{v_\perp} = \frac{2\pi m_e}{q|B|}. \tag{8.22}$$

Note that t_c is independent of $v_\perp$. Thus, d_c is the same for all electrons, regardless of their initial values of $v_\perp$. With suitable choices of magnetic and electrostatic field strengths and tube dimensions, it is possible to relay the image as it emerges from the photocathode accurately back to the anode.

Within the limitations discussed in the preceding section, electrostatic electron optics can produce results similar to those illustrated here with magnetic optics. Because the mathematics for this case is relatively complicated, we will not treat it in detail; see Csorba (1985) for details.

Continuing with the example of magnetic electron optics, if we assume that the emission velocity of the electron from the cathode is negligible then

$$v_\parallel = -\frac{q}{m_e}\mathscr{E}t. \tag{8.23}$$

To determine the design of a tube where the electrons conduct one spiral in the magnetic lens, we can substitute $\mathscr{E} = -V/d_c$ in equation (8.23) (where V is the voltage between the cathode and anode), integrate the result as in equation (8.21), and substitute for t_c from equation (8.22) to obtain

$$d_c = \left(\frac{2m_e}{q}V\right)^{1/2}\frac{\pi}{B}. \tag{8.24}$$

A standard application of the Biot–Savart law yields the magnetic field in an ideal solenoid (for example, far from edge effects) in the form

$$B = \frac{\mu_0 I_S N}{L}, \tag{8.25}$$

where I_S is the solenoid current, N is its number of turns of wire, and L is its length. Equations (8.24) and (8.25), together with the characteristics of the photocathode and phosphor, determine the first-order design and performance of the image intensifier, as illustrated in the example at the end of this chapter.

Because the first four types of image tube discussed above all have a phosphor as their output, they have a number of common characteristics. Imaging areas are typically a few to a hundred square centimeters, with resolution elements about 50 μm on a side. The time response is determined by the decay of the emission from the phosphors, which can have exponential time constants of several milliseconds.

Some phosphor output tubes show substantially lower detective quantum efficiencies than would be deduced from the properties of the photocathode alone (Cromwell, 1986). The detective quantum efficiency of the tube can be measured in terms of the counting statistics of output pulses produced by detection of single photons; in some of these cases it is only about 50% of the quantum efficiency of the photocathode. There appear to be two different causes: (1) in some tubes, the phosphor is too thin and there are voids which produce no output when struck by an energetic electron; and (2) the grains of some phosphor types appear to have dead spots where there is an inadequate density of luminescence centers, so even if an electron strikes the phosphor its signal may be lost (R. H. Cromwell, private communication, 1991). Among commonly used phosphors, this latter problem is particularly prevalent with those of type P11 and P20.

Phosphors emit over a very broad solid angle. It is difficult to design a conventional optical system that captures a large fraction of this output. This problem can be counteracted by using fiber optic exit bundles to conduct the light without allowing it to spread. A single fiber optic is a thin rod made of a low index glass core clad with a high index glass. A photon introduced into this rod at a small angle relative to the rod axis is relayed down the rod through total internal reflection off the high index cladding. Fiber optic plates are manufactured by bonding fiber optics into close packed bundles. Although the net transmission efficiency of these plates is only 50–60%, they are used as faceplates for many image intensifiers. For example, we have already mentioned that they accommodate the field curvature of Gen I tubes to provide flat entrance and exit windows. For all tube types, the output of the fiber optic bundle can be received by a photographic plate or electronic detector in contact with the flat surface of the fiber bundle, providing efficient transfer of the image to the recording medium. A variety of electronic detectors can be used in this way, as discussed below.

Since image intensifiers are often stacked to increase the net system gain, it has been found convenient to contact the exit faceplate of one tube to the entrance of another through fiber optic bundles. When tubes are cascaded in this manner, there is an inevitable degradation of resolution due to the convolution of the finite resolutions of each of the components, including misalignment of fibers from one plate to its mate. Output spot sizes from such chains of intensifiers are typically 80–100 μm. When coupling by conventional optics is unavoidable, many intensifier stages are cascaded to overwhelm the optical inefficiencies with gain. Such systems have degraded resolution and increased distortion compared with systems having fewer stages.

In a chain of image intensifiers, the resolution is determined by a variety of factors as discussed above; however, in a single tube, the image spreading is dominated by the aberrations in the electron optics and by scattering in the output phosphor. The resolution of image intensifiers is conventionally quoted in line pairs per millimeter. As discussed in Chapter 2, this quantity is relatively easy to measure, but the MTF provides a more complete description of the imaging properties of the detector. Figure 8.11 shows the MTF of an electrostatically focused tube, divided into contributions from the phosphor screen and the electron optics (along with their product) to give the net performance of the tube. Empirically, it is found that the image is roughly Gaussian in cross-section, and the modulation transfer function is therefore also roughly Gaussian. However, the MTF can vary substantially over the face of the tube. For an electrostatic device, the electron optics at the center of the tube have small aberrations, and the resolution is dominated by the phosphor. The aberrations increase off-axis and quickly dominate the performance; at this point, since the optics are radially symmetric, there can

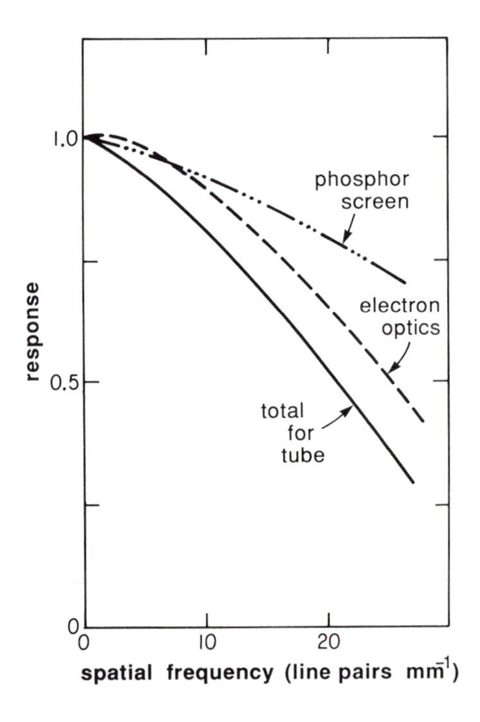

Figure 8.11. The MTF curves of the phosphor viewing screen and electron optics of an image intensifier and their combination for the entire device, after Stark, Lamport, and Woodhead (1969).

be a significant difference between the MTF in the radial and tangential directions.

As with other detector arrays, image intensifiers have fixed pattern noise. The primary causes are: (1) the granularity in their phosphors, (2) the large scale nonuniformities in both the photocathode and the phosphor, and (3) the pattern in which the optical fibers are arranged on the input and output. A further complication in flat fielding arises because of source induced background signal. We have mentioned with regard to proximity focused tubes the possibility that the output light will reach the photocathode and produce a signal. Although the problem is most severe for this tube type because of the small gap between photocathode and phosphor, it can also occur for the other image intensifier types that have phosphor output screens. Since signal induced backgrounds need not be uniform over the photocathode, they can violate the assumptions discussed in Section 7.3.1 that allow simple and accurate correction of fixed pattern noise. In situations requiring accurate flat fielding, signal induced background signals can limit the performance of image intensifiers significantly.

In addition to random photon noise and noise from thermally generated electrons, noise pulses occur in all types of image intensifier when residual gas atoms in the tube acquire a charge and are accelerated onto the phosphor or other output medium to produce ion events. For best performance at very low light levels, it is usually necessary to test and select commercial image tubes which have particularly low ion event rates. Tubes with high ion event rates can be used as second and third stages in image tube stacks; in this application, the ion events become unimportant because of the lower gain for them compared with that for photon events in the first stage. An unavoidable source of large events is the output from cosmic rays and other energetic radiations that strike the photocathode or phosphor.

A broad variety of techniques has been developed to record and store the outputs of image intensifiers. A number of them are described below, selected either because of historical interest or because they are well adapted to certain applications. After the first two, they all give electronic outputs that can be digitized and stored and manipulated in a computer. Many of the electronic tubes used for image tube readout also function directly as light detectors, but they have relatively high capacitances on their sensing electrodes, which lead to high read noise and make them intrinsically unsuitable for low light level detection without the gain in image brightness provided by the intensifier stage.

(a) Photographic plates

Photographic plates in close contact with the output face of a conventional image intensifier provide one of the simplest recording schemes. It is virtually required that the image tube have a flat exit window such as can be provided by a fiber optic exit bundle. The outputs of image intensifiers can also be coupled to plates with lenses. Plates provide a large number of pixels and efficient storage of the results, with low implementation costs. As with any use of plates, the approach has poor time response, dynamic range, and linearity, and quantitative extraction of intensity information is laborious and imprecise.

(b) Electronographic tube

An electronographic tube can greatly improve the linearity and dynamic range of the output compared with that of an intensifier/photographic plate system. The improvements result because a single photoelectron can be accelerated to an energy where it creates a number of developable grains when it enters the recording emulsion, thereby averaging out the vagaries of the photographic process as applied to photons. Departures from linearity can be held to less than 1–2% over a dynamic range of 10^4 to 10^5. A serious difficulty arises in extracting this information from the emulsion, since the maximum signal for a dynamic range of 10^5 corresponds to a density of 5, that is, a transmission of the plate of only 10^{-5}. Such small transmissions are hard to measure accurately with conventional plate measurement techniques.

A disadvantage with these devices is the difficulty in removing an exposed emulsion and replacing it with a fresh one for a new exposure. Some early devices required that the glass envelope be broken and the tube destroyed to remove the exposed emulsion (weep). More recent versions use vacuum windows of thin mica that allow about 75% of the electrons to pass through to an emulsion carefully pressed against them. These windows are fragile, and they limit the size of the tube output to no more than about 20 cm^2.

(c) Multiple anode intensifiers

When an electrical rather than a visual output is desired, an image intensifier can dispense with the phosphor screen and register the electron pulses on some suitable anode. For example, if an individual anode is provided for each microchannel in an intensifier with a microchannel plate, the full imaging information is retained. The implementation of this approach, however, would be undesirably complex for reading out a large number of microchannels. When photon rates are sufficiently low, a variety of encoding schemes can be employed (see Lampton, 1981) that use a more modest number of anodes to provide positional information. This type of image tube is most suitable for recording single photon events, and is used widely in the ultraviolet and X-ray spectral regions. The highest performance implementation of this concept is to use overlying grids of vertical and horizontal wires at the output of the microchannel plate combined with output circuitry that determines the address in each grid of an electron pulse that deposits charge on the wires. This device is termed a multi-anode microchannel array (MAMA) detector and is described by Cullum (1988), Morgan and Timothy (1988), and Timothy (1988).

(d) Image dissector

An image dissector can be used to record the output of an image intensifier. Because the image dissector is not capable of any internal integration, to avoid loss of signal the scan rate must be sufficiently high that the image intensifier phosphor can provide the integration; that is, each point in the output image must be sampled at least once per phosphor decay time. This requirement places a number of restrictions on the design of the image tube and the supporting digitizing and computing equipment that receives and processes the output of the dissector. With large signals, the phosphor can impose significant limitations in dynamic range and granularity; these devices are most useful at low illumination, where single photon events can be measured individually.

(e) CCD

A charge coupled device can also read out an image intensifier, again with the phosphor-imposed performance limitations discussed above. Obviously, the useful number of pixels in the detection system cannot exceed those of the CCD. Where long integrations (and low time resolution) are acceptable, and cooling can be provided to suppress dark current, modern CCDs produce better performance across the visible and near infrared when operated bare. However, an intensified CCD can operate at the photon noise limit even at very low light levels without cooling (albeit with reduced quantum efficiency compared with a bare CCD). In addition, the increase in CCD read noise with rapid readout rates can make an intensifier stage necessary to reach the photon-noise limit when high time resolution is needed, either for detection of single photon events or because of the variability of the object under observation.

(f) Reticon and Digicon

Diodes provide highly linear readouts with good dynamic range and low dark current. The disadvantage of these devices is the high read noise associated with the relatively large diode capacitance, frequently increased further by the design of the readout electronics. However, with sufficient intensifier gain these devices can count single photon events. One type of readout, usually termed a Reticon after the manufacturer of the most successful version, consists of linear arrays of diodes that are switched by transistors sequentially onto the input of an amplifier. An image intensifier output phosphor is optically coupled to the diode array. The Digicon (also a tradename) places the output diodes inside the vacuum of the tube and lets the accelerated photoelectrons fall directly on them. In this way, it circumvents the shortcomings posed by a phosphor output.

(g) Vacuum tube or television-type imaging detectors

A variety of readouts are based on television-type imaging detectors, illustrated generically in Figure 8.12. All these devices are read out by an electron beam, which is generated as follows. Free electrons are produced by heating a cathode inside the vacuum envelope of the tube. An enclosure is placed around the cathode with a small aperture, through which the electrons are emitted. Electron optics are used to form an image of the electron-emitting aperture onto a surface, called the target, that stores the light image in an electronic form. Because the beam of electrons has very high electrical conductivity, it effectively shorts the electrical signals on the target to the beam-generating cathode; the resulting current in the target is sensed to provide an output containing the image. Deflection coils allow this spot to be scanned over the target to read out the entire image. Because the target electrode of all these devices has a relatively large capacitance, there is a relatively low voltage swing for a given stored charge, and hence even very quiet output amplifiers produce a relatively large read noise in electrons. However, if coupled to an image intensifier, a television-type imager can readily count individual photons. Further discussion can be found in Csorba (1985).

The conventional vidicon uses a photoconductive target of antimony trisulphide (Sb_2S_3). The faceplate side of the target is a transparent electrode that is held positive. When the electron beam is swept over the target, it charges the backside to a slightly negative potential, placing a net charge differential across the target capacitance. The absorption of photons in the

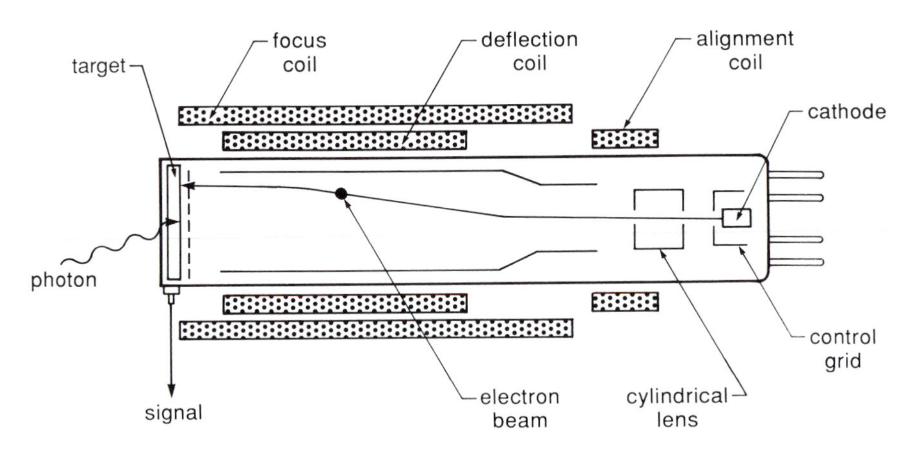

Figure 8.12. Cross-section of a conventional vidicon.

Sb_2S_3 generates electron–hole pairs; the electrons are swept to the faceplate electrode, and the holes are repelled toward the vacuum side of the target. The target capacitance is therefore locally discharged. When the electron beam strikes this portion of the target, it supplies charge to re-establish the potential on the capacitance. An equal charge flows in the faceplate electrode, where it is sensed to measure the image strength at that position.

Particularly at low light levels, conventional vidicons show an undesirable degree of 'lag'; that is, they have a memory of the image that takes a number of sweeps of the electron beam to erase, so when the illumination pattern changes on the tube there is a delay before the old image completely disappears. Much of this lag occurs because of traps in the bulk photoconductor that retain a portion of the charge over a number of read cycles. In addition, these vidicons show substantial 'blooming' (the spread of the bright portions of an image into neighboring pixels due to the lateral migration of charge). These problems can be reduced by fabricating the target as a PIN diode; the large field in the intrinsic portion of the diode causes the charges to be swept quickly to the surfaces of the target with reduced trapping and lateral migration. Such targets are fabricated in PbO; the tubes using them are called lead-oxide vidicons, or Plumbicons, after a particularly successful version.

The bandgap of PbO is 1.9 eV, so lead-oxide vidicons have red response only to 0.65 μm. Improved red response can be achieved with a silicon target (ST) vidicon. The bulk of the target in an ST vidicon is n-type silicon; an array of p-type implants in the backside of the target creates the diodes for collection of charge. Response extends to the silicon cutoff wavelength, 1.1 μm and the silicon target is extremely robust against damage by over-illumination. However, the smaller bandgap results in relatively large dark currents, so these vidicons need to be cooled for satisfactory operation at low light levels.

To sense very faint images, the tubes described above need to be coupled optically to the output phosphor of an image intensifier. This coupling can either be accomplished by re-imaging the output phosphor onto the faceplate of the readout with lenses or by providing both tubes with flat fiber optic faceplates that can be contacted directly. In either case, performance shortcomings are imposed by the phosphor: limited dynamic range; granularity; reduction in detective quantum efficiency. These problems can be avoided in a manner similar to the electronographic camera or the Digicon. A photocathode is evaporated on the faceplate of the tube, and the photoelectrons it produces are accelerated and focused by electron optics onto the tube target. If the target is similar to that for an ST vidicon, such a device is termed a silicon intensified target (SIT) vidicon. Figure 8.13 illustrates its operation. A second important example of this class of device

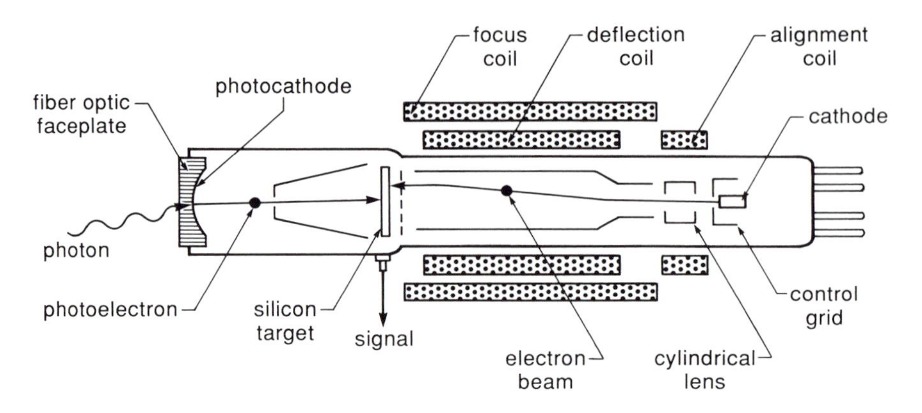

Figure 8.13. Cross-section of a silicon intensified target (SIT) vidicon.

is a secondary electron conduction (SEC) tube. The SEC tube is similar in concept to the SIT except that its target is a thin layer of KCl, with a thin aluminum coating to act as the positive electrode. The large bandgap in the KCl suppresses dark current, and even at room temperature on-target integrations of up to an hour are feasible. On the other hand, the KCl target is relatively easily damaged by over-illumination.

At low photon rates, the output of an image intensifier can be read out quickly enough by an electronic detector (100–1000 times per second) that the events due to individual photons can be distinguished. The relatively high noise that results from reading out the electronic detector quickly is overcome by the high gain of the image intensifier, so each photon event results in many photons being received by the readout device. For successful realization of this concept, elaborate follow-up computing is required. For example, with the readout rate set to less than a phosphor decay time, multiple detections of the same event will occur in successive outputs; the computer must recognize them and identify them with a single detection. Frequently, computational methods are also used to locate each photon event to within a fraction of its diameter by a centroiding algorithm. Typically, an input photon can be located to about 0.1 of the full width at half maximum (FWHM) of the output spot if the event rate is low enough that multiple events do not confuse the centroiding algorithm. An approximate maximum rate is 0.05 events per pixel per read cycle. When used in this way, the effective resolution of the tube is significantly increased over that obtained by integrating a long exposure on the output device before reading it out.

8.4 Example

We will consider a magnetically focused image intensifier as an illustration. We will use a GaAs-based photocathode, and we will design the tube to have a gain of a factor of 100. We assume a photocathode diameter of 40 mm and a distance from cathode to anode/phosphor of 60 mm.

First, consider the design of the photocathode. Typical properties of GaAs are as follows:

reflection coefficient:	0.16
optical absorption coefficient:	$2 \times 10^4\,\mathrm{cm}^{-1}$
carrier recombination lifetime:	$10^{-9}\,\mathrm{s}$
diffusion coefficient:	$150\,\mathrm{cm}^2\,\mathrm{s}^{-1}$
electron escape probability:	0.4

We note that the absorption length $= 1/(2 \times 10^4\,\mathrm{cm}^{-1}) = 0.5\,\mu\mathrm{m}$. The diffusion length is

$$L = (D\tau)^{1/2} = (150\,\mathrm{cm}^2\,\mathrm{s}^{-1}\,10^{-9}\,\mathrm{s})^{1/2} = 3.87\,\mu\mathrm{m}$$

Comparing, we see that the absorption length is much shorter than the diffusion length, as required to make a good photocathode; the high absorption allows us to treat the quantum efficiency approximately as given in the extreme represented by equation (5.24). If we take the absorption to go exponentially, that is, as in equation (2.3), we can show that the photocathode with the highest product of absorption factor and diffusive-limited quantum efficiency has a thickness of $1.5\,\mu\mathrm{m}$, where $\eta = \eta_{\mathrm{ab}} \times 0.929b = 0.883b$. We estimate b as the product of (one minus the reflection) of the photocathode and the electron escape probability, or $b = 0.336$. Therefore, $\eta = 0.297$.

To obtain a gain of 100, every electron emitted from the photocathode must produce on average $100/\eta = 337$ output photons. If we simplify the calculation by assuming that the phosphor emits light only at $0.55\,\mu\mathrm{m}$, then each photon requires an energy of 2.25 eV. If the phosphor efficiency is 25%, the electrons must be accelerated to an energy of $337 \times 2.25\,\mathrm{eV}/0.25 = 3039\,\mathrm{eV}$, that is, the potential between cathode and anode should be 3039 V. There is no allowance in this estimate for possible dead spots in the phosphor or other losses. To be prudent, we will design the tube for a potential of 5000 V.

To obtain a uniform magnetic field in the focusing region, we will design the coil to be twice the diameter of the photocathode, that is, 80 mm, and twice the length between cathode and anode, that is, 120 mm. Equation (8.24) allows us to estimate the required field strength:

$$B = \left(\frac{2m_{\mathrm{e}}}{q} V\right)^{1/2} \frac{\pi}{d_{\mathrm{c}}} = 1.602 \times 10^{-2}\ \mathrm{Wb\,m}^{-2},$$

where $d_c = 60$ mm and $V = 5000$ V. A reasonable number of turns for the coil is 1000, so we can compute the required current from equation (8.25),

$$I_S = \frac{LB}{\mu_0 N} = 1.53\ \text{A}.$$

A reasonable wire diameter in the coil is 1mm (giving a coil eight windings deep with no allowance for insulation). The total length of wire is $2\pi rN = 2\pi(0.044\ \text{m}) \times (1000) = 276.5\ \text{m}$. The conductivity of copper at room temperature is $6.5 \times 10^7\ (\Omega\ \text{m})^{-1}$. From equation (3.1), we can estimate that the resistance of the coil winding is $5.42\ \Omega$, and the power dissipated in the coil is $I_S^2 R = 12.7$ W. This power must be removed from the tube; if it is to be operated at a reduced temperature to reduce dark current from the photocathode, the cooling system must also contend with the heat conducted into the tube from the ambient temperature surroundings. Therefore, a substantial cooling system is required.

8.5 Problems

8.1 Using bandgap diagrams, discuss why n-type GaAs cannot be used effectively for a negative electron affinity photocathode.

8.2 Suppose you are making a measurement using a photomultiplier with very large dynode gain (so differences due to dynode gain can be ignored). The signal you are measuring corresponds to an average of 100 pulses per second. In addition, there is an excess noise process that produces an average of 1 pulse per second that is 100 times as large as a photon pulse. All the events arrive with a Poisson distribution in time (that is, the uncertainty in the number of pulses received in a time interval is the square root of the expected number). Compare the signal to noise that will be achieved in 100 seconds both by current measurement and by pulse counting.

8.3 A photomultiplier with cutoff wavelength of 1 μm has a dark current of 10^{-6} A when operated with a net gain of 10^6 at a temperature of 300 K. Assume that the effective quantum efficiency is 0.1 and that you want to operate the tube in pulse counting mode and to be degraded from background limited sensitivity by no more than 10% with a background of 100 photons s^{-1} onto the photocathode. Assuming the dark current is described accurately by the Richardson–Dushman equation (8.1), determine the temperature at which you need to operate the tube.

8.4 Assume the output of the phosphor of an image intensifier is Lambertian. Suppose that the gain of the intensifier is a factor of 50 and that its output is coupled to another intensifier by an f/2 relay lens. What is the effective gain of the intensifier/lens combination as seen by the second intensifier? Assuming a photocathode quantum efficiency of 10%, comment on the usefulness of this combination.

8.5 Consider a detector system consisting of two identical image intensifiers with fiber optic inputs and outputs, stacked and read out by a CCD in contact with the output of the second image tube. For an input point image, one of the image tubes without fiber optics would give an output image of gaussian cross-section with a FWHM of 30 μm. The faceplate fibers are 25 μm in diameter. The CCD pixels are 25 μm in size; the diffusion length and the pixel thickness are both 10 μm. Compute the MTF of the system (use equations (7.21) and (7.22)). Taking the spatial frequency at which the MTF = 0.5 as a measure of the resolution, compare the achievable resolution of the image tube system used in a photon centroiding mode with that of the CCD alone.

9

Bolometers

We now turn to thermal detectors, the second major class of detector listed in Chapter 1. Unlike all detector types described so far, these devices do *not* detect photons by the excitation of charge carriers directly. They instead absorb the photons and convert their energy to heat, which is detected by a very sensitive thermometer. The energy that the photons deposit is important to this process; the wavelength is irrelevant, that is, the detector responds identically to signals at any wavelength so long as the number of photons in the signal is adjusted to keep the absorbed energy the same. Thus, the wavelength dependence of responsivity is flat and as broad as the photon absorbing material will allow. Because the absorber is decoupled from the detection process, it can be optimized fully, and quantum efficiencies can be as high as 90–100%. Semiconductor bolometers are the most highly developed form of thermal detector for low light levels and are the detectors of choice for many applications, especially in the infrared and submillimeter spectral ranges. Nonetheless, these devices do have their fair share of disadvantages. They must be constructed carefully to ensure that they are isolated from the thermal surroundings, and the techniques typically used to construct them do not lend themselves to efficient development of large arrays; an array of 100 detector elements is quite ambitious. In addition, these detectors must be operated at extremely low temperature ($\leqslant 1$ K) for best sensitivity.

9.1 Basic operation

To illustrate the operation of a bolometer, consider the simple thermal model in Figure 9.1. A detector is connected by a weak thermal link of thermal conductance G (units of watts per kelvin) to a heat sink at temperature T_0. Assume there are no additional paths for heat loss. The detector is absorbing a constant power P_0, which raises its temperature by an amount T_1 above

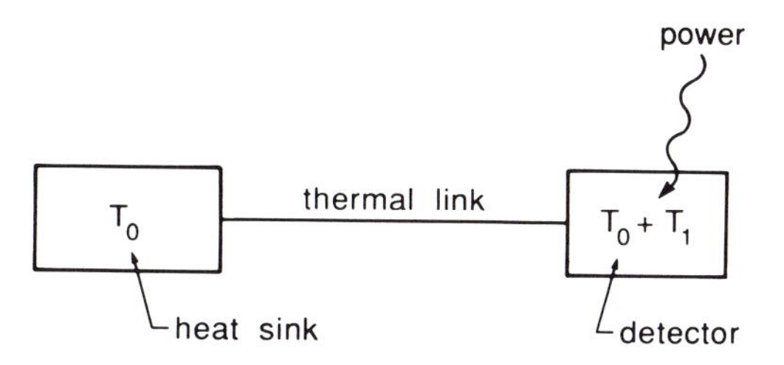

Figure 9.1. Thermal model of a bolometer.

the heat sink. The definition of G is then

$$G = \frac{P_0}{T_1}. \tag{9.1}$$

Now we introduce an additional, variable power component, $P_v(t)$, which we will assume is deposited by absorbed radiation. The temperature of the detector element will change following the time dependence of $P_v(t)$:

$$\eta P_v(t) = \frac{dQ}{dt} = C\frac{dT_1}{dt}, \tag{9.2}$$

where η is the fraction of power absorbed by the detector (that is, the quantum efficiency) and C is the heat capacity, which is defined by $dQ = C\,dT_1$ and has units of joules per kelvin. The total power, $P_T(t)$, being absorbed by the detector is

$$P_T(t) = P_0 + \eta P_v(t) = C\frac{dT_1}{dt} + GT_1. \tag{9.3}$$

Suppose $P_v(t)$ is a step function at $t=0$ such that $P_v = 0$ for $t<0$ and $P_v = P_1/\eta$ for $t \geqslant 0$. The solution to equation (9.3) is then

$$T_1(t) = \begin{cases} \dfrac{P_0}{G}, & t<0 \\ \dfrac{P_0}{G} + \dfrac{P_1}{G}\left(1 - e^{-t/(C/G)}\right), & t \geqslant 0. \end{cases} \tag{9.4}$$

By inspection, the thermal time constant of the detector is therefore

$$\tau_T = \frac{C}{G}. \tag{9.5}$$

For times long compared to τ_T, T_1 is proportional to $(P_0 + P_1)$. If T_1 can be measured, this device can measure the input power.

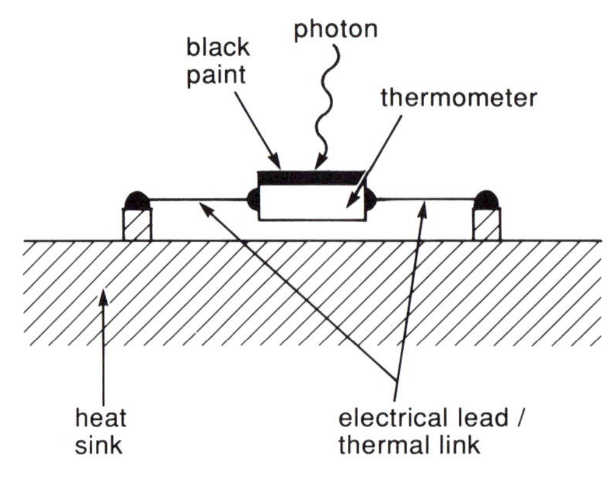

Figure 9.2. Simple bolometer.

A simple form of bolometer is illustrated in Figure 9.2. It consists of a small chip of silicon or germanium that has been doped both to have a resistance suitable for coupling the signals to a low noise amplifier and to have a large temperature coefficient of resistance. The detector chip is suspended between electrical contacts by very thin leads that are soldered or glued to the chip. These leads supply the thermal link to the heat sink and also carry a current to generate a voltage that can be measured to determine the electrical resistance of the detector. A high input impedance amplifier monitors the voltage across the bolometer, which varies in response to changes in its resistance and therefore its temperature, thereby measuring any changes in the input power. If necessary, the detector is covered with black paint to enhance its absorption of photons. A further description is provided by Low (1961).

9.2 Detailed theory

9.2.1 Electrical properties

From the discussion above, it is apparent that the performance of a semiconductor bolometer is based on the temperature dependence of its electrical properties. This dependence is described by the temperature coefficient of resistance,

$$\alpha(T) = \frac{1}{R} \frac{\mathrm{d}R}{\mathrm{d}T}. \tag{9.6}$$

Using equations (3.1), (3.5), (3.51), and (3.52), and assuming that the mobility is independent of temperature, the resistance of a piece of semiconductor can be shown to be of the form

$$R = R_0 T^{-3/2} e^{B/T}, \tag{9.7}$$

where R_0 and B are constants. Thermal detectors of modest performance and operating well above absolute zero can use thermometer elements described by equation (9.7). We shall see, however, that high performance bolometers need to be operated at very low temperatures. In this case, the thermally excited conduction (upon which the derivation of equation (9.7) depends) is very small. Even if the semiconductor material could be sufficiently controlled to produce the large resistances implied by equation (9.7), the resulting detectors would have little advantage over lower resistance ones because the low operating temperatures reduce Johnson noise below the level of other noise mechanisms even without using high impedances. Since the first stage amplifiers used with bolometers must be operated at higher temperatures than the detectors themselves, it would be difficult to couple extremely high impedance detectors to these amplifiers without picking up excess noise in the wiring between detector and amplifier.

To obtain appropriate electrical properties at very low temperatures (<5 K), the semiconductor material must be doped more heavily than assumed in equation (9.7) so that the dominant conductivity mode is hopping. That is, the impurity atoms are sufficiently close together in the crystal that their electron wave functions overlap slightly, and there is a chance for the electrons to shift from one atom to the next without first entering the conduction band. This mechanism freezes out relatively slowly; the resistance is given by an expression of the form:

$$R = R_0 \, e^{[\Delta/T]^{\xi}}, \tag{9.8}$$

where $\xi \approx 1/2$ (which value we will use in the following). Equation (9.8) is only applicable for $T \ll \Delta$, where Δ is a temperature of the order of 4–10 K. This relationship gives

$$\alpha(T) = -\frac{1}{2}\left(\frac{\Delta}{T^3}\right)^{1/2} \tag{9.9}$$

(with $\xi = 1/2$). An alternative expression, applicable at intermediate values of T, is determined from an empirical fit to the bolometer resistance and is of the form

$$R = R_0\left(\frac{T}{T_0}\right)^{-A}, \tag{9.10}$$

where typically $A \approx 4$. We then have

$$\alpha(T) = -\frac{A}{T}. \tag{9.11}$$

Note that, in both of these cases, $\alpha < 0$ and has strong temperature dependence.

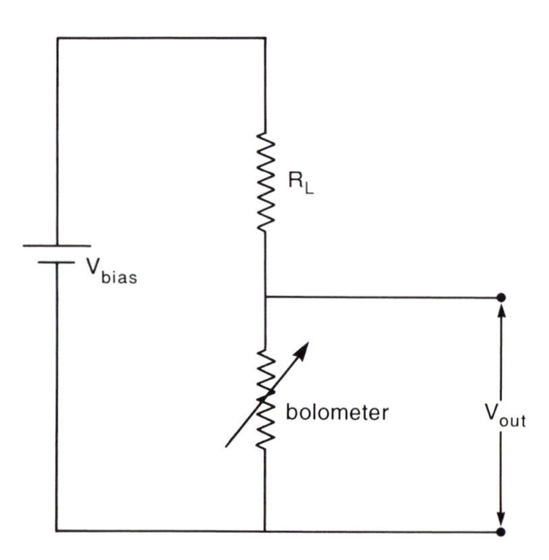

Figure 9.3. Bolometer biasing and readout circuit.

It is normally adequate to use the circuit in Figure 9.3 (which is similar to Figure 6.4) to determine the bolometer resistance. We have already discussed the shortcomings of this circuit that arise from the need to operate photodetectors at very large resistance to minimize the Johnson noise. Since there are compelling reasons to operate bolometers at extremely low temperatures, Johnson noise for bolometers is sufficiently small even with modest detector resistance, and the previous arguments no longer apply. As a result of the modest resistances, for many applications the amplifiers that sense the output of this circuit need not be cooled. However, for practical reasons having to do with controlling noise introduced by vibrations of the wiring between the detector and a warm amplifier, the first amplifier stage of most systems uses a transistor placed close to the detector and cooled as far as possible without compromising its operation.

The analysis of the circuit in Figure 9.3 is simplified by setting the load resistance, R_L, to be much larger than the detector resistance so the current through the detector is, to first order, independent of the detector resistance. If R_L is too small and the bolometer controls the current through the circuit, the circuit is subject to a runaway instability in which additional current drives down the bolometer resistance (because of the large negative coefficient of resistance), which allows even more current to flow. This arrangement frequently results in the demise of the bolometer through overheating. For protection, a large value of R_L is almost always used, and an analysis under that assumption does not suffer any serious loss of generality in terms of practical applications. The general treatment can be found in Mather (1982).

9.2.2 Time response

In this section and the following two, we will derive the time response, responsivity, and noise characteristics of a bolometer in the circuit in Figure 9.3, making the simplifying assumption just discussed. Let $P_I = I^2 R(T)$ be the electrical power dissipated in the detector resistance $R(T)$ by the current I. Because R is a function of temperature, we must modify equation (9.3):

$$P_T(t) = C\frac{\mathrm{d}T_1}{\mathrm{d}t} + GT_1 - \frac{\mathrm{d}P_I}{\mathrm{d}T}T_1. \tag{9.12}$$

However,

$$\frac{\mathrm{d}P_I}{\mathrm{d}T} = I^2\frac{\mathrm{d}R}{\mathrm{d}T} = \alpha I^2 R = \alpha P_I. \tag{9.13}$$

Substituting this expression into equation (9.12) and rearranging, we obtain

$$C\frac{\mathrm{d}T_1}{\mathrm{d}t} + (G - \alpha P_I)T_1 = P_T(t). \tag{9.14}$$

By comparing this result with equation (9.3) and its solution, we find that the time response is again exponential, but with an electrical time constant of

$$\tau_e = \frac{C}{G - \alpha P_I}. \tag{9.15}$$

Since $\alpha < 0$, the electrical time constant for a semiconductor bolometer is shorter than the thermal time constant τ_T defined in equation (9.5). This result is due to electrothermal feedback, which is discussed by Jones (1953) and Mather (1982). We will discuss it in more detail when we derive the noise performance of these detectors.

Because the time response is exponential, the response of the bolometer can be described as in equations (3.21) through (3.27). In particular, the frequency response is obtained by substituting τ_e for τ_{RC} in (3.25), that is,

$$S(f) = \frac{S(0)}{[1 + (2\pi f \tau_e)^2]^{1/2}}, \tag{9.16}$$

where $S(0)$ is the low frequency responsivity and both it and $S(f)$ have the units of volts (output signal) per watt (input photon power).

9.2.3 Responsivity

To derive the responsivity, let $\mathrm{d}R$, $\mathrm{d}T$, and $\mathrm{d}V$ be the changes in the resistance, temperature, and voltage, respectively, across the bolometer that are caused by a change in absorbed power, dP. Then, continuing to assume that R_L is large, we find that

$$dV = I\,dR = I(\alpha R\,dT) = \alpha V\,dT. \tag{9.17}$$

At low frequency ($t \gg \tau_e$), we obtain $dT = dP/(G - \alpha P_I)$ from equations (9.3), (9.4), and (9.14). Substituting this expression into equation (9.17), we get $dV = \alpha V dP/(G - \alpha P_I)$, yielding a responsivity in volts per watt of

$$S_E = \frac{dV}{dP} = \frac{\alpha V}{G - \alpha P_I}. \tag{9.18}$$

This responsivity can be determined entirely from the electrical properties of the detector and is therefore termed the electrical responsivity.

The electrical properties of a bolometer, however, are not always available in terms of α and G, and dismantling the detector to determine these parameters usually voids the warranty. What can be determined for an operating detector is an I–V curve (or load curve), which has the form shown in Figure 9.4 for a bolometer with $\alpha < 0$. The load curve is measured by using a high impedance

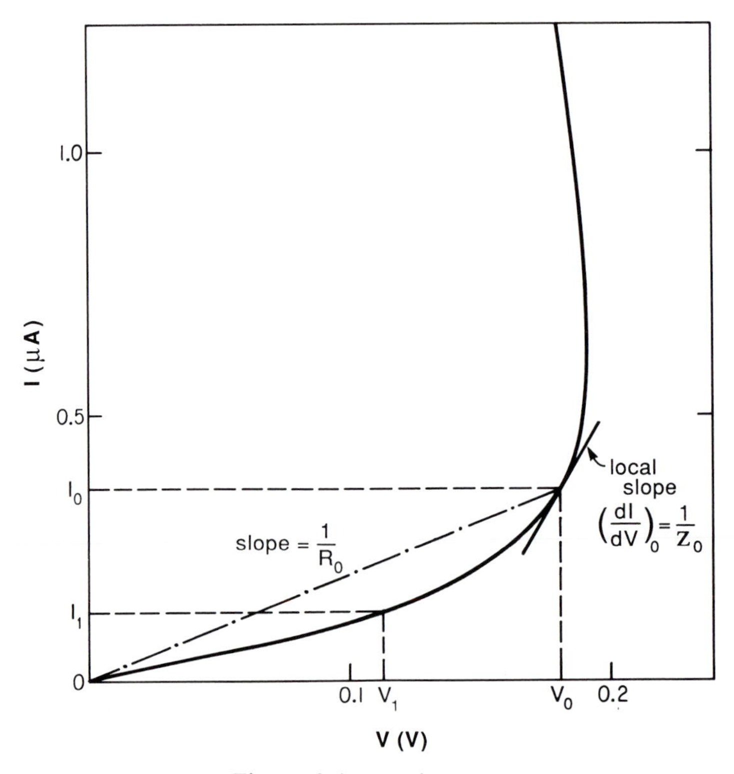

Figure 9.4. Load curve.

voltmeter (possibly in series with an FET input DC-coupled amplifier) to measure the voltage across the detector as a function of the current through it. The parameters governing the bolometer performance can be determined from a series of measurements centered on the load curve.

The circuit in Figure 9.3 is used to establish an operating point for the bolometer on the I–V curve by adjusting R_L and the bias voltage to establish the desired current through the detector. A generic example is indicated by the voltage V_0 in Figure 9.4. We are interested in determining the responsivity (and other parameters) of the detector at this operating point. The bolometer resistance is $R_0 = V_0/I_0$. We designate the slope of the load curve at any operating point as $1/\mathrm{Z}$, or

$$\mathrm{Z} = \frac{\mathrm{d}V}{\mathrm{d}I}, \tag{9.19}$$

which is different from the resistance because of the nonlinearity of the load curve. We can rewrite Z as follows:

$$\mathrm{Z} = R\frac{\mathrm{d}(\log V)}{\mathrm{d}(\log I)} = R\frac{\left[\dfrac{\mathrm{d}(\log P)}{\mathrm{d}(\log R)} + 1\right]}{\left[\dfrac{\mathrm{d}(\log P)}{\mathrm{d}(\log R)} - 1\right]}. \tag{9.20}$$

This expression can be simplified by defining the parameter $\mathrm{H} = G/\alpha P$ and noting from equations (9.1) and (9.6),

$$\frac{\mathrm{d}(\log P)}{\mathrm{d}(\log R)} = \frac{1}{I^2}\frac{\mathrm{d}P}{\mathrm{d}R} = \frac{1}{I^2}\frac{\mathrm{d}P}{\mathrm{d}T}\frac{\mathrm{d}T}{\mathrm{d}R} = \frac{G}{\alpha I^2 R} = \frac{G}{\alpha P} = \mathrm{H}. \tag{9.21}$$

Substituting this result into equation (9.20) and solving for H, we have

$$\mathrm{H} = \frac{\mathrm{Z} + R}{\mathrm{Z} - R}, \tag{9.22}$$

which can be determined readily from the load curve.

If we substitute $G = \alpha \mathrm{H} P$ (from equation (9.21)) into equation (9.18) and then use the form of H obtained in equation (9.22), the electrical responsivity takes the form

$$S_E = \frac{V}{P(\mathrm{H} - 1)} = \frac{1}{2I}\left(\frac{\mathrm{Z}}{R} - 1\right). \tag{9.23}$$

The power, P, in equation (9.23) is the electrical energy dissipated in the detector at the operating point, and the other quantities are also measured at this point.

If the temperature coefficient of resistance for the bolometer material is known, the combination of equations (9.21) and (9.22) permits the determination of the thermal conductance, G, from the load curve. Expressing equation (9.15) in these parameters, we obtain

$$\tau_e = \frac{\tau_T}{1 - \frac{1}{H}} = \tau_T \left[\frac{Z+R}{2R} \right], \tag{9.24}$$

which can be solved for τ_T $(=C/G)$ to determine the heat capacity of the detector in terms of τ_e. (The value of τ_e is obtained by determining the bolometer response to a varying signal; refer to equation (9.16).)

To convert S_E to the responsivity to incident radiation, or 'radiant responsivity', we must allow for the possibility that only a fraction, η, of the incident energy is absorbed by the bolometer, where η is the quantum efficiency. We then have

$$S_R = \frac{\eta}{2I}\left(\frac{Z}{R} - 1\right), \tag{9.25}$$

again in volts per watt. It is interesting to compare this expression with the responsivity of both a photoconductor (given in equation (3.13)) and a photodiode (given in equation (5.2)); note that the bolometer response is independent of the wavelength of operation (as long as the quantum efficiency is independent of wavelength). Figure 9.5 illustrates this behavior.

9.2.4 *Noise and noise equivalent power (NEP)*

A correct derivation of the bolometer noise must account for the electrothermal feedback mechanism already encountered in discussing the time response. The bolometer is, of course, subject to Johnson noise characterized by a noise

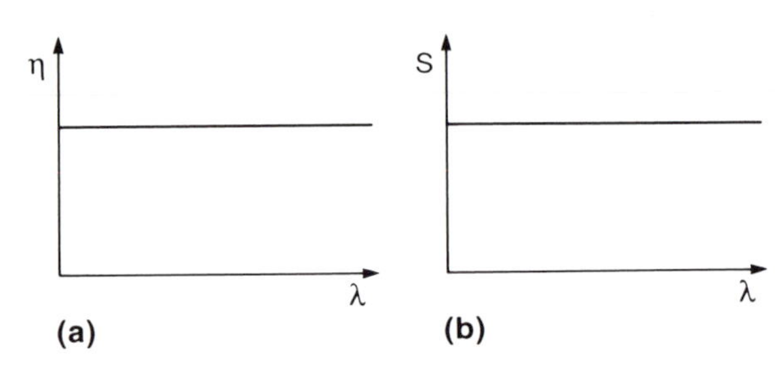

Figure 9.5. (a) Quantum efficiency and (b) responsivity for a bolometer.

voltage $V_J = \langle I_J^2 \rangle^{1/2} R$, where $\langle I_J^2 \rangle = 4kT\,df/R$ from equation (3.42). The Johnson noise dissipates power in the detector. When V_J is added to the bias voltage, the power dissipated in the detector increases, and, because of the negative temperature coefficient of resistance, the detector resistance decreases, thus reducing the net change in voltage across the detector. Similarly, when V_J opposes the bias voltage, the power dissipated in the detector is decreased, and the resistance increases, decreasing the net voltage change. Since the detector response opposes the ohmic voltage changes resulting from Johnson noise, the observed noise should be less than predicted by equation (3.42) with a fixed value of resistance.

We can simplify a quantitative analysis of this behavior by ignoring the change in operating conditions produced by the Johnson noise voltage. This assumption is equivalent to taking $I \gg \langle I_J^2 \rangle^{1/2}$ and $R_L \gg R$. Consider a bolometer with no signal and no noise. Then we can take the output voltage to be $V_0 = IR + I^2 R S_E$, where the second term represents the response of the detector to the power dissipated by the current established by the bias voltage. If we add Johnson noise, the output voltage becomes

$$\begin{aligned} V_0 &= IR + \langle I_J^2 \rangle^{1/2} R + \langle (I + I_J)^2 \rangle R S_E \\ &\approx \text{const.} + \langle I_J^2 \rangle^{1/2} R + I \langle I_J^2 \rangle^{1/2} R S_E, \end{aligned} \tag{9.26}$$

where we have made use of the assumption $I \gg \langle I_J^2 \rangle^{1/2}$. The noise, V_N, is the variable component of (9.26). Substituting for the responsivity from equation (9.23),

$$\begin{aligned} \langle V_N^2 \rangle^{1/2} &\approx \langle I_J^2 \rangle^{1/2} R + \frac{\langle I_J^2 \rangle^{1/2} R}{2} \left(\frac{Z}{R} - 1 \right) \\ &\approx \left(\frac{R+Z}{2} \right) \left(\frac{4kT\,df}{R} \right)^{1/2}. \end{aligned} \tag{9.27}$$

For example, if $Z = 0$, electrothermal feedback reduces the noise by a factor of two relative to Johnson noise for a fixed resistance, whereas a detector operated with a small bias voltage (for example, V_1 in Figure 9.4) will have $Z \approx R$ and nearly the full Johnson noise for a fixed resistance.

Mather's (1982) general derivation shows that no correction is needed in equation (9.27) but that there is a correction factor of $(R_L + R)/(R_L + Z)$ to the time constant in equation (9.24), that is,

$$\tau_e = \tau_T \left[\frac{Z + R}{2R} \right] \left(\frac{R_L + R}{R_L + Z} \right). \tag{9.28}$$

The NEP for a bolometer under Johnson-noise-limited conditions is

$\langle V_N^2 \rangle^{1/2}/(S_R(df)^{1/2})$. Using equation (9.25), it can be written as

$$\mathrm{NEP_J} = (4kTR)^{1/2} \frac{I}{\eta} \left| \frac{Z+R}{Z-R} \right|$$

$$= (4kTP)^{1/2} \frac{|H|}{\eta} = \left(\frac{4kT}{P} \right)^{1/2} \frac{G}{\eta |\alpha|}, \tag{9.29}$$

where we have used $H = G/\alpha P$. We have disregarded the frequency dependence of $\mathrm{NEP_J}$; to include this effect, equation (9.29) should be multiplied by the denominator in equation (9.16). The temperature dependence of $\mathrm{NEP_J}$ can be obtained by substituting for α from equation (9.9) or (9.11):

$$\mathrm{NEP_J} \sim \begin{bmatrix} GT^2 & \text{for } \alpha \sim T^{-3/2} \\ GT^{3/2} & \text{for } \alpha \sim T^{-1}. \end{bmatrix} \tag{9.30}$$

This expression makes clear the advantages in operating the bolometer at very low temperatures. We shall soon show that the temperature dependence of G can increase these advantages. In addition, equation (9.29) implies that if possible the detector should be biased such that Z is negative. Such values can be obtained, but it is usually found that excess noise is produced by the relatively large current that must flow through the detector to maintain the operating point.

In addition to Johnson noise, a second type of fundamental noise for a bolometer is thermal noise due to fluctuations of entropy across the thermal link that connects the detector to the heat sink. More detailed derivations can be found in van der Ziel (1976), and Mather (1982); here we will determine the magnitude of this noise through an analogy with the treatment of Johnson noise in Section 3.2.5.

We construct a thermal circuit diagram of the bolometer as shown in Figure 9.6, which is analogous to the electrical circuit used for the discussion of Johnson noise (see Figure 3.8). Here the thermal link is replaced by an equivalent resistance, R_T, and the ability of the bolometer to store thermal energy is represented by a capacitance, C_T. As in the case of Johnson noise, the energy that is stored fluctuates under thermodynamic equilibrium. The resulting fluctuating temperature of the bolometer is represented by T_N (analogous to V_N). By analogy with the electrical case (equation (3.41)), the energy E_T stored in C_T should be

$$E_T = \frac{1}{2} C_T T_N^2, \tag{9.31}$$

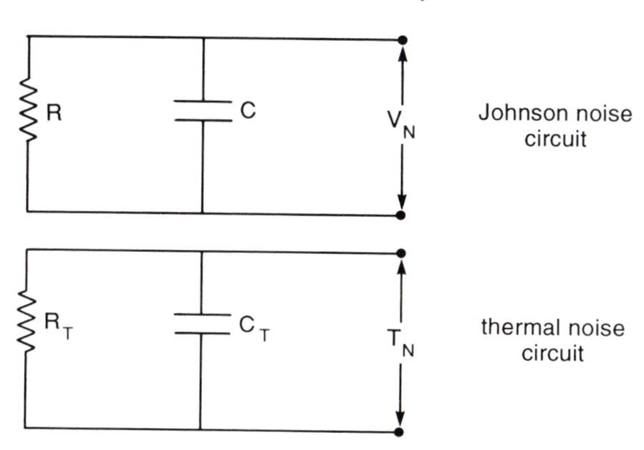

Figure 9.6. Analogy between Johnson noise and thermal noise.

and the dissipation of energy in R_T occurs at a rate

$$P_T = \frac{T_N^2}{R_T}. \tag{9.32}$$

From these relationships, we see that C_T has units of joules per kelvin squared and R_T has units of kelvin squared per watt. We also expect the circuit to have exponential time response with a time constant of $R_T C_T$. From equation (9.5), this relationship gives us

$$R_T C_T = \frac{C}{G}, \tag{9.33}$$

where C is the heat capacity of the bolometer and G is the thermal conductance to the bath. Applying the dimensional constraints stated just after equation (9.32) (and being careful not to confuse thermal capacitance C_T with heat capacity C), we find that

$$C_T = \frac{C}{T} \tag{9.34}$$

and

$$R_T = \frac{T}{G}. \tag{9.35}$$

From equation (3.42), we know that $\langle V_J^2 \rangle = (4kTR)df$, or in the present case by substituting $\langle T_N^2 \rangle$ for $\langle V_J^2 \rangle$ and R_T for R and using equation (9.35),

$$\langle T_N^2 \rangle = \frac{4kT^2 \, df}{G}. \tag{9.36}$$

Now let a signal power, P_v, fall on the bolometer; it will produce a temperature change of

$$\Delta T_S = \frac{\eta P_v}{G}. \tag{9.37}$$

We compute the value of P_v that will give unity signal to noise against thermal fluctuations by setting $\Delta T_S = \langle T_N^2 \rangle^{1/2}$, yielding

$$P_v = \frac{(4kT^2G\,df)^{1/2}}{\eta}, \tag{9.38}$$

and

$$\mathrm{NEP_T} = \frac{(4kT^2G)^{1/2}}{\eta}. \tag{9.39}$$

Again, there will be a frequency dependence as discussed after the derivation of $\mathrm{NEP_J}$ in equation (9.29).

The third fundamental performance limit for a bolometer is photon noise. Bolometers are not subject to G–R noise, so the NEP can be calculated analogously to equations (8.3) through equation (8.5). For the bolometer,

$$\langle I_N^2 \rangle = \frac{2(hc)^2 \varphi \eta\, df\, S_E^2}{(\lambda R)^2} \tag{9.40}$$

and

$$S(\mathrm{AW}^{-1}) = \frac{\eta S_E}{R}. \tag{9.41}$$

So the NEP is

$$\mathrm{NEP_{ph}} = \frac{hc}{\lambda}\left(\frac{2\varphi}{\eta}\right)^{1/2}. \tag{9.42}$$

The total NEP of a bolometer is given by the quadratic combination of the NEPs from various noise sources, for example:

$$\mathrm{NEP} = (\mathrm{NEP_J^2} + \mathrm{NEP_T^2} + \mathrm{NEP_{ph}^2} + \ldots)^{1/2}. \tag{9.43}$$

The NEP will be a function of the frequency of operation of the bolometer. The frequency dependence of the detector responsivity is given in equation (9.16), and the parameters which control it are discussed in detail in the following section. The corresponding modifications in the expressions for NEP – equations (9.29), (9.39), and (9.42) – are straightforward.

9.3 Bolometer construction and operation

The theoretical analysis we have just completed gives an accurate description of the bolometer performance achieved in practice – that is, bolometers whose construction and operating parameters are known accurately perform near to theoretical expectations. Bolometer development therefore concentrates on construction techniques, use of improved materials, measurement of material parameters, and utilization of low temperatures. These issues are illustrated in the following discussion.

The NEP at a given temperature is improved by reducing the thermal conductance, as shown by equations (9.29) and (9.39). The time response of the bolometer is given by equations (9.5) and (9.15) and is improved by reducing the heat capacity and increasing the thermal conductance of the detector. Therefore, both the heat capacity and the thermal conductance are critical in bolometer optimization. Heat capacity can be minimized by the choice of construction materials as well as by minimizing the volume of materials with high specific heat. Construction techniques have been developed that allow control of the thermal conductance over a very wide range. We will find that the temperature dependencies of the heat capacities and thermal conductances can significantly increase the benefits of operating the bolometer at very low temperatures.

9.3.1 Heat capacity

For pure semiconductor material at low temperature, an approximation to the specific heat (heat capacity per unit volume) is given by the Debye theory for heat capacity of a crystal lattice:

$$c_{\mathrm{v}}^{\mathrm{lat}}=\frac{12\pi^4\mathcal{N}_{\mathrm{A}}k}{5}\left(\frac{T}{\Theta_{\mathrm{D}}}\right)^3, \tag{9.44}$$

where $\mathcal{N}_{\mathrm{A}}$ is Avogadro's number times density over atomic weight and Θ_{D} is the Debye temperature. (For a discussion of this relationship and others to come, see, for example, Goldsmid, 1965). As examples, for germanium, $\Theta_{\mathrm{D}}=374$ K and $c_{\mathrm{v}}\sim 2.7\times 10^{-6}T^3$ J K^{-4} cm^{-3}, while for silicon, $\Theta_{\mathrm{D}}=645$ K and c_{v} is a factor of four times less when compared with germanium. For metals (and heavily doped semiconductors), c_{v} includes contributions both from the lattice and from free electrons. The electronic specific heat is

$$c_{\mathrm{v}}^{\mathrm{e}}=\gamma_{\mathrm{e}}T=\gamma_{\mathrm{e}}T\,(\rho\ \mathrm{mole}^{-1}), \tag{9.45}$$

where the parameter γ_{e} depends on the metal but is of order 10^{-3} J mole^{-1} K^{-2}, and the second equality converts the units of γ_{e} from mole^{-1} to volume^{-1}

by multiplying by the moles per unit volume, that is, ρ mole^{-1}. For brass ($\rho = 8.5\,\mathrm{g\,cm^{-3}}$, 1 mole = 64 g), $\gamma_e \sim 1.3 \times 10^{-4}\,\mathrm{J\,cm^{-3}\,K^{-2}}$. In general, the specific heat includes lattice and electronic contributions:

$$c_v = DT^3 + \gamma_e T \rightarrow \gamma_e T \text{ as } T \rightarrow 0, \tag{9.46}$$

where D is a constant from, for example, equation (9.44).

Although the bolometer chip itself is a semiconductor, there is usually metal and/or heavily doped semiconductor in the leads and in the contacts between the chip and the leads. Thus,

$$C = c_v^{\mathrm{lat}} V_{\mathrm{lat}} + c_v^{\mathrm{e}} V_{\mathrm{e}}, \tag{9.47}$$

For $T > 1$ K, V_{lat} is approximately the volume of semiconductor and V_e is the effective volume of the metal in the leads and contacts; at lower temperatures the electronic contribution to the specific heat of the semiconductor will become important, as indicated by condition (9.46). Silicon is preferred over germanium for bolometers operating above about 1 K because of its reduced heat capacity. For low temperature (that is, 0.3 K or 0.1 K) detectors, the electronic contribution dominates and the performance of the two semiconductors is similar.

The effect of the heat capacity of the leads is reduced because they do not experience the full temperature variation seen by the thermometer chip: although they are attached at one end to the thermometer, where they experience its full temperature excursion, at the other end the leads are attached to the cold sink where there is no temperature change. This effect can be taken into account in equation (9.47) by adjusting the volume of material attributed to the leads in computing V_e. The metal in the leads can be minimized by making them of dielectric material carrying thin metal films or ion implants for electrical conduction.

9.3.2 Thermal conductance

The time response, responsivity, and NEP depend on the conductance of the thermal link, G. We assume that this parameter is dominated by the behavior of the metal in the leads, although G may also be dominated by dielectrics in the leads or by additional supports used to hold the bolometer in place. The thermal conductivity of metals is given roughly by

$$k_e \approx \left[\frac{\pi^2 k^2}{3q^2}\right] \sigma T, \tag{9.48}$$

where the expression in brackets is the Lorentz number ($\approx 2.45 \times 10^{-8}\,\mathrm{W\,\Omega\,K^{-2}}$)

and σ is the electrical conductivity. At low temperatures, the value of the Lorentz number tends to decrease, in some cases by as much as an order of magnitude. This behavior can be allowed for by taking $k_e \sim T^a$, $a \geqslant 1$. Assuming that the bolometer is mounted on two leads of length L and cross-sectional area A, the conductance is

$$G = 2\frac{A}{L}k_e. \tag{9.49}$$

Thermal conductances useful with bolometers operating at 1 to 4 K range from about 10^{-7} to 10^{-3} W K^{-1}, whereas bolometers at 0.1 K utilize predominantly dielectric leads with $G < 10^{-10}$ W K^{-1}. This wide range of thermal conductances permits the frequency response and NEP of the bolometer to be tailored for a variety of applications.

With further understanding of the behavior of C and G, we can revisit the frequency response of bolometers. Substituting equations (9.47) and (9.49) into equation (9.5), we find that the time constant $\tau \sim T^2$ as long as the semiconductor lattice dominates the heat capacity of the bolometer, and τ is roughly independent of T when the metallic parts dominate.

Where the heat capacity is dominated by the lattice, bolometers built for differing operating temperatures in which G is adjusted to keep the same frequency response will have $G \propto T^2$. Substituting in equation (9.30), the Johnson-noise-limited NEP goes as $T^{7/2}$ for a given bolometer with $\alpha = -A/T$ as in equation (9.11); with the same assumptions the thermal noise limited NEP goes as T^2. If the heat capacity is dominated by free electrons, the frequency response is unaffected by reducing the temperature and the NEPs go as in equations (9.30) and (9.39).

As we have seen, the theoretically predicted improvement of NEP with reduced temperature varies considerably depending on the construction of the detector. As an empirical example, currently available silicon bolometers with similar sensitive areas ($\sim 1000\,\mu m^2$) and time response (~ 6 ms) have NEPs of $\sim 5 \times 10^{-15}$ W Hz$^{-1/2}$, $\sim 3 \times 10^{-16}$ W Hz$^{-1/2}$, and 0.5–1×10^{-17} W Hz$^{-1/2}$, respectively, at 1.5, 0.3, and 0.1 K. These values suggest NEP $\propto T^{(2\text{ to }2.5)}$; a scaling as $T^{5/2}$ is often used as an approximate relation.

9.3.3 Electrical properties

Doped silicon and germanium make good thermometer elements for bolometers. Very flexible control over the electrical properties of the element can be achieved by adjustments of the concentration of majority impurity and the compensation. Useful silicon material can be obtained from the semiconductor industry, or doping can be by ion implantation. In the case of germanium, highly uniform doping can be achieved by exposing high purity material to a flux of neutrons in a nuclear reactor. The dominant doping occurs when the $^{70}_{32}Ge$ atoms capture neutrons and decay to $^{71}_{31}Ga$; other reactions produce As and Se dopants (Palaio *et al.*, 1983). Particularly with silicon, low noise performance requires that contacts to the semiconductor be made through regions that are heavily doped by ion implants (for example, Moseley, Mather, and McGammon, 1984).

9.3.4 Quantum efficiency

Although the semiconductor chip plays an essential role as a thermometer for the detector, in some cases, it is also a perfectly adequate absorber of the incident radiation. For example, with silicon bolometers, where the material is heavily doped, the impurity concentration in the chip can also provide for efficient absorption of infrared photons. In many cases, however, the absorption properties of the chip may be inadequate to yield good quantum efficiency. The simplest solution, and one frequently employed for bolometers designed to operate in the near and middle infrared (say from 1 to 100 μm) is to enhance the absorption of the bolometer material by coating the thermometer chip with black paint. Either of these approaches requires that the thermometer be as large as the sensitive area of the detector. In addition, absorbing coats of black paint can significantly add to the heat capacity of the bolometer and degrade its performance.

A thermometer chip of adequate size to be used also as an absorber in the far infrared and submillimeter regions would contribute too much heat capacity, and composite bolometers are used instead, as illustrated in Figure 9.7. The thermometer chip is glued to a blackened plate of sapphire or

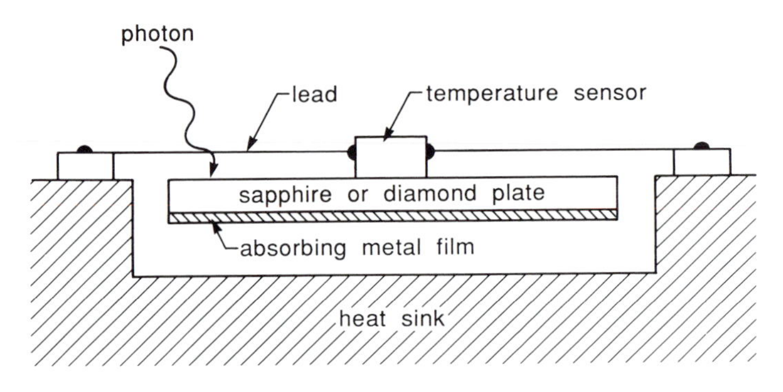

Figure 9.7. Composite bolometer.

diamond; since the heat capacity of sapphire is about 2% that of germanium (or ~8% that of silicon) and that of diamond is about ten times smaller still, the absorbing plate can be relatively large without affecting the bolometer time constant. At these wavelengths, the use of black paint as an absorber would add substantially to the heat capacity of the detector and degrade its performance.

The impedance of free space is $\zeta_0 = (\mu_0/\varepsilon_0)^{1/2} = 377\,\Omega$, where μ_0 is the absolute magnetic permeability and ε_0 is the permittivity of free space. An efficient absorber that has low volume and hence little heat capacity is a thin layer of metal adjusted in thickness to have a resistance of 377 ohms per square of surface area, thus matching the impedance of free space. (Note that a given film thickness will have the same resistance for any square surface area.) The theoretical foundation of this approach can be found in Stratton (1941, p. 511) or Born and Wolf (1975, p. 628), and it is given in practical form for this application by Clarke *et al.* (1977).

The latter reference considers the case of a metal film deposited on the back of a dielectric substrate such as sapphire or diamond (see Figure 9.7). Assuming normal incidence, the transmittivity and reflectivity at the first (unmetallized) surface are (compare to equation (2.4)):

$$t_1 = \frac{4n}{(n+1)^2} \tag{9.50}$$

and

$$r_1 = \frac{(n-1)^2}{(n+1)^2}, \tag{9.51}$$

respectively, where n is the index of refraction of the dielectric. The transmittivity, reflectivity, and absorptivity at the second (metallized) surface are

$$t_2 = \frac{4n}{(n+1+x)^2}, \tag{9.52}$$

$$r_2 = \frac{(1-n+x)^2}{(1+n+x)^2}, \tag{9.53}$$

and

$$a_2 = \frac{4nx}{(1+n+x)^2}, \tag{9.54}$$

respectively, where $x = \zeta_0/R_0$, and R_0 is the resistance per square area of the metal film. Although in general there can be complex interference phenomena in the dielectric plate, a simple and efficient case occurs when $r_2 = 0$ (that is, $x = n - 1$): (1) there are no multiple reflections, and (2) the absorptivity is independent of wavelength. For the case of a sapphire dielectric, $n \sim 3$, this case gives $t_1 = 0.75$, $a_2 = 0.667$, and a net absorption efficiency of $t_1 a_2 = 0.50$. Thus, the quantum efficiency using such an absorber would be 50%.

9.3.5 Arrays

Because of the necessity of having a thermally isolating suspension, bolometers lend themselves easily to one-dimensional arrays but can be built only with difficulty in close-packed two-dimensional arrays. The construction techniques for these detectors do not lend themselves to extremely large arrays, but 30–100 detectors is a feasible number.

Many types of bolometers are assembled one at a time, making it tedious to construct even one-dimensional arrays. The development of bolometers that are etched out of a single piece of silicon and that have the thermometer element and conducting leads created by ion implants will make arrays of this type of detector far more readily available (for example, Kelley *et al.*, 1988). This type of bolometer is illustrated in Figure 9.8.

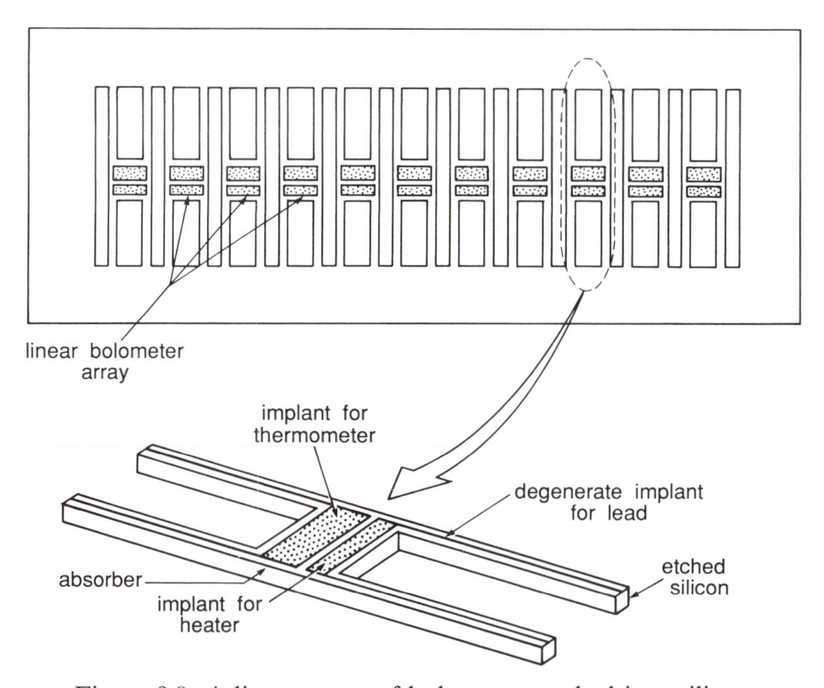

Figure 9.8. A linear array of bolometers etched into silicon.

9.3.6 Operating temperature

Because good performance of a bolometer depends strongly on operating it at a very low temperature, the cryostat or refrigerator is a critical part of the detection system. Four of the more common approaches used to cool bolometers are: (1) a helium 4 dewar operated at ambient pressure ($T = 4.2\,\mathrm{K}$); (2) a similar dewar but pumped to reduce the pressure over the helium 4 ($1 < T < 2\,\mathrm{K}$); (3) a closed cycle helium 3 refrigerator ($T \approx 0.3\,\mathrm{K}$); and (4) an adiabatic demagnetization refrigerator ($T \approx 0.1\,\mathrm{K}$).

The simplest refrigeration method is to cool the bolometer to 4.2 K in a helium dewar similar to that shown in Figure 9.9. Similar dewars are usually more than adequate to meet the cooling requirements for other types of detector. These dewars depend on the thermal insulation of a vacuum, which is augmented by a variety of design features. One of the most important of these features is a shield that blocks thermal radiation from the outer case.

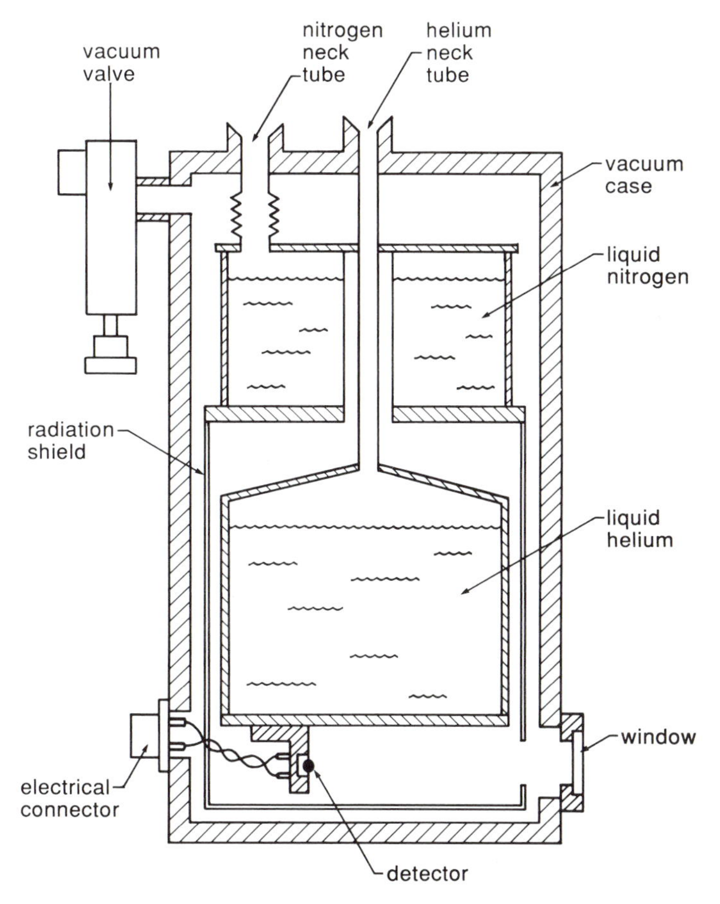

Figure 9.9. Helium 4 dewar. After a design produced by Infrared Laboratories, Inc.

In the illustration, this shield is cooled by a vessel containing liquid nitrogen (LN2), but simpler and lower performance dewars can cool this shield with a heat exchanger that dumps the heat from the shield into the gas emerging from the primary liquid container up the dewar neck tube. The surfaces of this shield, the inner surface of the vacuum case, and the outer surface of the primary liquid vessel are all coated with materials that have very low emissivity, so that radiative coupling of heat from one to another is suppressed. The dewar necks and supporting structure must also be constructed of low conductivity materials, and should have a large ratio of length to cross-sectional area to suppress thermal conduction along them (see equation (9.49)). Useful discussions of dewars and other cryogenic issues can be found in Scott (1959) and White (1987).

When operating with a high performance bolometer, the inner vessel of the dewar is filled with liquid helium 4, and a mechanical pump is attached to the dewar neck through a flexible hose. The pump is used to lower the pressure over the liquid helium; at the relatively easily obtainable vapor pressure of 3 mm, the bath temperature is 1.3 K. With careful control of the heat leak paths into the helium bath, it is feasible to bring the pressure down to 0.5 mm, at which the temperature of the bath is 1.1 K. One of the most pernicious sources of heat leak is superfluid helium creeping up the neck of the dewar against the force of gravity. Since superfluid helium is an almost perfect thermal conductor, a portion of the neck is shorted out thermally, and excess heat is conducted into the bath. This heat increases the rate at which helium boils off, thus increasing the load on the pump and placing a lower limit on the achievable temperature.

The isotope helium 3 does not have a superfluid state at the temperatures of interest. Unfortunately it is scarce, and the losses that occur in the helium 4 system just described could not be tolerated with helium 3. However, helium 3 can be used in a closed cycle refrigerator, such as that illustrated in Figure 9.10; this device would be mounted on the helium-cooled work surface of a dewar like the one shown in Figure 9.9. In this type of system, the helium is liquified by closing a mechanical clamp called a heat switch that sinks the helium 3 vessel to a bath of pumped helium 4 at $T < 2$ K. The switch is then opened to isolate the helium 3 vessel and allow it to cool further. An activated charcoal pump is sunk to the helium 4 bath through another heat switch. As the helium 3 evaporates, it is adsorbed onto the charcoal grains, which maintain a low vapor pressure on the helium 3 and reduce its temperature to $\sim$0.3 K. When the liquid helium 3 has all evaporated and been adsorbed, the heat switch for the charcoal pump is opened, and a heater is turned on to drive the helium 3 out. At the same time, the switch sinking the helium 3

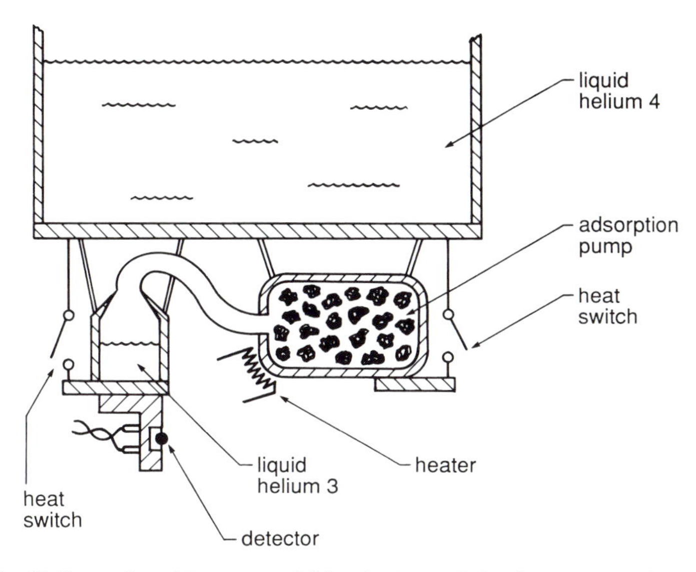

Figure 9.10. Helium 3 refrigerator. This device might be mounted on the helium work surface of a dewar similar to that in Figure 9.9.

vessel to the helium 4 bath is closed to liquify the helium 3 in the vessel. When this process has been completed, the cycle can be repeated.

Even lower operating temperatures can be reached by using refrigerators cooled by adiabatic demagnetization of a paramagnetic salt. These devices use a superconducting coil to magnetize the salt; simultaneously, a heat switch is closed to dump heat from the capsule containing the salt to a helium 4 bath at $T < 2\,K$. After the salt is magnetized, the current through the coil is turned off and the heat switch is opened. The salt can then absorb energy, which acts to randomize the magnetic alignment of its molecules and hence to demagnetize it. If sufficiently isolated from other heat sources, the salt cools to below 0.1 K. After substantial demagnetization of the salt has occurred, the refrigerator cycle must be repeated to remagnetize the salt and permit further operation at 0.1 K.

Low temperature refrigerators are reviewed by Radebaugh (1983).

9.3.7 Other operational considerations

Bolometers have excess low frequency noise with a $1/f$ behavior, which can arise from such causes as $1/f$ noise in the readout, from contact noise in the bolometer, or from fluctuations in the temperature of the cold station. Typically, $1/f$ noise dominates the NEP at frequencies below a few hertz. It is possible to suppress this type of noise significantly by using carefully matched pairs of detectors in a bridge circuit and with a rapidly modulated bias voltage (Rieke *et al.*, 1989). This gain is made at the expense of the complication in obtaining and operating matched bolometers and with a modest penalty in noise for frequencies above the $1/f$ regime.

Because the bolometer load curve is a function of the radiant power that the detector is absorbing, the device can be nonlinear. A large radiant power can significantly heat the thermometer chip and reduce its resistance, shifting the load curve toward larger currents for a given voltage. The result is a decrease in responsivity, leading to inherent nonlinearity. However, as long as the signal power is much smaller than the total power dissipated in the bolometer, a linear approximation to the responsivity can be quite accurate. Given the relatively large power provided by the bias current, this situation usually prevails.

However, large changes in dissipated power can occur during normal operations, for example when the background power on the detector is changed by changing the bandpass filter that defines the wavelength range of operation. As a result, both responsivity and time response can vary with operating conditions for a given detector system. To predict the detector performance accurately, the load curve and other parameters must be measured at or corrected to the appropriate background level.

9.4 Other thermal detectors

9.4.1 Hot electron bolometer

A bolometer for the submillimeter- and millimeter-wave spectral regions can be implemented quite differently from those described so far, by making use of properties of n-type InSb. In this material the impurity levels are merged with the conduction band even at very low temperatures; as a result, the ionization energy can be very small, of order 0.001 eV, and the bolometer has a sea of free electrons. Photons incident on the detector are absorbed by free electrons, which are thereby raised above their thermal equilibrium energies and become hot electrons. Eventually, an equilibrium will be reached between de-excitation and photo-induced effects. Since the hot electrons in InSb do not interact strongly with the lattice to de-excite, they can accumulate to a significant density. For InSb, the hot electrons strongly affect the mobility and hence the conductivity (for example, equations (3.5) and (3.6)); the photon signal can be monitored by measuring the detector resistance.

The parameters of these bolometers are largely set by the properties of the InSb (for example, Kinch and Rollin, 1963; Putley, 1977); nonetheless, they are well suited for certain applications (one of which is described in Chapter 11). Voltage responsivities are 100 to 1000 V W^{-1}, the thermal conductance, G, is about 5×10^{-5} W K^{-1}, and the heat capacity of the electron sea $C \sim 3/2nkv$, where n is the carrier concentration and v is the volume of the detector. A typical detector might be 2 mm on a side ($v \sim 10^{-2}$ cm^3) and have $n \sim 5 \times 10^{13}$ cm^{-3}; then $C \sim 10^{-11}$ J K^{-1} and the detector time constant (from equation (9.5)) is 2×10^{-7} s. In this example operating at 4 K, the thermally limited NEP is 2×10^{-13} W Hz$^{-1/2}$, from equation (9.39) and assuming unity quantum efficiency. The detector is very fast and of only modest sensitivity compared with the bolometers we have discussed previously (see also the example below).

The quantum efficiency of these detectors depends on the absorption efficiency of the free electrons; at low frequencies, the absorption is large and wavelength independent, and the quantum efficiency is near unity; at high frequencies, the absorption efficiency goes as λ^2. The crossover between these behaviors is at $\lambda \sim 1.6$ mm; at 1 mm, the absorption coefficient $a = 22$ cm^{-1}, and at $\lambda = 100$ μm, $a = 0.3$ cm^{-1}. Consequently, the detectors become inefficient at wavelengths shorter than about 300 μm.

9.4.2 Room temperature detectors

Because of the temperature dependencies of performance derived above, thermal detectors operating near room temperature are not suitable for truly low level detection. However, because of their extremely broad spectral response, a variety of thermal detectors find wide application in laboratory situations and other circumstances where relatively bright illumination is provided.

The role of the semiconductor thermometer chip described above can be taken by a thermometer based on the thermoelectric effect. When two dissimilar metals are contacted to make a thermocouple, the mismatch in their Fermi levels results in a temperature-dependent electromotive force being developed across the joint. If a similar junction is held at a reference temperature, a circuit connecting the two junctions will develop a voltage:

$$\Delta V = \alpha \Delta T. \quad (9.55)$$

Here, α is the Seebeck coefficient (of order 50 μV °C^{-1}, and available in tabular form for a variety of metal pairs), ΔT is the temperature difference between the two metallic joints, and ΔV is the voltage. The response can be increased by connecting a number of thermocouples in series to make a thermopile detector. For more information, consult Stevens (1970) or Pollock (1985).

Pyroelectric detectors are based on certain specialized materials (for example, triglycerine sulfite (TGS), strontium barium niobate, lithium tantalate). To create a detector, a sample of one of these materials is placed between two electrical contacts. The molecules of these materials have a charge asymmetry which can be aligned by applying an electric field to the contacts at elevated temperature, and which is retained as a net electrical polarization after the sample has been cooled and the field has been removed. To prepare it for use, the detector material is 'poled' by putting it through this series of steps. Afterwards, when the crystal changes temperature and the crystal lattice therefore expands or contracts, a change in the electrical polarization occurs which changes the charge on the contacts. As a result, a current will flow in an external circuit:

$$I = p(T)A\frac{\mathrm{d}T}{\mathrm{d}t}, \tag{9.56}$$

where A is the detector area and $p(T)$ is the pyroelectric coefficient (which can be of the order of $1\times10^{-8}\,\mathrm{C\,cm^{-2}\,K^{-1}}$). Further information on pyroelectric detectors can be found in Putley (1970). Unlike most other types of thermal detector, pyroelectric devices can be constructed in large scale two-dimensional arrays in a straightforward manner (see, for example, Watton, Manning, and Burgess, 1982).

A Golay cell is another thermal detector consisting of an hermetically sealed container filled with gas and arranged so that expansion of the gas under heating by a photon signal deflects a mirror. The mirror position is sensed by reflecting a beam of light off it to a suitable detector.

9.5 Example: design of a bolometer

We will now describe the design and performance of a hypothetical bolometer operating at 1.5 K. Its thermometer element is a cube of germanium that is 0.53 mm on a side. It has brass leads 25 μm in diameter and 0.5 cm long. Assume the germanium is doped with gallium at a level that gives the element a resistance of $1\times10^{6}\,\Omega$ at 1.5 K, and that the temperature dependence of its resistance is described by equation (9.10). We will also assume that the detector is blackened to give a quantum efficiency of 0.9, but that its other properties are unaffected by the blackening agent. Let the bolometer be operated in a circuit similar to that in Figure 9.3, with $R_L = 10^7\,\Omega$ and $V_b = 1.5$ V.

The low temperature electrical conductivity of brass is $\sigma = 3.1\times10^{5}$ $(\Omega\,\mathrm{cm})^{-1}$. From equation (9.48), we can calculate that the thermal conductivity is $k_e = 0.0114\,\mathrm{W\,K^{-1}\,cm^{-1}}$. Substituting into equation (9.49), we get a thermal conductance of

$$G = 0.22 \times 10^{-6}\ \mathrm{W\ K^{-1}}.$$

For Ge:Ga, $A \approx 4$ in equation (9.10), so at the operating temperature of this detector, $\alpha = -2.67\ \mathrm{K^{-1}}$. The load curve of the detector can be computed from equation (9.1) (rewritten in the form $T - T_0 = I^2 R/G$) and from equation (9.10). The current and voltage are then related through

$$I = \frac{V}{R_0}\left[1 + \frac{IV}{GT_0}\right]^4,$$

which can be solved for specific (I, V) points by choosing a value of I (or V) and iterating to obtain V (or I). This procedure yields the load curve plotted in Figure 9.4.

From equation (9.29),

$$\mathrm{NEP_J} = 8.3 \times 10^{-19}\left(\frac{1}{P}\right)^{1/2}\ \mathrm{W^{3/2}\ Hz^{-1/2}},$$

where $P = I^2 R$ is the electrical power dissipated in the detector and $\mathrm{NEP_J}$ is the Johnson-noise-limited NEP. To first order, the current through the bolometer is regulated by the load resistor, so

$$I \approx \frac{1.5\ \mathrm{V}}{10^7\ \Omega} = 1.5 \times 10^{-7}\ \mathrm{A}.$$

Using this current and the appropriate values of G, T_0, and R_0 in the load curve equation above, we find that the voltage drop across the bolometer is 0.12 V, so a more accurate estimate of the current is

$$I = \frac{(1.5 - 0.12)\ \mathrm{V}}{10^7\ \Omega} = 1.4 \times 10^{-7}\ \mathrm{A}.$$

Therefore, the power dissipated in the detector is

$$P = (1.4 \times 10^{-7}\ \mathrm{A})(0.12\ \mathrm{V}) = 1.7 \times 10^{-8}\ \mathrm{W}.$$

We then obtain

$$\mathrm{NEP_J} = 6.4 \times 10^{-15}\ \mathrm{W\ Hz^{-1/2}}.$$

In theory, $\mathrm{NEP_J}$ can be made smaller by increasing P, but excess current-induced noise can thwart this trend fairly quickly. From equation (9.39), the thermal-noise-limited NEP is

$$\mathrm{NEP_T} = 5.8 \times 10^{-15}\ \mathrm{W\ Hz^{-1/2}},$$

and the net NEP at low background (that is, where photon noise makes a negligible contribution) is the quadratic combination of NEP_J and NEP_T, or

$$NEP = 8.6 \times 10^{-15}\ \mathrm{W\,Hz^{-1/2}}.$$

From equation (9.44) and the discussion following it, the heat capacity of the germanium chip is

$$C^{sc} = (2.7 \times 10^{-6}\ \mathrm{J\,K^{-4}\,cm^{-3}})(1.5\ \mathrm{K})^3(0.053\ \mathrm{cm})^3$$
$$= 1.4 \times 10^{-9}\ \mathrm{J\,K^{-1}}.$$

From equation (9.45), the heat capacity of the two leads together is

$$C^m = (1.3 \times 10^{-4}\ \mathrm{J\,K^{-2}\,cm^{-3}})(1.5\ \mathrm{K})(2 \times 0.5\ \mathrm{cm})\pi\left(\frac{25 \times 10^{-4}\ \mathrm{cm}}{2}\right)^2$$
$$= 9.6 \times 10^{-10}\ \mathrm{J\,K^{-1}}.$$

To first order, the temperature variation of the leads can be taken to be 0.5 times that of the germanium chip, so we can allow for their heat capacity by adding half of it to that of the chip. In addition, the contacts between the leads and chip will add heat capacity, which we arbitrarily assume to be $1.5 \times 10^{-10}\ \mathrm{J\,K^{-1}}$, giving an effective value of $C = 2 \times 10^{-9}\ \mathrm{J\,K^{-1}}$. From equation (9.15), the detector time constant is

$$\tau_e = \frac{2 \times 10^{-9}\ \mathrm{J\,K^{-1}}}{(0.22 \times 10^{-6}\ \mathrm{W\,K^{-1}}) - (-2.67\ \mathrm{K^{-1}})(1.7 \times 10^{-8}\ \mathrm{W})}$$
$$= 0.0075\ \mathrm{s},$$

giving a frequency response up to $1/2\pi\tau_e = 21$ Hz (see equation (9.16)).

In general, the details of the bolometer construction are not known as precisely as we have assumed them to be; as a result, the detector performance must be deduced in part from laboratory measurements of its load curve, frequency response, and responsivity.

9.6 Problems

9.1 Prove that the resistance of a square of metal film depends only on the thickness of the film and not on its area.

9.2 Evaluate the performance of the bolometer of the example at $T = 0.3$ K. Assume the doping of the thermometer chip is adjusted to give a nominal resistance of $10^6\ \Omega$ at 0.3 K, but that the other aspects of the detector construction are unchanged.

9.3 Discuss the applicability of D^* as a figure of merit for bolometer performance.

9.4 Discuss the advantages and disadvantages of operating a bolometer with a transimpedance amplifier.

9.5 Consider a dewar where liquid helium cools apparatus that absorbs the energy passing from the room (at 20°C) through a 3 cm diameter hole (with full view of the room from one side). How long will it take to boil away one liter of helium? (The cooling capacity of liquid helium is $2562\ \mathrm{J\ L^{-1}}$.)

9.6 Consider a bolometer similar to the one in the example whose load curve is measured in a circuit like Figure 9.3 with a variable bias supply and $R_L = 10^9\ \Omega$. The results are given in the following table. Assuming $\alpha = -2.67$, show that G for this bolometer is $0.25 \times 10^{-6}\ \mathrm{W\ K^{-1}}$. If the frequency response with $V_{out} \sim 0.12$ V extends to 16 Hz, derive the heat capacity of the bolometer.

V_b (V)	V_{out} (mV)	V_b (V)	V_{out} (mV)	V_b (V)	V_{out} (mV)
10	10.0	120	105.1	230	158.6
20	19.9	130	111.7	240	161.8
30	29.7	140	117.9	250	164.8
40	39.3	150	123.7	260	167.5
50	48.7	160	129.1	270	170.1
60	57.8	170	134.2	280	172.5
70	66.6	180	139.0	290	174.7
80	75.1	190	143.5	300	176.8
90	83.2	200	147.7	310	178.7
100	90.9	210	151.6	320	180.4
110	98.2	220	155.2	330	182.1

9.7 Consider the bolometer of Problem 9.6, operated at 1.5 K and with $\eta = 0.9$. Imagine that it has excess current noise at the operating frequency in the amount $\langle I_{1/f}^2 \rangle^{1/2} = 5 \times 10^{-8} I_{bolo}\ \mathrm{Hz}^{-1/2}$. For $I_{bolo} = 5$, 10, 15, 20, 25, and 30×10^{-8} A, derive the zero-background NEP and show that the optimum bias condition sets $I_{bolo} \sim 2 \times 10^{-7}$ A.

10

Visible and infrared coherent receivers

Coherent receivers are the third and last general category of detector listed in Section 1.2. These devices mix the electromagnetic field of the incoming photons with a local oscillating field to produce a signal at the difference, or beat, frequency. Unlike the output from the incoherent detectors discussed so far, this signal directly encodes the spectrum of the incoming signal over a range of input frequencies and also retains information about the phase of the incoming wavefront. As a result, these receivers are easily adapted for spectroscopy; in addition, their outputs can be combined to reconstruct the incoming wavefront, making interferometry between different receivers possible. A dramatic application of this latter capability is the use of intercontinental baseline radio telescope interferometers to achieve milliarc-second resolution in astronomy. Very weak signals can be detected because mixing the local field with the incoming photons allows the signal to be amplified in some situations independently of many noise sources. Coherent receivers monopolize radio applications; they are not used as widely at infrared frequencies because of their narrow spectral bandwidths, small fields of view, and inability to be constructed in simple, large format spatial arrays.

10.1 Basic operation

In general, any device that measures the field strength of the incoming photon – that is, that has the potential to measure and preserve phase information directly – is a coherent receiver. Heterodyne receivers are the most important class of such device. They function by mixing signals of different frequency; if two such signals are added together, they 'beat' against each other. The resulting signal contains frequencies only from the original two signals, but its *amplitude* is modulated at the difference, or beat, frequency (see Figure 10.1). Heterodyne receivers measure this amplitude. Further discussion can be found in Torrey and Whitmer (1948).

In Figure 10.1, two oscillating fields have been mixed, and the resulting field is indicated by the solid line. If this field is measured by a linear device, two results are possible: (1) if the time constant of the device is short, the output will simply follow the solid curve in Figure 10.1, which contains power only at the original frequencies; and (2) if the time constant is long, the device will not respond because on average the output signal contains equal positive and negative excursions. In neither case will the output contain any component of power at the beat frequency. For heterodyne operation, the mixed field must be passed through a nonlinear circuit element or *mixer* that converts power from the original frequencies to the beat frequency. In the submillimeter- and millimeter-wave (and radio) regions this element is a diode or other nonlinear electrical circuit component. For visible and infrared operation, the nonlinear element is a photon detector, sometimes also termed a photomixer.

The action of the nonlinear element in converting the power to the beat frequency is illustrated in Figure 10.2. We have already argued that if the element has a linear *I–V* curve (see Figure 10.2(a)), then the conversion efficiency is zero. Similarly, any mixer having a characteristic curve that is an odd function of voltage around the origin will have zero conversion efficiency if operated at zero bias; the cubic curve in Figure 10.2(b) is an

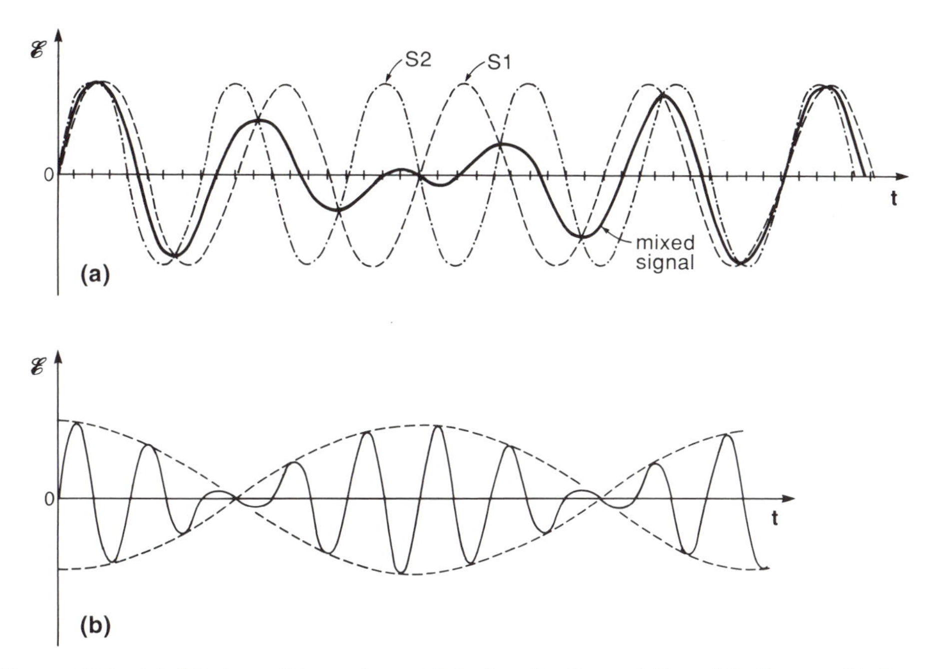

Figure 10.1. (a) Mixing of two sinusoidal signals, S_1 and S_2, with the result shown by the heavy line. (b) The mixed signal on a compressed timescale to illustrate the beating at the difference frequency.

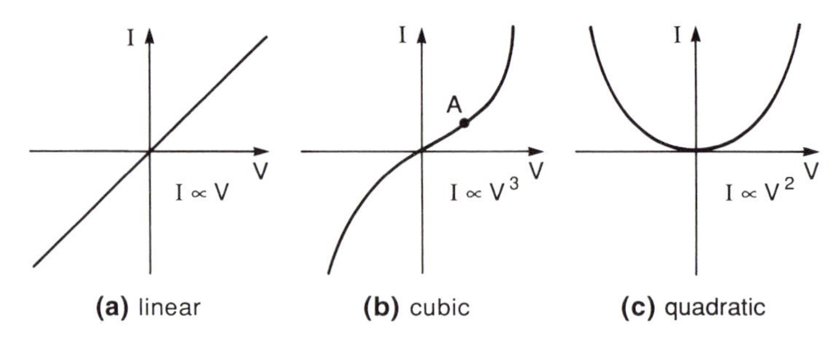

Figure 10.2. *I–V* curves of three hypothetical mixer elements.

example. In this case, some conversion will occur if the mixer is biased above zero. For example, if the operating point is set at A, the average change in current will be greater for positive than for negative voltage swings. Much greater efficiency is achieved, however, with a characteristic curve that is an even function of voltage – for example the quadratic curve in Figure 10.2(c). Moreover, if $I \propto V^2$, the output current is proportional to the square of the signal amplitude. The amplitude is proportional to the field strength in the incoming photon, and since the signal power goes as the square of the field strength, a quadratic *I–V* curve for the mixer produces an output proportional to power, which is usually what we want to measure (that is, $I \propto V^2 \propto \mathscr{E}^2 \propto P$, where $\mathscr{E}$ is the strength of the electric field). Hence, we prefer to use 'square law' devices as fundamental mixers because their output is linear with input power.

Even if we cannot find an electronic device with an *I–V* curve exactly like the one shown in Figure 10.2(c), we can assume that our mixer has a characteristic curve that can be represented by a Taylor series,

$$I(V)=I(V_0)+\left(\frac{\mathrm{d}I}{\mathrm{d}V}\right)_{V=V_0}\mathrm{d}V+\frac{1}{2!}\left(\frac{\mathrm{d}^2I}{\mathrm{d}V^2}\right)_{V=V_0}\mathrm{d}V^2 + \frac{1}{3!}\left(\frac{\mathrm{d}^3I}{\mathrm{d}V^3}\right)_{V=V_0}\mathrm{d}V^3+\cdots, \tag{10.1}$$

where V_0 is the voltage at the operating point. As we have seen, the first two terms are of no importance for fundamental mixers ($I(V_0)$ is a DC current, and the second term gives a net response of zero). The fourth and higher terms can usually be ignored as long as $\mathrm{d}V\ (=V-V_0)$ is small; the device is then a good square law mixer as long as $\mathrm{d}^2I/\mathrm{d}V^2$ is reasonably large.

Assume that we are using a square law mixer, and let it be illuminated with a mixture of two sources of power, one a signal at frequency ω_S and the other at ω_{LO}. Let the power at ω_{LO} originate from a source within the instrument called the local oscillator (LO), and assume it is much stronger than the weaker, unknown signal at ω_S. We also specify that $\omega_S > \omega_{LO}$. Then the mixed signal will be amplitude modulated at the intermediate frequency $\omega_{IF} = |\omega_S - \omega_{LO}|$. The signal at ω_{IF} contains spectral and phase information about the signal at ω_S as long as the LO signal at ω_{LO} is steady in frequency and phase and the difference between ω_S and ω_{LO} is neither so large that it exceeds the maximum frequency response of our equipment nor so small that it falls below its low frequency cutoff. The signal has been downconverted to a much lower frequency than ω_S or ω_{LO}, in the frequency range where it can be processed by conventional electronics. As a result, it is relatively easy to extract the spectrum of the source or measure the phase of the incoming photons.

Note that there is no way of telling in the mixed signal whether $\omega_S > \omega_{LO}$ or $\omega_{LO} > \omega_S$. Because we have lost the initial information regarding the relative values of ω_S and ω_{LO}, many of the derivations of receiver performance will assume that the input signal contains two components of equal strength, one above and the other below the LO frequency ω_{LO} (see Figure 10.3). Since the signal at ω_{IF} can arise from a combination of true inputs at $\omega_{LO} + \omega_{IF}$ and $\omega_{LO} - \omega_{IF}$, it is referred to as a double sideband signal. When observing continuum sources, the ambiguity in the frequency of the input signal is usually a minor inconvenience. When observing spectral lines, however, the unwanted 'image' frequency signal can result in serious complications in calibration. Submillimeter- and millimeter-wave receivers can be operated single sideband if the image is suppressed by tuning the mixer or with a narrow bandpass 'image rejection' filter that is placed in front of the receiver; this filter blocks the photons at image frequencies before they are mixed with the LO signal. In general, visible or infrared receivers must operate double sideband.

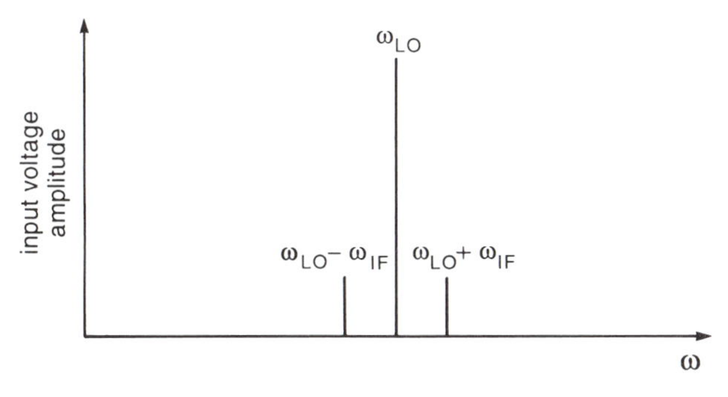

Figure 10.3. The ambiguity in the frequency of the input signal after heterodyne conversion to the intermediate frequency.

The output of the mixer is passed to an intermediate frequency amplifier and then to a rectifying and smoothing circuit called a detector. This terminology differs from that used for incoherent devices, where the detector is the device that receives the photons and converts them to an electrical signal. Rather than a single detector, a 'backend' spectrometer often receives the output of the IF amplifier. The spectrometer could consist of a number of filters tuned to different frequencies with detectors on their outputs, a digital correlator, or an acousto-optic spectrometer (AOS) such as a Bragg cell, where the frequencies are converted to ultrasonic waves that disperse a monochromatic light beam onto an array of visible-light detectors. Kraus (1986) describes these spectrometer types in further detail.

10.2 Visible and infrared heterodyne

10.2.1 Mixer

At the highest frequencies at which heterodyne receivers are used, a continuous wave (CW) laser is used as the local oscillator. The laser light and the signal are combined by a beam splitter, sometimes called a diplexer (see Figure 10.4). The output is mixed in a photon detector – a photoconductor, photodiode, photomultiplier, or bolometer. Because a photomixer responds to power, or field strength squared, it is a square law mixer. Reviews of the operation of this receiver type can be found in Keyes and Quist (1970), Teich (1970), Blaney (1975), and Kazovsky (1986).

The operation of visible and infrared heterodyne mixers is described by the theories derived in Chapters 3, 4, 5, 8, and 9, so we will make use of the

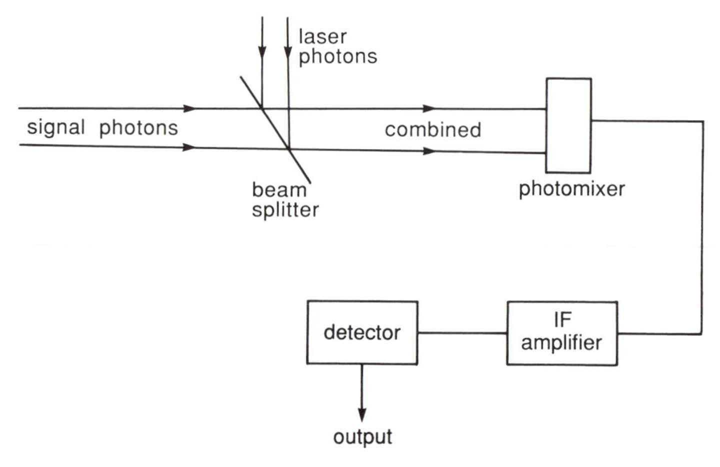

Figure 10.4. An infrared or visible heterodyne receiver.

formalisms in those chapters here. Because the mixer must follow the heterodyne signal, it must respond to far higher frequency (up to 1 GHz) than we considered in the earlier discussions. Compared with typical operation as incoherent detectors, where time constants of milliseconds to seconds are acceptable, heterodyne use demands careful optimization for high frequency response.

Assume we have such a mixer arranged as in Figure 10.4, that the two photon beams falling on the mixer are well collimated, and that both are linearly polarized in the same direction. The combined electric field in that direction is

$$\mathscr{E}_{\mathrm{T}}(t)=\mathscr{E}_{\mathrm{L}}(t)+\mathscr{E}_{\mathrm{S}}(t), \tag{10.2}$$

where $\mathscr{E}_{\mathrm{L}}(t)$ and $\mathscr{E}_{\mathrm{S}}(t)$ are the electric fields of the laser and source, respectively.

The electric and magnetic fields of any electromagnetic wave can be written as

$$\mathscr{E}(t)=\mathscr{E}_0 e^{-\mathrm{j}\omega t}=\mathscr{E}_0\cos\omega t-\mathrm{j}\mathscr{E}_0\sin\omega t \tag{10.3a}$$

and

$$\mathscr{H}(t)=\mathscr{H}_0 e^{-\mathrm{j}\omega t}=\mathscr{H}_0\cos\omega t-\mathrm{j}\mathscr{H}_0\sin\omega t, \tag{10.3b}$$

where $\mathscr{E}_0$ and $\mathscr{H}_0$ are the complex amplitudes of the fields. The electric and magnetic field amplitudes are related by

$$\mathscr{E}_0=\left(\frac{\mu}{\varepsilon}\right)^{1/2}\mathscr{H}_0, \tag{10.4}$$

where μ is the magnetic permeability and ε is the dielectric permittivity of the medium.

We want to determine the current that is generated in the mixer as a result of the electric field $\mathscr{E}_{\mathrm{T}}(t)$. The rate at which energy passes through a unit area normal to the propagation direction of the wave is given by the real part of the Poynting vector:

$$S(t)=\mathrm{Re}[\mathscr{E}(t)]\,\mathrm{Re}[\mathscr{H}(t)]\hat{\imath}. \tag{10.5}$$

The direction of S is given by the unit vector $\hat{\imath}$ and is perpendicular to the directions of both $\mathscr{E}(t)$ and $\mathscr{H}(t)$. The time average of the Poynting vector is[1]

$$\langle S\rangle_{\mathrm{t}}=\langle\mathrm{Re}[\mathscr{E}(t)]\,\mathrm{Re}[\mathscr{H}(t)]\rangle_{\mathrm{t}}\hat{\imath}=\frac{1}{2}\mathrm{Re}[\mathscr{E}(t)\mathscr{H}^*(t)]\hat{\imath}. \tag{10.6}$$

Using equation (10.4) to substitute for $\mathscr{H}^*(t)$ in equation (10.6), the power falling on a mixer of area A (normal to the propagation direction of the radiation) and averaged in time over a cycle of the input frequencies is

$$\langle P(t)\rangle = \frac{A}{2}\left(\frac{\varepsilon}{\mu}\right)^{1/2} |\mathscr{E}(t)|^2. \tag{10.7}$$

To express the time dependence explicitly, we replace $\mathscr{E}(t)$ and $\mathscr{E}^*(t)$ in $|\mathscr{E}(t)|$ in the expression above with $\mathscr{E}_T(t)$ and $\mathscr{E}_T^*(t)$ from equations (10.2) and (10.3a) to get

$$\begin{aligned}\langle P(t)\rangle &= \frac{A}{2}\left(\frac{\varepsilon}{\mu}\right)^{1/2} [(\mathscr{E}_{OL}e^{-j\omega_L t} + \mathscr{E}_{OS}e^{-j\omega_S t}) \\ &\quad \times (\mathscr{E}_{OL}^* e^{j\omega_L t} + \mathscr{E}_{OS}^* e^{j\omega_S t})] \\ &= \frac{A}{2}\left(\frac{\varepsilon}{\mu}\right)^{1/2} [|\mathscr{E}_{OL}|^2 + |\mathscr{E}_{OS}|^2 \\ &\quad + \mathscr{E}_{OS}\mathscr{E}_{OL}^* e^{-j(\omega_S-\omega_L)t} + \mathscr{E}_{OS}^*\mathscr{E}_{OL}e^{j(\omega_S-\omega_L)t}].\end{aligned} \tag{10.8}$$

The photocurrent, $I(t)$, that is generated by this power can be obtained, for example, from equation (3.16) (for a photoconductor) or equation (5.1) (for a photodiode). Equation (3.11) is used to write φ in terms of P. If the mixer in Figure 10.4 is a photoconductor then equation (3.16) gives

$$I(t) = \frac{\eta q G}{h\nu} P(t), \tag{10.9}$$

where η is the quantum efficiency of the mixer, G is its photoconductive gain, and $\nu = \omega/2\pi$ is the frequency of the incoming photons. Substituting for $P(t)$ from equation (10.8), we obtain

$$I(t) = I_L + I_S + 2(I_L I_S)^{1/2}\cos[(\omega_S - \omega_L)t + \phi], \tag{10.10}$$

where the DC current from the laser is

$$I_L = \frac{\eta q G A}{2h\nu}\left(\frac{\varepsilon}{\mu}\right)^{1/2} |\mathscr{E}_{OL}|^2, \tag{10.11}$$

and a similar expression holds for the current from the source, I_S. The relative phase between $\mathscr{E}_S$ and $\mathscr{E}_L$ is

$$\phi = \arctan\left[\frac{\mathrm{Re}[\mathscr{E}_{OS}]\,\mathrm{Im}[\mathscr{E}_{OL}] - \mathrm{Im}[\mathscr{E}_{OS}]\,\mathrm{Re}[\mathscr{E}_{OL}]}{\mathrm{Re}[\mathscr{E}_{OS}]\,\mathrm{Re}[\mathscr{E}_{OL}] + \mathrm{Im}[\mathscr{E}_{OS}]\,\mathrm{Im}[\mathscr{E}_{OL}]}\right]. \tag{10.12}$$

Thus, the photocurrent contains a component oscillating at the intermediate frequency, $\omega_{IF} = |\omega_S - \omega_L|$. In principle, this current also contains components at ω_S and ω_L, but since in this case these frequencies may be 10^{12} to 10^{14} Hz, these components appear as the cycle-averaged currents I_S and I_L, which are

DC in nature. The IF current (the third term in equation (10.10)) is the heterodyne signal and has a mean-square-amplitude of

$$\langle I_{\mathrm{IF}}^2 \rangle_t = 2 I_L I_S; \tag{10.13}$$

recall that $\langle \cos^2 \omega t \rangle = 1/2$.

It is important to note that the signal strength in equation (10.13) depends on the LO power. As a result, many forms of noise can be overcome by increasing the output of the local oscillator. The ability to provide an increase in power while downconverting the input signal frequency is characteristic of quantum mixers, such as the photomixers discussed here. The conversion gain is defined as the IF output power that can be delivered by the mixer to the next stage of electronics (sometimes called the exchangeable power) divided by the input signal power. The maximum power transfer occurs when the input impedance of the next stage has been matched to the output impedance of the mixer. In general, this condition results in half of the total power appearing at the input of the following electronics, so for a photoconductor mixer the gain is:

$$\begin{aligned}\Gamma_c &= \frac{\text{deliverable IF signal power}}{\text{input signal power}} \\ &= \frac{(\eta q G/h\nu)^2 P_S P_L R}{P_S} \\ &= \left(\frac{\eta q}{h\nu}\right)(G V_b),\end{aligned} \tag{10.14}$$

where we have made use of equations (3.13), (10.9) and (10.13), and assumed $P_L \gg P_S$. For example, at 10 μm with $\eta = 0.5$, $G = 0.5$, and a detector bias voltage $V_b = 5$ V, we get $\Gamma_c = 10$. In practice, the gain in output power with increasing LO power is limited by saturation in the mixer. Nonetheless, the gain of quantum mixers can be useful in overcoming the noise contributions of the post-mixer electronics and in providing high signal to noise from the receiver. The gain provided by quantum mixers can be contrasted with the behavior of classical mixers, which downconvert the frequency but provide no increase in power. Chapter 11 provides further discussion of this distinction.

The phase, ϕ, in the heterodyne signal is a measure of the phase of the input signal (assuming that the phase of the local oscillator – the laser in this case – is stable). Thus, the heterodyne signal can be used directly with another, similar signal to reconstruct information about the wavefront. This attribute allows signals from two different telescopes to be combined coherently, as if the telescopes were part of a single large instrument, making it possible to do wide-baseline spatial interferometry.

10.2.2 Post-mixer electronics

The heterodyne signal derived above is a low level, high frequency AC current; to be compatible with simple post-mixer electronics, it needs to be amplified and converted into a slowly varying DC (or nearly DC) voltage that is proportional to the time-averaged input signal power.

Because of the low power in the IF signal, a critical component in a high performance receiver is the amplifier that follows the mixer and boosts the IF signal power. High speed amplifiers can be constructed with MOSFETs, but the best performance in this application is usually obtained with high electron mobility transistors (HEMTs) built on GaAs (Streetman, 1990).

The HEMT is based on the metal–semiconductor field effect transistor (MESFET), which is a cross between the JFET and MOSFET discussed in Chapter 6. To make a MESFET, one starts with a substrate of GaAs doped with chromium. The Fermi level is then close to the middle of the bandgap, giving the material relatively high resistance. An n-doped layer is grown on this substrate to form the channel, with contacts for the source and drain and a gate formed as a Schottky diode between them on this layer (see Figure 10.5). The electron flow between source and drain in this channel can be regulated by the reverse bias on the gate; as with the JFET, with an adequately large reverse bias the depletion region grows to the semi-insulating layer and pinches off the current. Because this structure is very simple, MESFETs can be made extremely small, which reduces the electron transit time between the source and drain and increases the response speed.

From the definition of mobility (equation (3.6)), for a given field the electron drift velocity between drain and source will be proportional to the mobility of the material in the channel. It is for this reason that high speed MESFETs are built in GaAs rather than in silicon. However, as the size of the transistor is decreased, the n-type doping in the channel must be increased leading to

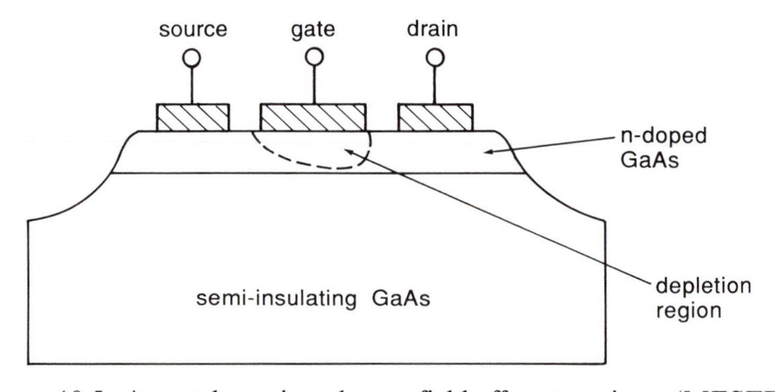

Figure 10.5. A metal–semiconductor field effect transistor (MESFET).

decreased mobility due to impurity scattering. This problem can be overcome by forming a heterojunction; refer to the discussion of quantum well detectors in Section 5.4, particularly to Figures 5.13 and 5.14. If the MESFET is grown on a thin layer of GaAlAs heavily doped n-type and deposited over an undoped layer of GaAs, the Fermi level in the GaAsAl can lie above the bottom of the conduction band in the GaAs. Consequently, the electrons collect in the GaAs just below the interface with the GaAlAs as shown in Figure 10.6; because this material is undoped, the mobility is high. The resulting high electron mobility transistor can have response extending to $\approx 10^{11}$ Hz. One factor which limits the high frequency response is the physical size of the gate and the associated parasitic capacitances. Gates smaller than 1 μm are now being manufactured.

The lowest noise is achieved with HEMTs operated near 1.5 GHz with bandwidths of 1 GHz or less. Despite their ability to operate at much higher frequencies and bandwidths, in low noise receivers they are usually used near these minimum noise conditions. Therefore, the range of frequencies accepted by the IF stage is typically $\Delta f_{IF} \approx 10^9$ Hz (or less if limited by the frequency response of the mixer).

The conversion to a slowly varying output can be done by a 'detector' stage that rectifies the signal and sends it through a low-pass filter. We would like the circuit to act as a square law detector because $\langle I_{IF}^2 \rangle$ is proportional to I_S (see equation (10.13)), which in turn is proportional to the power in the incoming signal.

A suitable circuit for this purpose is shown in Figure 10.7. Here C_B is used to block DC components of the signal such as I_L and I_S in equation (10.10). If the diode is held near zero bias with $I \ll I_0$, it will act as a good square law

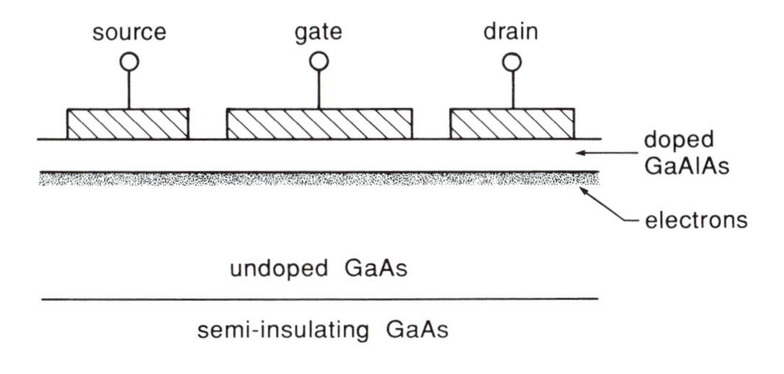

Figure 10.6. A high electron mobility transistor (HEMT).

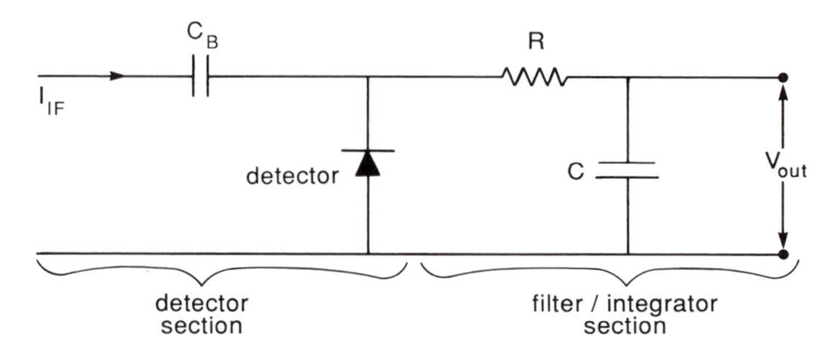

Figure 10.7. The detector stage.

detector. To demonstrate, solve the diode equation (5.33) for the voltage and expand in terms of I/I_0:

$$V = \frac{kT}{q}\ln\left(1 + \frac{I}{I_0}\right)$$
$$= \frac{kT}{q}\left[\frac{I}{I_0} - \frac{1}{2}\left(\frac{I}{I_0}\right)^2 + \frac{1}{3}\left(\frac{I}{I_0}\right)^3 - \frac{1}{4}\left(\frac{I}{I_0}\right)^4 + \cdots\right]. \quad (10.15)$$

As previously discussed, the first term in the expansion will have zero conversion efficiency, and, near zero bias, the third term will have negligible efficiency. For $I \ll I_0$, the fourth and higher terms are also negligible. Thus, the dominant behavior is square law as long as the current is kept small and the operating point is held near zero bias.

The *RC* circuit on the output of the detection stage acts to integrate the output with a time constant $\tau = RC$. It therefore smooths the output to provide V_{out} in a form that is easily handled by the circuitry that follows.

Sometimes it is desirable to carry out a variety of operations with the IF signal itself before smoothing it with a circuit similar to that just discussed. For example, imagine that the input signal contains components over a range of frequencies. Then the heterodyne signal might be sent to a bank of narrow bandpass electronic filters with a smoothing circuit on the output of each filter. The frequencies present at the input to the filter bank are limited to the bandwidth of the IF amplifier or of the photomixer (typically ~1 GHz). The filters can be tuned to divide this IF signal into components at a range of frequencies over the IF stage bandpass. Because the intermediate frequency goes as $|\omega_S - \omega_L|$, the frequencies in the heterodyne signal correspond to a similar range of frequencies in the source centered on the signal and image frequencies. The filter bank therefore provides the spectrum of the source under study.

Another possibility, which can be used in combination with a filter bank, is to introduce a second mixer to convert the heterodyne signal to a frequency range in which it can be processed more efficiently by other electronics. In this situation, the image signal from the second mixer can be easily suppressed by blocking it with an electronic filter placed ahead of the mixer in the signal processing train of electronics.

10.2.3 Local oscillator

At the high frequencies of the infrared and visible spectral regions, the only form of local oscillator with reasonably high power output is a continuous wave laser. Thus, to achieve good sensitivity the receiver must work within $\pm\Delta f_{IF}$ of the frequencies of suitable lasers. Stable, fully tunable lasers are not available over the accessible frequency range of the receivers; the receivers can only operate at a limited number of frequencies near high power laser lines. Many lines can be produced by a molecular gas laser operating with various gases that are pumped with another laser operating at higher frequency; nonetheless, many gaps remain in the frequency coverage. An extensive summary of potentially useful laser lines is given by Douglas (1989). Button, Inguscio, and Strumia (1984) provide individual descriptions of the most important laser types, and DeTemple (1987) and Lehecka *et al.* (1990) review the overall operation of these devices.

10.2.4 Signal to noise and fundamental detection limits

The derivation of the detection limits of a heterodyne receiver is significantly different from similar derivations for incoherent detectors. As pointed out after the derivation of equation (10.13), we need to distinguish between types of noise: (1) that are independent of the LO-generated current, I_L, and (2) that depend on I_L. In principle, the first category can be eliminated at least for quantum mixers by using a local oscillator with sufficient power to raise the signal strength out of the noise (there are, however, frequently practical problems in doing so). Thus, the second category alone contains the fundamental noise limits for heterodyne receivers. Two types of fundamental noise need to be considered: (1) noise in the mixer from the generation (and recombination if the mixer is a photoconductor) of charge carriers by the LO power; and (2) noise from thermal background detected by the system.

In considering the generation–recombination (G–R) noise from the LO, we assume that the LO power has been increased to provide $I_L \gg I_S$ and $I_L \gg I_B$, where I_B is the signal in the mixer from background photons. From equation

(10.9), the current generated by the LO in a photoconductive mixer is

$$I_L = \frac{\eta q G P_L}{h\nu}, \tag{10.16}$$

where P_L and ν are the LO power and frequency, respectively, G is the photoconductive gain of the mixer, and η is its quantum efficiency. This equation is valid for a photodiode mixer if we set $G=1$. (To keep things reasonably simple, we will not consider bolometer mixers in the following discussion.) We define the frequency bandwidth over which the mixer operates to be Δf_{IF}, the IF bandwidth. A general expression for the noise current can then be obtained from equations (3.36) (substituting an a for the 2 in the numerator), (3.40), and (10.16) to be

$$\langle I_{G-R}^2 \rangle = 2aqI_L G\Delta f_{IF} = \frac{2aq^2\eta G P_L \Delta f_{IF}}{h\nu}. \tag{10.17}$$

As we have seen, $a=2$ for a photoconductor, whereas for a photodiode mixer, we set $a=1$ and $G=1$ (this nomenclature should be clear from equations (5.1), (5.3), and (10.16)).

In considering the noise in the thermal background, we note from equation (10.13) that the background noise current is

$$\langle I_B^2 \rangle = 2I_L I_B, \tag{10.18}$$

where I_L is from equation (10.16) and similarly

$$I_B = \frac{\eta q G P_B}{h\nu}. \tag{10.19}$$

Here the background power is

$$P_B = \frac{1}{2} L_\nu(\varepsilon, T_B) A\Omega(2\Delta f_{IF}), \tag{10.20}$$

where L_ν is spectral radiance for an effective emissivity ε and a temperature T_B. The factor 1/2 accounts for the heterodyne receiver's sensitivity to only one polarization direction, A is the sensitive area of the receiver, and Ω is the solid angle over which it is illuminated. We have assumed a double sideband receiver, so background power is taken to be detected in two bands of width Δf_{IF} placed, respectively, above and below the LO frequency (see Figure 10.3); that is, the bandwidth $\Delta\nu$ for signals is $2\Delta f_{IF}$. This bandwidth is twice the frequency bandwidth in computing the G–R noise (equation (10.17)) because the two sidebands are merged into a single range of IF frequency.

The product $A\Omega$ is called the etendue of the system; it will be discussed at greater length in Section 10.3.3. We anticipate that discussion by taking $A\Omega \approx \lambda^2 = c^2/\nu^2$ (see equation (10.30) and the surrounding discussion) for heterodyne detection. Substituting into equation (10.18) from equations (10.16), (10.19), and (10.20), we have

$$\langle I_{\mathrm{B}}^2 \rangle = \frac{4\eta^2 q^2 G^2 \varepsilon P_{\mathrm{L}} \Delta f_{\mathrm{IF}}}{h\nu(e^{h\nu/kT_{\mathrm{B}}} - 1)}. \tag{10.21}$$

Equations (10.17) and (10.21) show how to calculate the two fundamental noise contributions. We next consider how these two noise mechanisms affect the signal-to-noise ratio achieved by the receiver. We continue to assume that we can make I_{L} arbitrarily large so that all noise mechanisms not dependent on I_{L} are negligible compared with the two discussed above. Then, since the output power goes as I^2, the instantaneous signal to noise at the output of the IF stage and at the IF bandwidth is

$$\left(\frac{S}{N}\right)_{\mathrm{IF}} = \frac{\langle I_{\mathrm{IF}}^2 \rangle}{\langle I_{\mathrm{G-R}}^2 \rangle + \langle I_{\mathrm{B}}^2 \rangle}, \tag{10.22}$$

where the signal and noise currents are from equations (10.13), (10.17), and (10.21). Substituting for these currents and then for I_{L} from equation (10.16) and I_{S} from a similar relation, we obtain

$$\left(\frac{S}{N}\right)_{\mathrm{IF}} = \frac{\eta P_{\mathrm{S}}}{h\nu \Delta f_{\mathrm{IF}} \left[\dfrac{a}{G} + \dfrac{2\eta\varepsilon}{e^{h\nu/kT_{\mathrm{B}}} - 1} \right]}, \tag{10.23}$$

where again $G = 1$ and $a = 1$ for a photodiode mixer and $a = 2$ for a photoconductor.

Note that the signal to noise derived in equation (10.23) is independent of P_{L} and can therefore be treated as the fundamental performance limit of a heterodyne receiver system at the IF output. This equation has two limiting cases. The first case, known as the quantum limit, occurs when the G–R noise from the LO power dominates; the second case is called the thermal limit, for which noise from the background emission dominates. The dividing point between these limits can be derived by setting the two terms in square brackets in equation (10.23) equal to each other and solving to obtain:

$$\frac{h\nu}{kT_{\mathrm{B}}} = \ln\left(1 + \frac{2\eta G\varepsilon}{a}\right). \tag{10.24}$$

Since η, G, and ε are all less than, but of the order of, one, and $a = 1$ or 2, the logarithmic term is of order unity. Therefore, the division between the two regimes can often be simplified by stating that for $h\nu \gg kT_B$, the quantum limit holds, while for $h\nu \ll kT_B$ we get the thermal limit.

A quantity useful for high speed systems such as those used in communications is the minimum detectable power, MDP_{IF}, required to give $S/N=1$ at the IF bandwidth:

$$MDP_{IF} = \frac{h\nu \Delta f_{IF}}{\eta}\left[\frac{a}{G} + \frac{2\eta\varepsilon}{e^{h\nu/kT_B}-1}\right]. \tag{10.25}$$

However, in most applications a detector stage is used to rectify and smooth the IF signal and we need to include this stage in the performance evaluation. As with other detection systems, the instantaneous signal-to-noise ratio is improved in proportion to the square root of the effective integration time. When a smoothing stage is used (such as the filter integrator circuit shown in Figure 10.7), the S/N increases in proportion to the square root of the ratio of effective integration times of the IF and smoothing circuits, that is,

$$\left(\frac{S}{N}\right)_{out} = \left(\frac{S}{N}\right)_{IF}\left(\frac{\tau_{RC}}{\tau_{IF}}\right)^{1/2}. \tag{10.26}$$

To convert integration times to frequency bandwidths, we should use the procedures in Chapter 3, for example, equations (3.37), (3.38), or (3.39). We can usually assume that the output of the IF stage is sampled uniformly over a time interval τ_{IF}, so $\tau_{IF}=1/2\Delta f_{IF}$, and that the integrator is a single stage RC circuit (as shown in Figure 10.7), so $\tau_{RC}=1/4\Delta f_{RC}$. Recalling that NEP goes inversely as the signal-to-noise ratio and refers to unity frequency bandwidth (that is, $\Delta f_{RC}=1$ Hz), the NEP in the quantum limit for a heterodyne receiver system, NEP_H is

$$\begin{aligned} NEP_H &= \frac{h\nu a}{\eta G}\Delta f_{IF}\left[\frac{\tau_{IF}}{\tau_{RC}(1\ \mathrm{Hz})}\right]^{1/2} \\ &= \frac{h\nu a}{\eta G}\Delta f_{IF}\left[\frac{2\Delta f_{RC}(1\ \mathrm{Hz})}{\Delta f_{IF}}\right]^{1/2} \\ &= \frac{h\nu a}{\eta G}(2\Delta f_{IF})^{1/2}. \end{aligned} \tag{10.27}$$

In the thermal limit,

$$NEP_H = \frac{2h\nu\varepsilon}{e^{h\nu/kT_B}-1}(2\Delta f_{IF})^{1/2}. \tag{10.28}$$

Although the result is obscured by our normalization of the NEP to unity bandwidth, inspection of (10.27) also shows the general result that the effective noise bandwidth of the receiver is proportional to the geometric average of the predetector and postdetector bandwidths, that is, to $(\Delta f_{IF}\,\Delta f_{RC})^{1/2}$ (for example, Robinson, 1962).

Note that the NEP_H is affected by the details of the time response of the IF and output stages (factors of order unity) through the conversion of effective integration time into frequency bandwidth. Equations (10.27) and (10.28) provide a specific example; if the IF stage had been assumed to have a frequency response like that of a single stage *RC* circuit ($\tau_{IF} = 1/4\Delta f_{IF}$), the NEP would be $\sqrt{2}$ times less than in these expressions.

Although we have concentrated on fundamental noise sources, noise that is independent of the LO power is also present. For example, the mixer detects the total background power, including that at frequencies outside the bandpasses determined by Δf_{IF}. This signal can contribute noise, particularly if the background is fluctuating rapidly. A second source of noise is Johnson noise in the mixer; it can normally be eliminated by cooling the mixer. A third possibility is that the local oscillator contributes noise through phase or amplitude instability or both. Also, the amplifiers used in the signal chain may dominate the noise. We will consider these nonfundamental noise sources in more detail in Section 10.3.5.

10.3 Performance attributes of heterodyne detectors

10.3.1 Bandwidth

The spectral bandwidth of a heterodyne receiver is determined by the achievable bandwidth at the intermediate frequency. The IF bandwidth, $\Delta f_{IF} = \Delta\omega_{IF}/2\pi$, can be limited either by the frequency response of the photomixer or by that of any circuitry that amplifies or filters the heterodyne signal. Some forms of infrared mixer impose severe limitations on the bandwidth. The *RC* time constants of photoconductor mixers are usually short because of the reduction in *R* by the LO power, but recombination times can limit the bandwidth. For example, the long recombination times in germanium limit the frequency response of far infrared germanium photoconductors to $<10^8$ Hz. InSb hot electron bolometers can be used as mixers in the submillimeter region, but their thermal time constants limit the achievable bandpasses to $\sim 10^6$ Hz. In more favorable cases, such as photodiode mixers, current technical limitations often hold Δf_{IF} to no more than a few times 10^9 Hz due to the *RC* time constant of the mixer. In the infrared, the bandwidth so defined is very small compared with the frequency of the signal photons; at 10 μm, $f = 3 \times 10^{13}$ Hz, and, even assuming a heterodyne bandwidth of 3×10^9 Hz, the spectral bandwidth is 0.01% of the operating frequency. Therefore, heterodyne receivers operating in this wavelength region have poor signal to noise on continuum sources. As a result, they are used primarily for measurements of spectral lines at extremely high resolution. In the submillimeter- and millimeter-wave regimes it becomes reasonable to detect continuum sources with a heterodyne system. (At 3 mm (10^{11} Hz) $\Delta f_{IF} = 3 \times 10^9$ Hz corresponds to a 6% bandwidth for a double sideband receiver.)

10.3.2 Time response

The time response of a heterodyne receiver can be as fast as the period of the IF signal, $1/f_{IF}$; that is, a few nanoseconds. It may also be limited by the time response of the mixer (see above) or of the detector.

10.3.3 Throughput

The signal photons cannot be concentrated onto the mixer in a parallel beam; even for a point source, they will strike it over a range of angles. The requirement that interference occurs at the mixer between the laser and the signal photons sets a requirement on the useful range of acceptance angle for the heterodyne receiver.

We will assume that the LO photon stream is perfectly collimated and strikes the mixer perpendicular to its face. Consider Figure 10.8, which shows

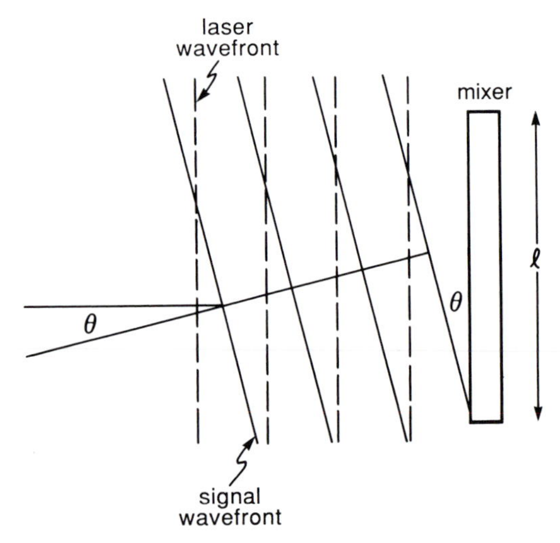

Figure 10.8. Required alignment of the local oscillator and signal wavefronts on the photomixer.

the general case of a signal wavefront striking the mixer face at a nonzero angle θ (θ is measured from the mixer face). If $\theta \neq 0$, there will be a phase shift across the mixer between the signal and laser fields; thus, the IF signals from different parts of the mixer will not add and may even begin to cancel. Full cancellation will occur when

$$\ell \sin\theta = \lambda \approx \ell\theta, \tag{10.29}$$

where ℓ is the length of the mixer (assumed square) and λ is the wavelength of observation. Taking the central ray of the signal also to strike the mixer perpendicularly, from equation (1.9), in the case of small θ, we get $\theta^2 \approx \Omega$, the solid angle over which signal interferes efficiently with the LO. We also have $\ell^2 = A$, the area of the mixer. Thus, we can rewrite equation (10.29) as

$$A\Omega \approx \lambda^2. \tag{10.30}$$

$A\Omega$, frequently called the etendue, is invariant in any aberration-free system. That is, optical elements can be placed in the beam to magnify or demagnify any focal plane of the system, but perfect geometric optics leave $A\Omega$ unchanged, and imperfect optics can only increase it. Therefore, equation (10.30) is also a constraint on the beam that can be accepted by a telescope or by any other optical system that concentrates the signal on the heterodyne mixer; this condition can be alternately expressed as

$$\Phi \approx \frac{\lambda}{D}, \tag{10.31}$$

where D is the diameter of the telescope aperture and Φ is the angular diameter of the field of view on the sky. This relationship defines the condition for diffraction-limited imaging; compare it to the Rayleigh criterion, which is just

$$\Phi = 1.22\frac{\lambda}{D}. \tag{10.32}$$

Thus, a coherent receiver should operate at the diffraction limit of the telescope (if the receiver only accepts a resolution element smaller than the diffraction limit of the telescope, there is a significant loss of energy from the source).

At infrared wavelengths, where atmospheric turbulence can cause significant image motion and broadening beyond the diffraction limit, the constraint on etendue can result in loss of signal because the receiver accepts only a fraction of the broadened image. The constraint is a much less serious shortcoming at longer wavelengths.

A second restriction, already mentioned, is that the interference that produces a heterodyne signal only occurs for components of the source photon electric field vector that are parallel to the electric field vector of the laser power. Since the laser light is polarized, only a single polarization of the source emission produces any signal.

The relation between wavelength and etendue in equation (10.30) and the constraints on polarization are manifestations of the 'antenna theorem', which applies to all heterodyne receivers. The combination of the constraint derived in equation (10.31) and the sensitivity to only one plane of polarization is sometimes collectively described by saying that heterodyne receivers are single mode detectors, or that they are sensitive to only one transverse mode of the radiation field.

10.3.4 Spatial arrays

We have seen that heterodyne receivers provide spectral information in a direct and powerful way. Although it would be relatively easy to construct a spatial array of mixers fed by a diplexer and local oscillator, every element in the array would require an IF amplifier and either a circuit similar to that shown in Figure 10.7 or some other device performing a similar function. These circuit elements are bulky and not amenable to automated production that uses integrated circuit technology. This situation is quite different from that of the infrared and CCD arrays discussed in Chapter 7, where the signals appear at widely spread time intervals and can be multiplexed to a simple set of output electronics. Because of their added complexities, large-scale heterodyne spatial arrays are not yet feasible. However, wideband digital correlators are under development that will provide compact and relatively inexpensive back ends, permitting the construction of modest sized spatial arrays.

10.3.5 Noise temperature

Although equations (10.27) and (10.28) are useful for describing the theoretical performance of heterodyne receivers in a given situation, they are inadequate for making general comparisons between heterodyne and incoherent detector response to continuum sources. If the LO power can be increased without limit, these equations imply that the NEP of heterodyne receivers is reduced (that is, the signal-to-noise ratio is increased) by narrowing the IF bandwidth. This behavior arises because the equations assume that all the source power continues to fall within this bandwidth, which clearly is a physically unreasonable assumption for a continuum source. To overcome this limitation, a thermal continuum source is introduced through a noise temperature, T_N, defined such that a matched blackbody at the receiver input at a temperature T_N produces $S/N=1$. The lower T_N, the fainter a source gives $S/N=1$, and the better is the performance of the receiver.

In the thermal limit, if the effective source emissivity $\varepsilon=1$, then by definition $T_N=T_B$. More generally, by substituting

$$P_S=L_\nu(T_N)A\Omega\,\Delta f_{IF} \tag{10.33}$$

into equation (10.23) and setting $S/N=1$, it can be shown that the double sideband noise temperature in the thermal limit at the output of the IF stage is

$$T_N=\frac{h\nu}{k}\frac{1}{\ln(\varepsilon-1+e^{h\nu/kT_B})-\ln\varepsilon}. \tag{10.34}$$

To derive a noise temperature in the quantum limit, we again start from equations (10.23) and (10.33). For the double sideband case, the quantum limit is

$$T_N=\frac{h\nu}{k\ln\left(1+\dfrac{2G\eta}{a}\right)}. \tag{10.35}$$

In an ideal case, we set G, η, and a to 1 to obtain

$$T_N\approx\frac{h\nu}{k}. \tag{10.36}$$

The quantum limit expressed in equation (10.36) can be justified in terms of the Heisenberg uncertainty principle, which states that the uncertainty, ΔP, in a measurement of power will be

$$\Delta P=\frac{h\nu}{\Delta t}, \tag{10.37}$$

where Δt is the observation time. Starting with equation (3.42) for Johnson noise in a resistor within a frequency bandwidth df and converting to power noise within a time interval Δt,

$$\langle P\rangle\Delta t=kT_N. \tag{10.38}$$

Setting $\Delta P\approx\langle P\rangle$, we obtain

$$T_N\sim\frac{h\nu}{k}. \tag{10.39}$$

It is often convenient to express the flux from a source as an antenna temperature, T_S, in analogy with the noise temperatures. This concept is particularly useful in the millimeter and submillimeter (and radio) regions, where the observations are virtually always at frequencies that are in the Rayleigh–Jeans regime ($h\nu \ll kT$). In this case, the antenna temperature is linearly related to the input flux density (see Problem 1.3):

$$\frac{P_S}{\Delta\nu} = 2kT_S, \tag{10.40}$$

where $\Delta\nu$ is the frequency bandwidth (for example, $2\Delta f_{IF}$ for a double sideband receiver) and we have assumed $A\Omega = \lambda^2$ from equation (10.30). To maintain the simple formalism in terms of noise and antenna temperatures, it is conventional to use a Rayleigh–Jeans equivalent temperature such that equation (10.40) holds by definition whether the Rayleigh–Jeans approximation is valid or not.

The concept of noise temperature provides a convenient means of quantifying the various sources of LO-independent noise, such as amplifier noise. The amplifier noise is usually expressed as the Johnson noise of an equivalent perfect resistor ('perfect' in the sense that it generates no excess noise) at a fictitious noise temperature; for example, assuming voltage noise can be ignored,

$$\langle I_A^2 \rangle = \frac{4kT_A \Delta f_{IF}}{R_A}, \tag{10.41}$$

where R_A and T_A are the amplifier input resistance and noise temperature, respectively, and $\langle I_A^2 \rangle^{1/2}$ is the amplifier noise current. A lower limit to the amplifier noise temperature is set by equation (10.39), since this relation is grounded in fundamental physics that must apply generally. For example, an amplifier operating at 1.5 GHz has a quantum limit of about 0.07 K; at this frequency, the best HEMT amplifiers currently have noise temperatures about 100 times this limit.

Assuming that the amplifier noise is dominant among the various LO-independent noise types, we can derive the LO power required to approach fundamental noise limits. We need $\langle I_A^2 \rangle < \langle I_{G-R}^2 \rangle + \langle I_B^2 \rangle$, or in the quantum limit (for which we take $\langle I_B^2 \rangle \to 0$),

$$P_L > \frac{2kT_A L h\nu}{q^2 \eta R_A a G^2}. \tag{10.42}$$

Here, L represents the loss ($L \geqslant 1$) in transferring the output of the mixer to the amplifier.

The noise temperature of a receiver system contains contributions from the mixer, amplifiers, thermal background, and other sources. The total noise can be expressed as the sum of these temperatures all referred to the system input; hence, the temperatures of individual components must be multiplied by the appropriate losses (or divided by the gains) that occur as the signals lose energy (or are amplified) in being passed from one system component to another. In short, the system noise is

$$T_N^S = T_M + L_1 T_A + L_1 L_2 T_3 + \cdots, \tag{10.43}$$

where T_M is the equivalent input noise temperature of the mixer, T_A is the equivalent input noise temperature of the amplifier, and T_3 is the noise temperature of some other system component. The L's express the effective losses (or the inverses of the gains) for each noise contribution. L_1 is the loss or inverse of the gain in passing the signal from the mixer input to the amplifier; assuming the mixer output is coupled efficiently to the amplifier, $L_1 = 1/\Gamma_c$, where Γ_c, is from equation (10.14) or an equivalent expression appropriate to other mixer types. L_2 is the loss or inverse of the gain in passing the signal from the input of the amplifier (including the inverse of the amplifier gain) to the input of the next component, and so on. Comparing with equations (3.45) and (10.41), we see that this formalism in terms of temperatures adds noise powers, which we showed was the correct dependence in equation (3.45) and the following discussion.

The achievable signal-to-noise ratio for a coherent receiver is given in terms of antenna and system noise temperatures by the Dicke radiometer equation:

$$\left(\frac{S}{N}\right)_c = \mathrm{K}\frac{T_S}{T_N^S}(\Delta f_{IF}\Delta t)^{1/2}, \tag{10.44}$$

where Δt is the integration time of the observation. This relation follows from the definitions of the antenna and noise temperatures (and that the noise temperature is referred to the receiver input), and from equations (10.27) and (10.28). As discussed following equation (10.28), details of the time response of the output stages may introduce minor corrections into the conversion to frequency bandwidth; for simplicity, these corrections are subsumed in the constant K, of order unity. In the following discussion, we set $\mathrm{K} = 1$.

Equations (10.40) and (10.44) give us the means to compare the performance of coherent and incoherent detectors, as long as we also keep in mind the bandwidth and single mode detection restrictions that we have already discussed. From equation (10.40) and the definition of NEP, the signal-to-noise ratio with an incoherent detector system operating at the diffraction limit is

$$\left(\frac{S}{N}\right)_i = \frac{2kT_S\Delta\nu(\Delta t)^{1/2}}{\mathrm{NEP}}. \tag{10.45}$$

Therefore, using equation (10.44), we obtain the ratios of signal to noise achievable with the two types of system under the same measurement conditions:

$$\frac{(S/N)_c}{(S/N)_i} = \frac{\mathrm{NEP}(\Delta f_{\mathrm{IF}})^{1/2}}{2kT_N^S\Delta\nu}. \tag{10.46}$$

To illustrate, consider a bolometer operating at the background limit and a heterodyne receiver in the thermal limit, both viewing a source through an etendue $= \lambda^2$ and against a background of unity emissivity and at a temperature of T_B. Assume we are observing at a frequency $h\nu \ll kT_B$. Then it can be shown that

$$\frac{(S/N)_c}{(S/N)_i} = \left[\left(\frac{1}{\eta}\right)\left(\frac{\Delta f_{\mathrm{IF}}}{\Delta\nu}\right)\left(\frac{h\nu}{kT_B}\right)\right]^{1/2}. \tag{10.47}$$

In this case, the incoherent detector yields better signal to noise unless $\Delta f_{\mathrm{IF}}/(\eta\,\Delta\nu) \gg 1$, for example where measurements are being made at spectral resolution significantly higher than the IF bandwidth and the incoherent detector must be operated in a very narrow band.

On the other hand, where the bolometer is detector noise limited and the heterodyne receiver operates at the quantum limit,

$$\frac{(S/N)_c}{(S/N)_i} = \frac{\mathrm{NEP}(\Delta f_{\mathrm{IF}})^{1/2}}{2h\nu\,\Delta\nu}. \tag{10.48}$$

Particularly if the spectral resolution $\nu/\Delta\nu$ is kept constant (as is typical), the transition from the case favoring incoherent to that favoring coherent detectors occurs abruptly with decreasing frequency in this case because the figure of merit in equation (10.48) goes as $1/\nu^2$. Further gains in spectroscopy result from the ability of a heterodyne receiver to obtain more than one spectral measurement at a time by dividing its IF output into a bank of narrowband electronic filters.

10.4 Test procedures

The underlying principles for testing heterodyne receivers are similar to those discussed in Chapter 6. In those discussions, we showed that considerable care is required to specify and allow for the geometry of illumination, if we want to obtain meaningful results. In contrast, for heterodyne receivers the antenna theorem defines an appropriate geometry.

A particularly simple determination of the noise temperature of a receiver can be made with two blackbody emitters, or loads, at different temperatures that are alternately placed over the receiver input. It is assumed that the temperatures of these loads are well determined and that their emissivities are close to unity. A 'Y-factor' is defined as

$$Y = \frac{V_{hot}}{V_{cold}}, \tag{10.49}$$

where V is the output voltage of the receiver and the subscripts indicate which load is over the input. Assuming that the receiver is linear, the output voltages should be proportional to the sum of the antenna temperatures of the loads and the noise temperature of the receiver, T_{rec}, or

$$Y = \frac{T_{hot} + T_{rec}}{T_{cold} + T_{rec}}. \tag{10.50}$$

Equation (10.50) can be solved for the receiver noise temperature,

$$T_{rec} = \frac{T_{hot} - Y T_{cold}}{Y - 1}. \tag{10.51}$$

10.5 Example

Consider a heterodyne receiver operating between 10 and 11 μm, viewing a background of 290 K, and using a HdCdTe diode photomixer with the following properties:

cutoff wavelength = 15 μm,
quantum efficiency = 40% at wavelengths short of 13 μm,
size = 100 μm diameter,
depletion region width at operating bias = 1 μm,
reverse bias impedance = 150 Ω (due to electron–hole generation in the depletion region, there is no well defined I_0, but a relatively constant effective resistance),
dielectric constant = 10.

The local oscillator is a CO_2 laser illuminating the photodiode by reflection off a diplexer with 10% reflectivity (the diplexer is built to transmit 90% of the incoming signal to maximize signal to noise). There are roughly 50 strong laser lines in the 10 to 11 μm region. Let the output of the photomixer go to an amplifier with an input impedance of 150 Ω, a gain of 100, a bandwidth of 3×10^9 Hz, and an output current noise of 33.2 μA Hz$^{-1/2}$. Viewing a 500 K source, the receiver output is 1.040 V, while viewing a 300 K source it is 1.000 V. Answer the following: (a) what is the IF bandwidth of the receiver;

(b) will the receiver operate in the thermal or quantum limit; (c) what is the noise temperature of the amplifier; (d) what laser power is required to overcome amplifier noise; (e) assuming the laser can be run at 20 times the value from (d), what is the conversion gain; (f) what is the expected system noise temperature; (g) what is the measured noise temperature, and is there excess noise; and (h) what is the probability that the receiver can operate at an arbitrary frequency in the 10 to 11 μm band?

(a) The IF bandwidth will be limited by the amplifier or by the RC time constant of the mixer. To estimate the RC time constant, we take

$$C = \frac{A\kappa_0\varepsilon_0}{w},$$

where w is the depletion width, A is the area of the diode, and κ_0 is the dielectric constant. Substituting, we find $C = 6.95 \times 10^{-13}$ F, $RC = 1.043 \times 10^{-10}$ s $= \tau$, and a cutoff frequency of

$$f_c = \frac{1}{2\pi\tau} = 1.5 \times 10^9 \text{ Hz}.$$

Since this frequency is less than the bandwidth of the IF amplifier, the achievable IF band will be limited to the photomixer response. In the following, we assume the IF amplifier bandwidth is also limited to this value to avoid excess noise.

(b) At 10.5 μm, $\nu = 2.855 \times 10^{13}$ Hz. At the ambient background temperature of 290 K, $h\nu/kT = 4.7 > 1$, so we expect to operate in the quantum limit. A more precise determination can be made from equation (10.23); setting $G = a = 1$, the quantum limit is obtained if

$$\frac{2\eta\varepsilon}{e^{h\nu/kT} - 1} \ll 1.$$

The left side of this expression is $\sim 0.007\varepsilon$, and since $\varepsilon \leqslant 1$, it is confirmed that we should be in the quantum limit.

(c) To convert the output noise of the amplifier to an input-referred noise, we divide by the gain to get the input-referred current noise $\langle I_A^2 \rangle^{1/2} = 3.32 \times 10^{-7}$ A Hz$^{-1/2}$. We also have $\Delta f_{IF} = 1.5 \times 10^9$ Hz and $R_A = 150$. From equation (10.41), we obtain

$$T_A = \frac{\langle I_A^2 \rangle R_A}{4k\,\Delta f_{IF}} = 200 \text{ K}.$$

(d) The LO power required for quantum limited operation can be obtained from equation (10.42). Since the output impedance of the diode matches the input impedance of the amplifier, $L=1$. For a photodiode, $G=a=1$, and we have $\eta=0.4$. Substituting, we need a local oscillator power of $\gg 0.068$ mW to overcome the amplifier noise. Since only 10% of the laser energy is reflected by the beamsplitter, the required laser power in the LO line is $\gg 0.68$ mW.

(e) In analogy with the derivation of equation (10.14), the conversion gain of the mixer is

$$\Gamma_c = \left(\frac{\eta q}{h\nu}\right)(I_{ph}R).$$

For a laser output of 13.6 mW, 1.36 mW is reflected onto the photomixer off the diplexer; with $\eta=0.4$, this signal produces a current of 2.88×10^{16} electrons s^{-1}, or 4.61 mA. We therefore find a conversion gain of 2.34.

(f) For the mixer, we obtain the quantum limit noise temperature from equation (10.38) as

$$T_N = \frac{h\nu}{k\ln(1+0.4)} = 2.97\frac{h\nu}{k} = 4073\text{ K}.$$

From (e), L_1 in equation (10.43) is $=1/2.34=0.427$. From equation (10.43) with $L_1=0.427$ and $T_A=200$ K, the system noise temperature is expected to be 4158 K.

(g) The output voltages give a Y-factor of 1.04 from equation (10.49). From equation (10.51), the receiver noise temperature is then 4700 K, giving an excess noise 4700 K $-$ 4158 K $=$ 542 K.

(h) The total frequency band between 10 and 11 μm is 2.73×10^{12} Hz. If each laser line allows a band of $\pm\Delta f_{IF}$ about the line, or a total of 3×10^9 Hz, the receiver can access a total frequency range of 1.5×10^{11} Hz or 5.5% of the 10 to 11 μm region. The probability of reaching an arbitrary frequency is 0.055.

10.6 Problems

10.1 Compare the signal to noise achievable at a wavelength of 300 μm on a continuum source with (1) a helium-3 cooled bolometer with an electrical NEP of 2×10^{-16} W Hz$^{-1/2}$ and a quantum efficiency of 0.53, operated through a spectral band of 30% of the center frequency; and (2) a heterodyne receiver with a single sideband noise temperature of 1500 K, an IF bandwidth of 3×10^9 Hz, and operated double sideband at the same frequency as the bolometer. At what spectral bandwidth would the two systems give equal S/N?

10.2 Derive equations (10.34) and (10.35).

10.3 Consider a square mixer of width ℓ illuminated by plane-parallel signal and LO waves. Assume that the LO is perfectly aligned on the mixer and the signal strikes it at an angle θ, in a direction parallel to the x axis of the mixer. Show that the current from a surface element of the mixer is

$$\mathrm{d}I(t) \approx \frac{\mathrm{d}x\,\mathrm{d}y}{\ell^2}\left[I_\mathrm{L} + 2(I_\mathrm{S} I_\mathrm{L})^{1/2}\cos\left(\omega t + \frac{2\pi \sin\theta}{\lambda}x + \phi''\right)\right],$$

where ϕ'' is a fixed phase shift. Use this result to prove the result of equation (10.29).

10.4 Compute the 'quantum limit' for an ideal incoherent detector. That is, for a photodiode or photomultiplier show that the minimum detectable power when dark current and Johnson noise can be neglected is

$$P_\mathrm{min} = \frac{h\nu\,\mathrm{d}f}{\eta},$$

where ν is the frequency of the signal photons, $\mathrm{d}f$ is the detector electronic bandwidth, and η is the quantum efficiency. Compare with the heterodyne detector.

Note

1 The second equality can be easily demonstrated. Referring to equations (10.3a) and (10.3b), and letting $\mathscr{E}_0 = a + jb$ and $\mathscr{H}_0 = c + jd$,

$$\begin{aligned}\mathrm{Re}[\mathscr{E}(t)]\,\mathrm{Re}[\mathscr{H}(t)] &= (a\cos\omega t + b\sin\omega t)(c\cos\omega t + d\sin\omega t)\\ &= (ac)\cos^2\omega t + (bd)\sin^2\omega t\\ &\quad + (ad + bc)\cos\omega t\sin\omega t.\end{aligned} \tag{10.6a}$$

Because $\langle\cos\omega t\sin\omega t\rangle_\mathrm{t} = 0$ and $\langle\sin^2\omega t\rangle_\mathrm{t} = \langle\cos^2\omega t\rangle_\mathrm{t} = 1/2$, we get

$$\langle\mathrm{Re}[\mathscr{E}(t)]\,\mathrm{Re}[\mathscr{H}(t)]\rangle_\mathrm{t} = \frac{1}{2}(ac + bd). \tag{10.6b}$$

Now we calculate the quantity in the second form of equation (10.6); it is

$$\begin{aligned}\frac{1}{2}\mathrm{Re}[\mathscr{E}(t)\mathscr{H}^*(t)] &= \frac{1}{2}\mathrm{Re}[(a + \mathrm{j}b)(\cos\omega t - \mathrm{j}\sin\omega t)\\ &\quad \times (c - \mathrm{j}d)(\cos\omega t + \mathrm{j}\sin\omega t)]\\ &= \frac{1}{2}(ac + bd),\end{aligned} \tag{10.6c}$$

which is equivalent to the result in equation (10.6b).

11
Submillimeter- and millimeter-wave heterodyne receivers

The general principles derived in Chapter 10 are equally valid for submillimeter- and millimeter-wave receivers. Performance attributes that limit the general usefulness of infrared heterodyne receivers, such as limited bandpass and diffraction limited throughput, cease to be serious limitations as the wavelength of operation increases. Heterodyne receivers are therefore the preferred approach for high resolution spectroscopy in the submillimeter spectral region, and their usefulness is expanded as the wavelength increases into the millimeter regime. At wavelengths longer than a few millimeters, they are used to the exclusion of all other kinds of detectors.

11.1 Basic operation

The operational principles of heterodyne receivers were described in Section 10.1, and the operation of the components that follow the mixer in a submillimeter- or millimeter-wave receiver is essentially identical to the systems discussed in Chapter 10. Such components can be used for amplification, frequency conversion, and detection. Often, much of the expense in a heterodyne receiver system is in the 'backend' spectrometer (for example, filter bank or correlator) and in other equipment that processes the IF signal. Because these items can be identical from one system to another, it is common to use a single set of them as 'back ends' with many different receiver 'front ends' that together can operate over a broad range of signal frequencies. In this chapter, we will focus our attention on the receiver components that must be changed from the infrared devices in Chapter 10 for operation in the submillimeter and millimeter, that is, on mixers and local oscillators.

There are two underlying reasons for the differences between the mixer design for submillimeter- or millimeter-wave and infrared receivers. The first is that high quality, fast photon detectors are not available at wavelengths longer than the infrared. As discussed in Chapters 3, 4, and 5, no high performance photodiodes or photoconductors are available with response at wavelengths longer than 200 μm. The long recombination times in germanium (see Table 3.1) result in relatively poor frequency response and low IF bandwidths for heterodyne systems based on the photoconductors available between 40 and 200 μm. Some work has been done in the submillimeter using bolometers as mixers, but they permit only very limited IF bandwidths.

The second reason is a more general result of the requirement for high frequency response. Efficient absorption of the energy of the incoming photons by a photomixer requires the mixer to have dimensions at least comparable to the photon wavelength. As electronic devices are made larger, their frequency response generally becomes worse. Consequently, as we approach longer wavelengths it becomes desirable to couple the energy into a mixer much smaller than the wavelength of the photons.

Submillimeter- and millimeter-wave mixers are therefore designed as optimized electrical components. The photon stream from the source is collected by a telescope or 'primary antenna' and concentrated onto a secondary antenna. The electric field of the photon stream creates an oscillating current in the secondary antenna, which conducts the current to the mixer. When the signal and LO power are combined, the oscillating current will beat in amplitude at their difference frequency as illustrated in Figure 10.1. The mixer is a suitably nonlinear electrical element that converts this amplitude beating into the IF signal. A block diagram of this arrangement is shown in Figure 11.1.

The secondary antenna often takes the form of a wire arranged within the receiver optics. Because of the restrictions expressed in the antenna theorem (see Section 10.3.3), the secondary antenna should be adjusted to receive only one mode of the photon stream because the fields due to the others will differ in phase and will tend to interfere with the heterodyne process. Since the bulk of the energy is in the longest wavelength, or fundamental mode, the antennas and their surrounding optics are designed to operate at this mode. This behavior accounts for the term 'single mode detector' introduced in the preceding chapter. The antenna wire is attached directly to the mixer; the efficiency of transfer of power to the mixer depends on the impedance matching of the mixer to the antenna.

At submillimeter wavelengths, the signal energy from free space may be concentrated onto the antenna wire by using reflectors around the wire to improve the efficiency and directivity of its antenna pattern. Good coupling can be achieved with a corner cube reflector such as the one illustrated in Figure 11.2. The secondary antenna contacts the mixer through a hole in the

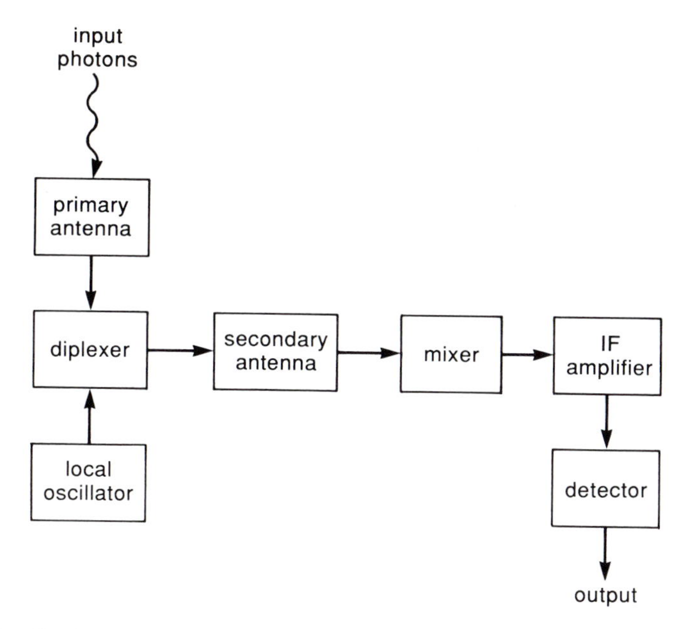

Figure 11.1. Block diagram of a millimeter- or submillimeter-wave heterodyne receiver.

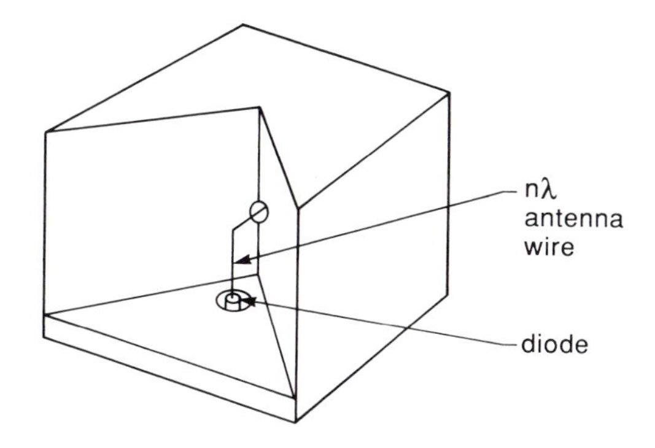

Figure 11.2. A corner reflector quasi-optical mixer mount.

bottom of the corner reflector. In general, the coupling schemes require resonant cavities or similar structures; consequently, their efficiency tends to be strongly wavelength dependent. Because of the heavy reliance on standard optical techniques and equipment (lenses and mirrors), the arrangements discussed in this paragraph are called quasi-optical feeds. They are described by Goldsmith, Itoh, and Stophan (1989).

At frequencies below ~800 GHz, the signal can be coupled to the mixer in a different way. The signal flux is directed into a feed horn that concentrates the energy into a waveguide. Details are given in Christiansen and Högbom (1985), Kraus (1986), Bahl (1989), and Walker *et al.* (1992). Waveguides can carry photon streams in a variety of modes corresponding to different oscillation patterns in the guide. To concentrate the fundamental mode onto the secondary antenna, waveguides are terminated with conducting plugs called backshorts. The position of these elements is adjusted to optimize the transfer of power to the secondary antenna. A suitably optimized waveguide usually couples energy to the secondary antenna more efficiently than does a quasi-optical feed.

At microwave and radio frequencies, waveguides can carry signals substantial distances without significant losses; however, the required fabrication tolerances in the millimeter- and submillimeter-wave regions are too stringent to achieve this advantage. There is active research in this area, and continued progress is anticipated in improving waveguide performance at high frequencies.

A variety of schemes are used to feed the energy from both the signal and the LO into the mixer optics. In many cases, the LO power is reflected off a diplexer that transmits the signal, an arrangement very similar to that used with infrared heterodyne receivers and illustrated in Figure 10.4. Other arrangements can be used with waveguides; an example, called a crossbar mixer, is illustrated in Figure 11.3. Two diode mixers are mounted in the signal waveguide with the electrical contact supplied by a wire across the guide. A second waveguide is used to couple the LO power into the mixers.

At very high frequencies, the local oscillator may be a continuous wave (CW) laser, whose output is combined with the source photon stream by

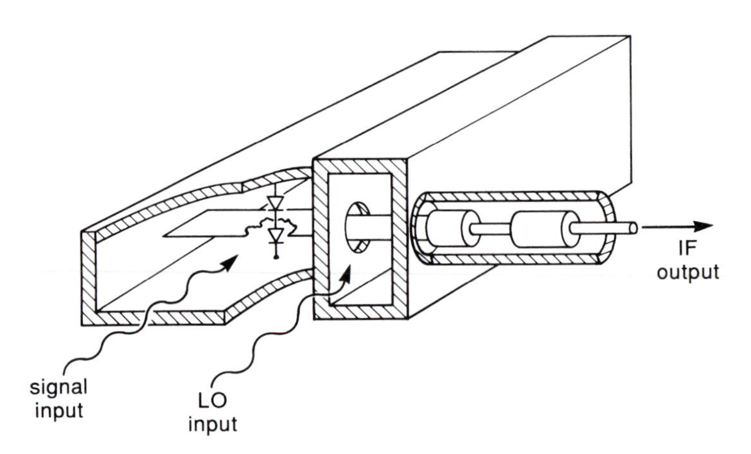

Figure 11.3. Crossbar mixer arrangement. The signal and LO power are brought to the mixer through separate waveguides.

optical elements and a beamsplitting diplexer. Laser LOs are discussed in Section 10.2.3; in general, they can provide large LO powers but allow operation only at a discrete set of frequencies; changing from one of these frequencies to another usually requires changing the laser gas, a cumbersome procedure. At submillimeter and millimeter wavelengths, a similar feed arrangement can be used with an electronic LO. The electronic LO can drive a wire antenna or waveguide feed horn that radiates its output into the optical system. Electronic oscillators have the advantage that they can usually be tuned with relatively little difficulty to the frequency of interest. At very high frequencies (above ~ 800 GHz), however, the power available from tunable oscillators is very small, and laser LOs are required to achieve adequately low system noise temperatures.

Despite the differences in mixers and LOs, the performance attributes discussed in Section 10.3 are applicable to submillimeter- and millimeter-wave receivers as well as to infrared ones. As pointed out in Chapter 10, the maximum frequency bandwidth of heterodyne receivers is often limited by the IF bandwidth, which is similar from one receiver to another; therefore, the fractional bandwidth changes inversely with operating frequency (assuming that the bandwidth is not limited by the mixer). Hence, these receivers can be used more effectively for low resolution spectroscopy and continuum detection as the operational frequency decreases from the infrared to the submillimeter- to the millimeter-wave spectral region. With decreasing frequency, it also becomes increasingly feasible for a receiver to be fed efficiently with a diffraction limited beam; the tolerances on optical components relax in proportion to wavelength, and the effects of atmospheric turbulence are decreased relative to the achievable angular resolution. The only infrared performance limitation discussed in Chapter 10 that remains as an equally serious inconvenience at longer wavelengths is the difficulty in making large scale spatial arrays of heterodyne receivers. Most of the discussions of fundamental performance limits and of the characterization and testing of receiver performance in Chapter 10 also apply to submillimeter- and millimeter-wave receivers with minor modifications.

11.2 Mixers

One measure of the performance of a mixer is the magnitude of the deviation of the device's I–V curve from linearity at the operating point, that is, a useful figure of merit is

$$\mathbb{R} \sim \frac{\mathrm{d}^2 I/\mathrm{d}V^2}{\mathrm{d}I/\mathrm{d}V}. \tag{11.1}$$

Therefore, strongly nonlinear circuit elements are used as mixers. The two most widely applied approaches are semiconductor diodes and superconductor–insulator–superconductor junctions, both of which are described below.

11.2.1 Diode mixers

Recall the discussion in Section 10.2.2, where we showed under what conditions a diode acts as a square-law mixer. A diode mixer is an example of a classical device. The electric field of the signal drives an oscillating current in a conductor; the action of the mixer is analogous to a voltage controlled switch which conducts whenever this current produces a voltage across the mixer that exceeds some threshold. Examination of Figure 10.1(b) demonstrates that the action of such a switch can generate a heterodyne signal suitable for detection and smoothing, but that the conversion gain can be no more than one and in practical cases conversion will occur only with loss of signal power.

This behavior contrasts with that of the photomixers discussed in Chapter 10. The photomixers are an example of quantum devices, where absorption of a photon frees a charge carrier. After its release, electric fields can do work on the charge carrier, making possible a mixer with conversion gain greater than unity, that is, the output power exceeds the power in signal photons incident on the mixer.

In theory, a junction diode similar to those described in Chapter 5 could be used as a mixer. However, junction diodes have frequency response limited to $\leqslant 1$ GHz by the recombination time required for charge carriers that have crossed the junction (for example, Sze, 1985). For submillimeter- and millimeter-wave receivers, it is necessary to use a device that has been optimized for high frequencies by restricting the junction area and removing the surrounding semiconductor through which charge must diffuse and recombine. Nonetheless, many of the parameters of diode mixers can be described in terms of the derivations in Chapter 5.

Suitable high frequency diodes can be produced at a contact between a metal and a semiconductor, for example, Schottky diodes as already discussed in Chapter 5. Figure 11.4 shows the band diagram of such a contact, where the work function of the metal, W_m, is larger than the electron affinity of the semiconductor, χ; the semiconductor is assumed to be doped n-type. Surface charges will accumulate at the contact to equalize the Fermi levels while maintaining the contact potential. The behavior of this device under forward and reverse biases is illustrated in Figure 11.5. The height of the potential barrier seen from the semiconductor side changes with the direction of the bias, whereas the height is virtually independent of bias as viewed from the

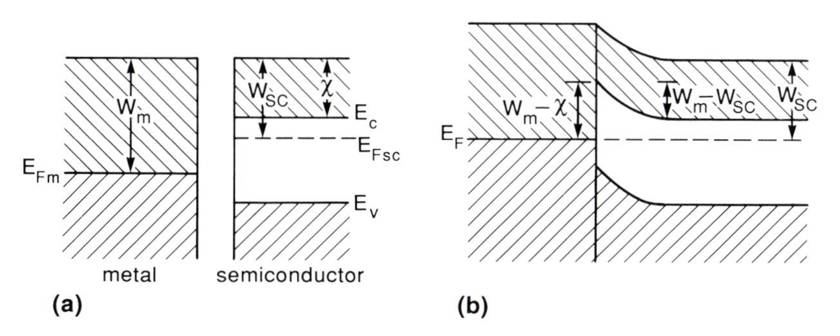

Figure 11.4. Band diagram for a Schottky diode. (a) shows the bands before contact and (b) shows the bands after contact and establishment of equilibrium.

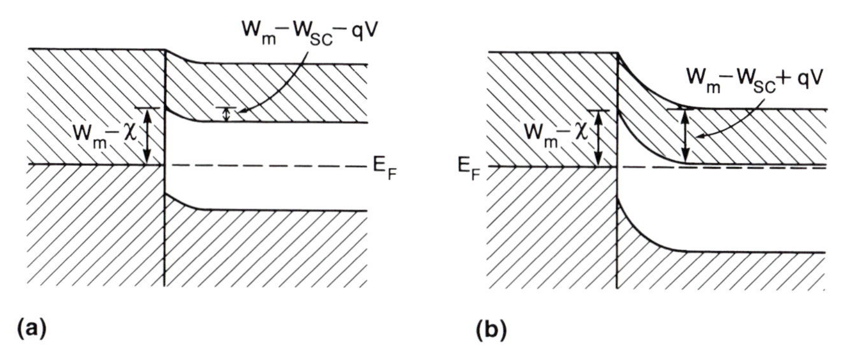

Figure 11.5. The diode action of a Schottky junction: (a) forward-biased; (b) back-biased.

metal side. This asymmetry produces the nonlinear *I–V* curve of the diode. A Schottky diode can also be made with a p-type semiconductor as long as $W_m < \chi + E_g$.

The *I–V* curve of a Schottky diode is described by the diode equation (5.33); any deviations from this behavior can frequently be treated by rewriting the expression as in equation (5.41):

$$I = I_0(e^{qV/mkT} - 1), \tag{11.2}$$

where m is called the slope parameter or 'ideality factor' and I_0 is the saturation current. In most realistic situations, the calculation of the saturation current is complex; however, in the ideal case where it arises purely through thermal emission, it is given by:

$$I_0 = AA^{**}T^2 e^{-q\phi_b/kT}, \tag{11.3}$$

where A^{**} is a constant called the modified Richardson constant, A is the diode area, and ϕ_b is the barrier height as seen from the metal side.

Other properties of Schottky diodes can be derived from our previous discussion of junction diodes by assuming that the metal has extremely large N_A if the semiconductor is n-type or extremely large N_D if the semiconductor is p-type. Then from equation (5.47), the thickness of the potential barrier is

$$w=\left[\frac{2\varepsilon}{qN}(V_0-V_b)\right]^{1/2}, \tag{11.4}$$

where N is the dopant concentration in the semiconductor, ε is its dielectric permittivity, and V_0 is the contact potential. The diode capacitance can be obtained from equations (5.48) and (11.4) to be

$$C_0=A\left[\frac{qN\varepsilon}{2(V_0-V_b)}\right]^{1/2}. \tag{11.5}$$

A large value of m reduces the sharpness of the knee in the diode curve, thus reducing the response of the mixer to the input field. Note that an increase in m is equivalent to an increase in the operating temperature (for the same I_0). Cooling the diode can improve its performance significantly as long as the ideality factor does not change. For a diode described by equation (11.2), the figure of merit in equation (11.1) is $\mathbb{R}\sim 1/mT$. The predicted improvement is seen down to about 70 K. Below this temperature, the ideality factor begins to increase, and below about 30 K there is little performance improvement with reduced temperature. One reason for this behavior is that quantum mechanical tunneling through the diode potential barrier has less temperature dependence than the thermally driven current over the barrier and therefore dominates the behavior at low temperatures. Since tunneling does not have as strong a voltage dependence as the thermal current, the ideality factor increases.

Schottky diodes were originally constructed by pressing a pointed metal wire, called a cat's whisker, against a piece of doped semiconductor. More consistent performance is obtained if the contact is made by depositing metal on the semiconductor in a carefully defined geometry, using photolithographic techniques from integrated circuit technology. Electrical contact to the metal film is made by pressing a thin wire against it.

The critical doping concentration for the Schottky diode need only be maintained in a thin layer of the semiconductor below the contact. The remainder of the semiconductor can be doped to minimize the resistance in series with the diode junction. A cross-section of such a diode is illustrated in Figure 11.6. Here an n-doped layer has been epitaxially grown on a heavily doped substrate and is in turn covered with an oxide layer. The metallization has been applied through a hole in the oxide layer. In the closely related

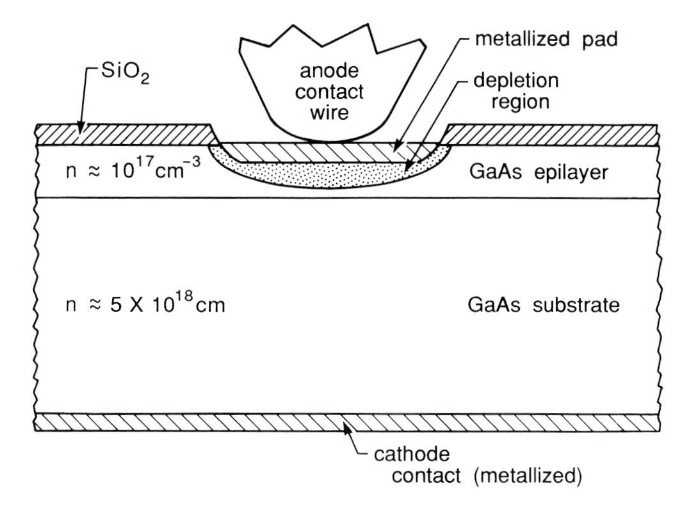

Figure 11.6. Cross-section of a Schottky diode mixer.

Mott diode, the thickness of the doped layer is adjusted so that the depletion region will just reach the conducting substrate at zero bias. Consequently, the contribution to the series resistance from the relatively low conductivity epitaxially grown layer is minimized.

A more thorough discussion of Schottky diodes can be found in Kelly and Wrixon (1980), Phillips and Woody (1982), Schneider (1982), Maas (1986), or Kollberg (1990). Archer (1986) discusses the practical considerations in using these devices in a complete receiver.

11.2.2 Superconductor–insulator–superconductor (SIS) mixers

The second important class of mixer uses the properties of superconductors to produce a strongly nonlinear circuit element. Superconductivity was successfully described by Bardeen, Cooper, and Schrieffer (1957) in what has become known as the BCS theory. A simplistic explanation is that the crystal lattice of the superconductor is deformed by one electron in a way that creates an energy minimum for a second electron of opposite spin, resulting in an attractive force between the two. The bound electrons are referred to as a Cooper pair. Cooper pairs can be broken into two quasiparticles[1] by the addition of an energy of at least 2Δ, where Δ is the binding energy per electron. This situation leads to the concept of a bandgap in the superconductor. The crystal distortions that lead to the binding energy of the Cooper pairs are not fixed to the crystal lattice but can move; if a Cooper pair is given momentum by an applied electric field, there is no energy loss mechanism except for that resulting from the breaking up of the pair, which itself is energetically unfavorable. Cooper pairs can form over relatively large distances. The value of this binding distance (also known as the coherence length) depends on the elemental composition, but it can be of the order of 0.1 to 1 μm. As a result of their large binding distances, all Cooper pairs in a superconductor interact with each other. In a quantum mechanical sense, they all share the same state and can be described in total by a single wave function. Therefore, they all participate in any imposed momentum and can conduct electric currents with no dissipative losses.

In many ways, the bandgap in superconductors is similar to that in semiconductors and insulators, although it is only a few milli-electron-volts wide. Besides the small gap width, the superconductivity gap is distinguished by the very large number of permitted energy states that lie just at the top of the 'valence' and bottom of the 'conduction' bands. This high density of states is possible because Cooper pairs have an integer net spin and can behave as identical particles without having to obey the exclusion principle that governs the behavior of normal electrons. Finally, the energy gap in a superconductor has a strong temperature dependence, varying from zero at the critical temperature (at which the material first becomes superconducting) to a maximum value at a temperature of absolute zero.

When two superconductors are separated by an insulator thinner than the binding distance of a Cooper pair, we have an SIS, or superconductor–insulator–superconductor structure. With no voltage applied across the junction, Cooper pairs can flow from one superconductor to the other to carry small currents (the Josephson effect) because the two superconductors share the same energy states. An applied voltage will shift the energy states, but the insulator will be effective in blocking the flow of additional current as long as the voltage difference is smaller than the energy gap. When the voltage just exceeds the gap, and if the insulator is thin enough, a significant current suddenly becomes possible via quasiparticle tunneling through the insulator.

The current that can be carried by the SIS structure can be understood in terms of the band structure of the superconductor and the process of normal electrons tunneling through an insulator. Referring to Figure 11.7, for simplicity we assume that the superconductors are close enough to absolute zero that the ground states are full and the excited states are empty. We are interested in the flow of electrons from left to right as the voltage across the device is increased (for example, from that in Figure 11.7(a) to that in Figure 11.7(b)). The current is proportional to {the density of filled states in the left side superconductor} times {the density of accessible empty states in the right side superconductor} times {the tunneling probability, P_T}.

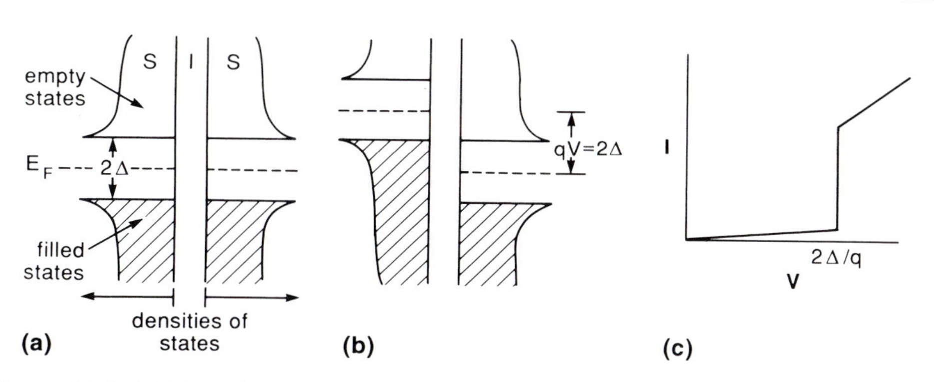

Figure 11.7. SIS junction: (a) shows the band diagram with no applied voltage; (b) shows the band diagram with just enough applied voltage to align the highest energy filled states on one side with the lowest energy empty ones on the other; (c) is the I–V curve.

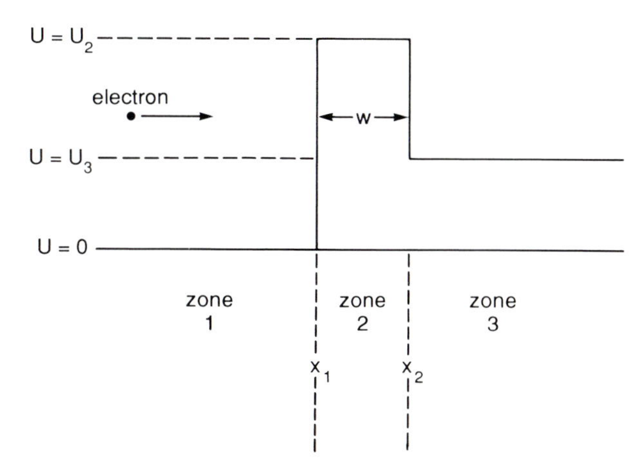

Figure 11.8. Quantum mechanical tunneling through a barrier.

In the simplest model, $P_T = 0$ for $V < 2\Delta/q$. For $V \geqslant 2\Delta/q$, P_T can be estimated from a simple one-dimensional quantum mechanical calculation. We refer to Figure 11.8 and consider an electron incident on the potential barrier from zone 1 to the left. We want to know the probability that the electron will tunnel through the potential barrier and appear in zone 3 on the right.

The behavior of the electron is described in quantum mechanics by the time-independent Schrödinger equation:

$$\left(-\frac{\hbar^2}{2m_e}\frac{\partial^2}{\partial x^2} + U(x)\right)\psi(x) = E\psi(x), \tag{11.6}$$

where $\hbar$ is Planck's constant divided by 2π, m_e is the electron mass, U is the potential energy, E is the total particle energy, and ψ is the wave function.

We will assume that $E < U_2$; otherwise the electron would be able to cross the barrier without tunneling. The solution of equation (11.6) in zone 1 is

$$\psi_1(x) = A\,e^{jk_1x} + B\,e^{-jk_1x}, \tag{11.7}$$

where

$$k_1 = \frac{(2m_eE)^{1/2}}{\hbar}. \tag{11.8}$$

Here, the term with coefficient A represents the electron incident onto the barrier from the left and that with coefficient B represents the electron reflected off the barrier. Similarly, the solution in zone 2 is

$$\psi_2(x) = C\,e^{\kappa x} + D\,e^{-\kappa x}, \tag{11.9}$$

where

$$\kappa = \frac{[2m_e(U_2 - E)]^{1/2}}{\hbar}, \tag{11.10}$$

and in zone 3 the solution is

$$\psi_3(x) = F\,e^{jk_3x}, \tag{11.11}$$

where

$$k_3 = \frac{[2m_e(E - U_3)]^{1/2}}{\hbar}. \tag{11.12}$$

Equation (11.11) is the wave function of an electron traveling to the right after tunneling through the barrier. A, B, C, D, and F are constants that can be related through the boundary conditions – namely, that both $\psi(x)$ and $\partial\psi/\partial x$ should be continuous from one zone to another. We wish to determine the relationship between wave function amplitudes for the input in zone 1 (amplitude $= A$) and the output in zone 3 (amplitude $= F$). This relationship can be shown to be

$$\left|\frac{F}{A}\right| = \frac{4k_1\kappa}{(k_1^2 + \kappa^2)^{1/2}(k_3^2 + \kappa^2)^{1/2}}\,e^{-\kappa w}, \tag{11.13}$$

where w is the width of the barrier and we have neglected terms of order $e^{-\kappa w}$ in comparison with those of order $e^{\kappa w}$. The probability of tunneling is just

$$P_T = \frac{k_3}{k_1}\left|\frac{F}{A}\right|^2. \tag{11.14}$$

The height of the potential barrier is determined by the bandgap of the insulating material and is typically of the order of 1 eV; substituting for $U_2 - E$ in equation (11.10), it can be seen from equations (11.13) and (11.14) that reasonably large tunneling probabilities ($P_T \sim 10^{-6}$) require barrier widths of $w \sim 10^{-9}$ m.

Although the above derivation has been carried out for tunneling by normal electrons, it turns out to be applicable (but not rigorous) for quasiparticles tunneling between superconductors. It is clear from the derivation that P_T is not steeply dependent on the particle energy as long as $E \ll U_2$ (that is, the energy of the particle is well short of allowing it to get over the barrier without tunneling). Therefore, the rate of tunneling will be controlled largely by the density-of-state terms – referring to Figure 11.7, the density of filled 'valence' states in the left side superconductor and of empty 'conduction' states in the right side superconductor. In particular, because the densities of permitted states tend to infinity at the edges of the bands, the tunneling current rises abruptly at the voltage difference across the device that just brings the top of the lower band in the left side superconductor to the energy level of the bottom of the upper band in the right side one. Figure 11.7(c) illustrates this behavior. The sharp inflection in the I–V curve provides the nonlinearity needed for a high-performance mixer.

A number of effects act to round off the abrupt onset of conductivity predicted in the above discussion. For example, as the temperature increases above absolute zero, the bandgap becomes smaller and its effect on the I–V curve becomes less pronounced. In addition, since the probability of excited states follows the Fermi function (equation (3.47)), as the temperature increases an increasing number of electrons will be lifted into the upper energy band, leaving vacancies in the lower one. Consequently, the probability increases with temperature that tunneling will occur between states created by thermal excitation, which is possible without imposing sufficient voltage to align the lower band on one side of the junction with the upper band on the other. As mentioned earlier, even at very low temperatures it is possible for Cooper pairs to tunnel across the insulator when the voltage is below $2\Delta/q$ through the Josephson effect. Finally, various other factors – for example, strain on the material, a dependence of the bandgap on direction in the crystal – can smear out the sharp density-of-state peaks at the superconducting bandgap. Nonetheless, a typical SIS mixer has a far sharper inflection in its I–V curve than does a Schottky diode; fundamentally, this behavior is possible because the superconducting bandgap is of order 1000 times smaller than the bandgaps in semiconductors. The sharp inflection in the I–V curves of SIS devices makes it possible for them to operate with far lower LO power than for Schottky diodes and other nonsuperconducting mixers. This behavior is particularly important at high frequencies, where it is difficult to obtain a large LO power from tunable oscillators.

Although the discussion in the preceding paragraph indicates some of the practical limitations, the mixer performance as measured by the figure of merit in equation (11.1) appears to improve without limit as the curvature of the *I–V* curve is made sharper. However, there is a fundamental limit to this behavior; for a sufficiently sharp inflection in the *I–V* curve, a single absorbed photon causes an electron to tunnel through the energy barrier. In this limit,

$$\mathbb{R} \sim \frac{q}{h\nu}, \tag{11.15}$$

(Hinken, 1989), where q is the charge on the electron and ν is the frequency of the absorbed photon; that is, an input power of $h\nu/\Delta t$ produces a current of $q/\Delta t$. Figure 11.9(a) shows how quantum-assisted tunneling occurs in an SIS junction and the effect on the *I–V* curve is illustrated in Figure 11.9(b).

An SIS mixer operating in a regime where quantum-assisted tunneling occurs becomes a quantum device. Therefore, it can produce conversion gains greater than unity (Tucker, 1979, 1980). The conditions to achieve this state are discussed by Hinken (1989). However, very low noise receivers can also use SIS mixers with gain <1 if quiet HEMT-based IF amplifiers are used with them.

SIS junctions can be made by depositing a layer of superconducting metal such as a lead alloy on an insulating surface (this process takes place in a vacuum), exposing it to air or a controlled oxygen atmosphere so that it oxidizes (to create the insulator layer), and then depositing a second layer of the metal on top of the oxide. It is found that junctions using pure lead for the superconductor are unstable during thermal cycling between room and

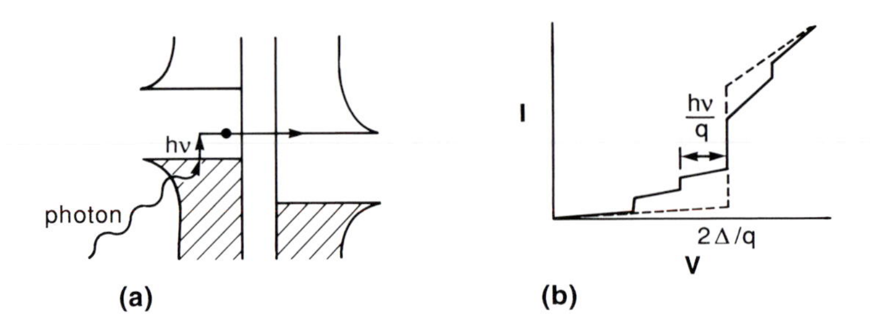

Figure 11.9. Quantum assisted tunneling in a SIS junction: (a) shows the band diagram and (b) shows the *I–V* curve.

operating temperature, but adequate stability can be achieved by the addition of small amounts of indium or bismuth. Standard photolithographic techniques are used to confine the overlap (junction) area to be small so that the frequency response of the finished device is acceptable. Junctions can also be made of niobium, where the insulator is Al_2O_3 or some other metal oxide deposited in a vacuum. Niobium junctions are now used to the exclusion of other types because the niobium is more chemically inert than lead, making the junction more durable. Additional discussion of the fabrication of these devices can be found in Dolan, Phillips, and Woody (1979), Richards *et al.* (1979), Huggins and Gurvitch (1985), and Gundlach (1989).

Our discussion of the method of operation of the SIS junction has been brief and nonrigorous. A more complete treatment of the background physics can be found in Solymar (1972), while Hinken (1989) provides a review of the operation as mixers. Further exploration of the theoretical performance of these devices can be found in Tucker (1979, 1980) and Tucker and Feldman (1985). Phillips and Woody (1982) and particularly Archer (1986) provide useful descriptions of practical considerations in using SIS junctions in receivers. The advantages of SIS mixers are discussed by Phillips and Woody (1982). The primary disadvantage is the necessity to operate at a temperature near absolute zero. The SIS mixer is now the standard for operation whenever maximum performance in the submillimeter- and millimeter-wave regions is worth the expense of providing very low temperature operation.

11.2.3 Other quasiparticle tunneling mixers

A variety of other mixers operate by quasiparticle tunneling; their operational principles are therefore similar to those just discussed in Section 11.2.2 for SIS junctions. Two examples are superconductor–insulator–normal metal, or SIN, junctions and closely related devices where the metal is replaced by heavily doped semiconductor; these superconductor–degenerate semiconductor junctions are known as super Schottky diodes.

11.2.4 Josephson junctions

Mixers can be built making use of the Josephson effect – the tunnelling of Cooper pairs. Although successful operation has been reported by a number of investigators, the devices are intrinsically noisy. In addition, they tend to be very fragile, with unstable and nonrepeatable behavior. Given the excellent performance of SIS mixers, Josephson junction mixers have found little favor.

11.2.5 Hot electron bolometers

The hot electron bolometer described in Chapter 9 is a relatively easily constructed detector for the submillimeter- and millimeter-wave regime that can be adapted for use as a mixer. In this case, the photons are absorbed directly by the mixer element, as was the case for the infrared heterodyne mixers in Chapter 10; in many ways, the bolometer mixer is more closely analogous to those devices than to the ones discussed in this chapter. To minimize the volume of the bolometer element, it is usually made long and thin and placed across a waveguide or other feed arrangement, much as a secondary antenna might be for some other mixer type. The speed of the bolometer is limited by the hot electron relaxation time of $\sim 10^{-7}$ s, so the achievable IF bandwidths are at best a few megahertz. To achieve adequate spectral range, it is common to sweep the LO frequency and to sample one frequency at a time, thus losing the spectral multiplex advantage available with faster mixers whose output can be divided spectrally by a bank of electronic filters or other dispersive device. Fortunately, the mixer achieves low noise temperatures with the use of modest ($<1\,\mu$W) LO power. A description of a hot electron bolometer receiver can be found in Phillips and Jefferts (1973).

11.3 Performance characteristics

11.3.1 Noise limits

Submillimeter- and millimeter-wave mixers are subject to fundamental noise limits similar to those already discussed for infrared heterodyne receivers. In this case, the quantum limit arises from the discreteness of electric charges driven through the mixer by the LO current shot noise. From equation (5.3), this noise current is

$$\langle I_S^2 \rangle = 2qI_L \Delta f_{IF}, \tag{11.16}$$

where I_L is the current from the local oscillator. The resulting quantum limit noise temperature is given for double sideband operation approximately as in equation (10.35) with $G = a = \eta = 1$, that is,

$$T_N \approx \frac{h\nu}{k}. \tag{11.17}$$

For single sideband operation,

$$T_N \approx \frac{h\nu}{k \ln 2}. \tag{11.18}$$

In the thermal limit, the noise temperature is given by equation (10.34).

Mixers and their attendant components are also subject to excess noise. The noise temperature of the mixer is defined in terms of an imaginary resistor R_N placed across the input to the mixer and having a resistance equal to the effective resistance of the mixer. For example, refer to the simple equivalent circuit of the Schottky diode shown in Figure 11.10, where C_0 is from equation (11.5), R_S is the series resistance of the semiconductor substrate on which the diode is grown, and R_0 is the nonlinear equivalent resistance of the diode. Then $R_N = R_S + R_0$. The mixer noise temperature, T_M, is the temperature of R_N such that its Johnson noise (equation (3.42)) would be equal to the observed mixer noise. Particularly for a cooled mixer, we can take $R_S \ll R_0$. From equation (11.2),

$$R_0 = \left(\frac{dI}{dV}\right)^{-1} = \frac{mkT}{qI_0(e^{qV/mkT})} \approx \frac{mkT}{qI}. \tag{11.19}$$

We then have

$$\langle I_S^2 \rangle = 2qI_L\, df = \langle I_J^2 \rangle = \frac{4kT_M\, df}{R} \approx \frac{4qIT_M\, df}{mT}, \tag{11.20}$$

or

$$T_M = \frac{\langle I_S^2 \rangle mT}{4qI\, df} = \frac{mT}{2}. \tag{11.21}$$

Equation (11.21) can be compared with equation (11.1) and the discussion following equation (11.5) to show that the noise temperature of the diode is inversely proportional to the degree of nonlinearity of the I–V curve of the device as represented by the figure of merit $\mathbb{R}$. The noise temperatures of the mixer and of other components can be added (after suitable multiplication by losses and division by gains) to compute an overall system noise temperature, as in equation (10.43).

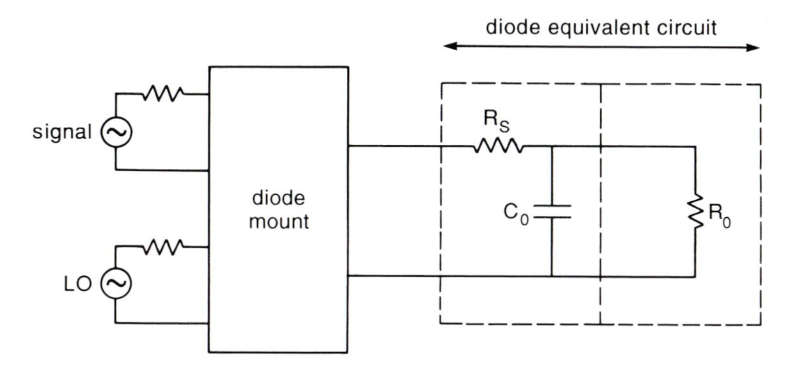

Figure 11.10. Simplified equivalent circuit for a diode mixer.

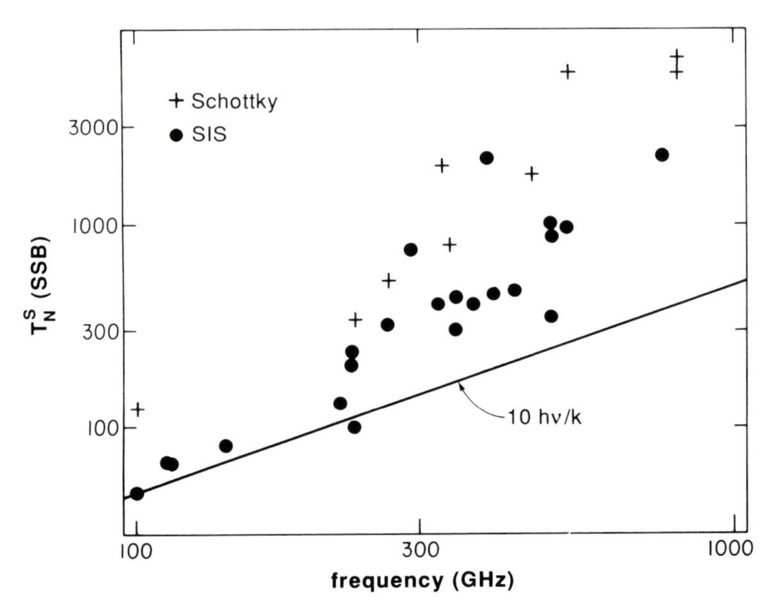

Figure 11.11. Achieved performance of high frequency heterodyne receivers compared with ten times the quantum limit.

In the millimeter-wave range, existing receivers approach the quantum limit to within a factor of ten; the achieved noise temperatures gradually increase relative to the quantum limit as the frequency is increased above ~200 GHz, as shown in Figure 11.11. At least in the millimeter-wave region, the mixers themselves can be produced with performance close to the quantum limit (see, for example, Mears *et al.*, 1991); most of the loss in performance comes from the difficulties in coupling the signal into the mixer. Therefore, continued improvement toward the quantum limit should be expected.

11.3.2 Frequency range

Again referring to Figure 11.10, the mixing process in a Schottky diode occurs in R_0; R_S and C_0 are parasitic elements that interfere with the operation of the mixer. As this circuit is drawn, it emphasizes that R_S and C_0 act as a low-pass filter on the mixer element and hence control the high frequency limit of the receiver. The cutoff frequency of the mixer is

$$f_c = \frac{1}{2\pi R_S C_0}, \tag{11.22}$$

which shows that we want to minimize both R_S and C_0. In addition, the series resistance R_S must be kept small to avoid resistive losses.

To minimize R_S, we see from equations (3.1), (3.5), and (4.5) that we would like to increase the doping concentration N. From equation (11.5), however, it is apparent that doing so increases the capacitance. Thus, the best we can do to control R_S is to select a semiconductor with high mobility (see equation (3.5)). The mobility of GaAs is over six times that of silicon for equivalent doping levels (see Table 3.1); hence, high frequency diodes are normally fabricated on GaAs (InP also has high mobility and is a useful alternative material to GaAs). Given the doping concentration, C_0 can be minimized by reducing the diode area. However, from equations (5.33) and (5.39), R_0 increases as the diode area, A, decreases. A compromise that also takes fabrication difficulties into account is to make diodes with diameters of about 2 μm, although smaller diodes (down to $\sim 0.5\ \mu$m) are used in the submillimeter. The performance of these mixers is seriously degraded above a few times 10^{11} Hz by their parasitic capacitances and the resulting RC time constants.

When Schottky diodes are not cooled, the thermal generation of charge carriers can reduce R_S sufficiently that the frequency response is limited by the electron transit time across the junction, rather than by the RC time constant of the mixer.

In the case of SIS mixers, the device leads are superconducting. Although they have no resistance to direct currents, they will have some dissipative losses at the high AC frequencies of interest; moreover, there are likely to be some normal conductors in the current path. Hence, we cannot set R_S quite to zero, but it will be much smaller than for a Schottky diode. These devices are therefore well adapted to high cutoff frequencies as long as the SIS junction area is small enough that it does not have an overly large capacitance (see Problem 11.2).

In addition to issues of fabrication and junction size, the superconducting material for the SIS junction must be selected to allow operation at the desired frequency. If an alternating current is applied at such a high frequency, f, that the photon energy, hf, exceeds 2Δ, then the current can break the Cooper pairs; the response of the mixer falls rapidly (as $1/f^2$) with increasing frequency above this limit. This energy limit is characterized by a frequency $f_u = 2\Delta/h$; for example, for lead ($\Delta = 2.7$ meV), $f_u = 6.5 \times 10^{11}$ Hz. Slightly higher frequencies can be reached with niobium junctions because of their 15% larger bandgap. Some very high frequency mixers operate the SIS junction over a voltage range of $\pm 2\Delta/q$, to achieve upper frequency limits approaching $2f_u$. These devices must be operated in a strong magnetic field to suppress Josephson tunneling near zero voltage (Walker *et al.*, 1992); otherwise, this effect imposes an unstable operating regime which dramatically increases the noise. Many mixers avoid the complication of applying magnetic fields and must operate below f_u. The bandwidths of these devices are further reduced by the necessity to avoid the low voltage regime of Josephson tunneling.

11.3.3 Conversion loss and gain

A critical parameter for a mixer is the conversion gain Γ_C; its inverse is the conversion loss, $L = 1/\Gamma_C$. If L is large, then equation (10.43) shows that an extremely quiet amplifier is required so that T_A does not dominate the system noise temperature. In this section, we expand on the previous discussion of this parameter, using the case of a Schottky diode mixer as an example.

Assume that the single sideband signal power into the mixer is P_S and the IF power out of the mixer from this input is P_{IF}. Then the conversion loss for the signal is

$$L_S = \frac{1}{\Gamma_{CS}} = \frac{P_S}{P_{IF}}. \tag{11.23}$$

The conversion loss is often quoted in decibels (db), where

$$L_S(\text{db}) = 10\log(L_S) = 10\log\left(\frac{P_S}{P_{IF}}\right). \tag{11.24}$$

A similar pair of definitions holds for the conversion loss for the image, L_I. If equal power is available from both signal and image, the double sideband conversion loss of the mixer, L, is given by

$$\frac{1}{L} = \frac{1}{L_S} + \frac{1}{L_I}. \tag{11.25}$$

To compute mixer conversion losses, it is necessary first to go back to the discussion of basic mixer performance (Section 10.1) and describe in further detail the behavior of a high frequency mixer. In Chapter 10, we discussed the simple situation where the output contains the input frequencies and the beat frequency. However, harmonics of the local oscillator, $n\omega_{LO}$ (where $n = 1, 2, 3, \ldots$), will also be present as well as the 'sum frequency' $\omega_{LO} + \omega_S$. All of these signals can beat with each other, producing a highly complex output signal as indicated in Figure 11.12. Only the IF signal is useful for the detection process; power from the incoming photon stream that is used in generating signals at the other frequencies is lost and reduces the conversion efficiency of the mixer.

To illustrate the calculation of the conversion loss of a Schottky diode, we follow Torrey and Whitmer (1948) (hereafter TW), to which the reader is referred for the full derivation. We will consider the conversion loss for the

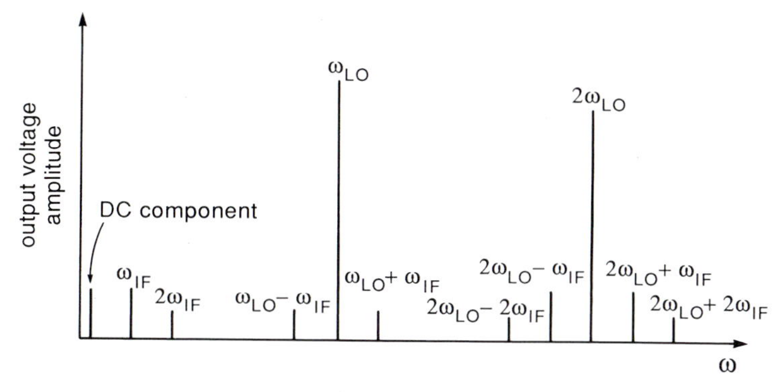

Figure 11.12. Signals present in a mixer.

single sideband case, L_S. At high operating frequencies, the mixer capacitance and other stray capacitances will tend to short-circuit the higher harmonics; see Figure 11.10 and equation (11.22). Therefore, the reduction in conversion efficiency through generation of unwanted signals can be simplified by considering only the strongest of these signals, the image sideband at $\omega_I = \omega_{LO} - \omega_{IF}$. This signal is generated as a result of the beat of the signal at ω_S with the second harmonic of the LO at $2\omega_{LO}$ as well as the beat between ω_{IF} and ω_{LO}. With this simplification, electrical signals need be considered at three frequencies, ω_S, ω_I, and ω_{IF}. We further assume that the signals at these frequencies are sufficiently small that the relevant voltages and currents can be related linearly, so the mixer can be regarded as a linear network with separate terminals for each of these three frequencies. The terminals then are connected to each other through the network by a set of complex impedances; for example, $Z_{S,I}$ is the impedance between the signal and image terminals.

If the complex current amplitudes at the three frequencies are designated by i_S, i_I,, and i_{IF} and the complex voltage amplitudes by e_S, e_I, and e_{IF}, then, by our assumption of linearity, the currents and voltages are related by three simultaneous linear equations:

$$\begin{aligned} i_S &= y_{S,S}\, e_S + y_{S,IF}\, e_{IF} + y_{S,I}\, e_I^*, \\ i_{IF} &= y_{IF,S}\, e_S + y_{IF,IF}\, e_{IF} + y_{IF,I}\, e_I^*, \\ i_I^* &= y_{I,S}\, e_S + y_{I,IF}\, e_{IF} + y_{I,I}\, e_I^*, \end{aligned} \tag{11.26}$$

where the y's are the complex admittances, defined as the inverse of the impedance, $y = 1/Z$, and the * indicates complex conjugation (see TW for an explanation of why the image current and voltage appear as complex conjugates). Although we will not pursue this point, it is often convenient to use the formalism of linear algebra for analysis of mixer performance, in which case equations (11.26) are expressed in matrix form:

$$\begin{vmatrix} i_S \\ i_{IF} \\ i_I^* \end{vmatrix} = Y \begin{vmatrix} e_S \\ e_{IF} \\ e_I^* \end{vmatrix}, \tag{11.27}$$

where Y is the admittance matrix,

$$Y = \begin{vmatrix} y_{S,S} & y_{S,IF} & y_{S,I} \\ y_{IF,S} & y_{IF,IF} & y_{IF,I} \\ y_{I,S} & y_{I,IF} & y_{I,I} \end{vmatrix}. \tag{11.28}$$

Without significant loss of generality, it is possible to simplify equations (11.26) and replace the nine complex admittances with five real conductances:

$$\begin{aligned} i_D &= g_{S,S}\, e_S + g_{S,IF}\, e_{IF} + g_{S,I}\, e_I^* \\ i_{IF} &= g_{IF,S}\, e_S + g_{IF,IF}\, e_{IF} + g_{IF,S}\, e_I^* \\ i_I^* &= g_{S,I}\, e_S + g_{S,IF}\, e_{IF} + g_{S,S}\, e_I^*. \end{aligned} \tag{11.29}$$

According to equations (11.29), the mixer can be treated as a resistor network as illustrated in Figure 11.13. TW show how all the relevant mixer equivalent network resistances can be measured in the laboratory. If we attach a source of alternating current with amplitude A and conductance g_A to the signal terminals of the mixer, then the available signal power, defined as the power delivered into a matched load, is

$$P_S = \frac{A^2}{8g_A}. \tag{11.30}$$

The power at the output will be

$$P_{IF} = \frac{i_{IF}^2}{8g_{IF}}, \tag{11.31}$$

and the conversion loss will be given, as in equation (11.23), as P_S/P_{IF}. TW show that

$$g_{IF} = g_{IF,IF} - \frac{2g_{S,IF}\, g_{IF,S}}{g_{S,S} + g_{S,I} + g_A}, \tag{11.32}$$

and

$$i_{IF} = \frac{A g_{IF,S}}{g_{S,S} + g_{S,I} + g_A}. \tag{11.33}$$

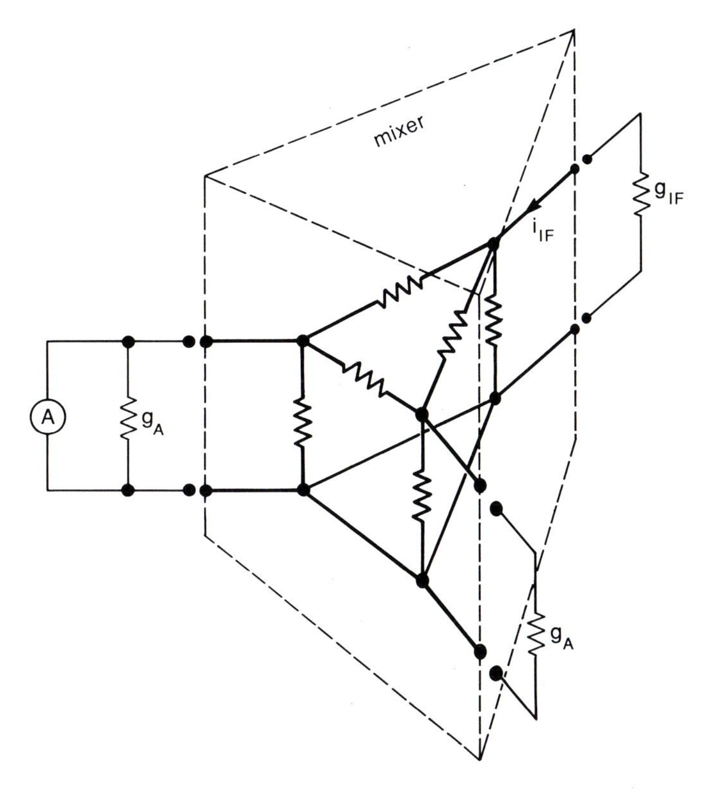

Figure 11.13. Linear network model of a diode mixer.

From equation (11.23), we then obtain

$$L_S = \frac{2g_{S,IF}}{g_{IF,S}} f(x), \tag{11.34}$$

where

$$f(x) = \frac{(x+a)(x+a-b)}{bx}, \tag{11.35}$$

and

$$a = 1 + \frac{g_{S,I}}{g_{S,S}}$$

$$b = \frac{2g_{S,IF}\, g_{IF,S}}{g_{S,S}\, g_{IF,IF}} \tag{11.36}$$

$$x = \frac{g_A}{g_{S,S}}.$$

So long as $0<b<a, f(x)$ has a unique minimum at

$$x_0=[a(a-b)]^{1/2}, \tag{11.37}$$

and the minimum conversion loss is

$$L_S(\text{min})=2\frac{g_{S,IF}}{g_{IF,S}}\frac{1+(1-b/a)^{1/2}}{1-(1-b/a)^{1/2}}. \tag{11.38}$$

The smallest loss occurs when $a=b$. It is usually true that $g_{S,IF}=g_{IF,S}$, in which case the loss can never be less than two:

$$L_S(\text{min})\geqslant 2\rightarrow 3\,\text{db}. \tag{11.39}$$

This loss occurs because half the power is converted to the image frequency (where it is of no use in the detection process) along with half to the IF. Because of this behavior, useful Schottky diode mixers always deliver *less* power to the IF stage than the signal power into them. With large LO powers, these mixers can be driven very close to the minimum 3 db loss. Nonetheless, unlike quantum devices (such as photo- and SIS mixers) that can provide gain, Schottky diode mixers are useful only because they reduce the signal frequency to a range where very low noise amplifiers and other circuitry can be used.

11.4 Local oscillators

High frequency local oscillators can be built around electronic components with negative resistance; the operation of such circuits will be illustrated through the case of the commonly used Gunn oscillator.

11.4.1 Gunn oscillator

A detailed description of the Gunn effect can be found in Streetman (1990). Consider the band diagram for n-type doped GaAs drawn in Figure 11.14. In previous cases, we have used bandgap diagrams which showed the maximum energy level in the valence band and the minimum energy level in the conduction band plotted versus a spatial dimension. In discussing indirect transitions in Chapter 1, we referred to the issue of band structure as a function of quantum mechanical wave vector; Figure 11.14 shows this behavior explicitly by plotting energy versus the electron propagation vector k (for example, equation (11.8)).

In Figure 11.14, the valence band is full, and, because the material is n-type, conduction will be through electrons excited into the conduction band. If the

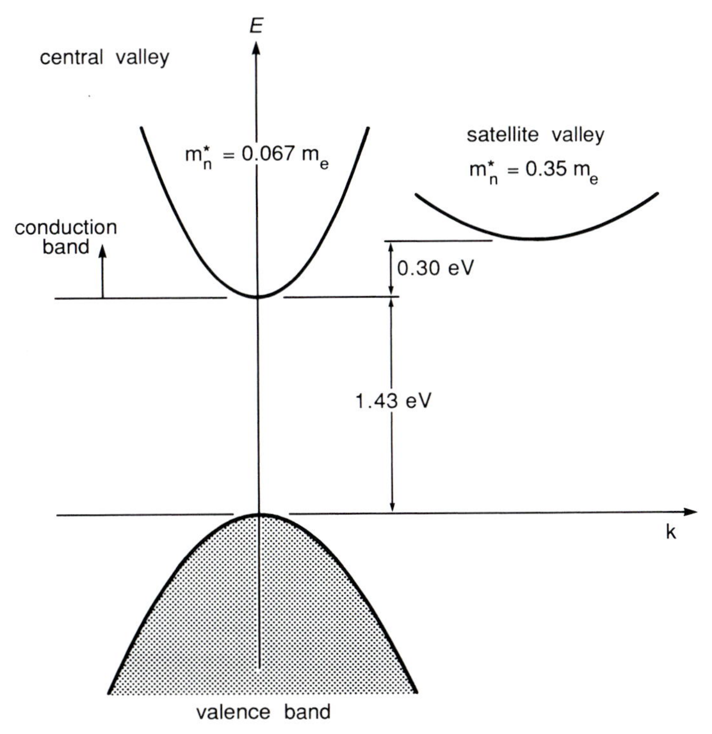

Figure 11.14. Band diagram of GaAs.

electric field is small (for this material, less than $\sim 3000\,\mathrm{V\,cm^{-1}}$), the conduction electrons will be excited into the central valley in the conduction band at $k=0$, and the material will behave as other semiconductors we have discussed previously. If, however, the electric field is high enough (greater than $\sim 3000\,\mathrm{V\,cm^{-1}}$), the electrons gain enough energy that they can be scattered into the satellite valley. The satellite valley differs from the central valley in two important ways. First, the effective density of states is about 70 times higher, so once the electrons reach the satellite valley they tend to stay there as long as the electric field is sufficiently large. Secondly, the electron mobility is much lower in the satellite than in the central valley. From equations (3.3) and (3.6), it can be seen that the current decreases when the mobility is reduced. Thus, if the electric field exceeds a critical value, the differential conductivity of the material, $\mathrm{d}J/\mathrm{d}\mathscr{E}$, is negative, as is its differential resistance (for relations among these quantities, see equations (3.1)–(3.6)).

If a space charge perturbation is imposed within a semiconductor, it normally dies out exponentially. We considered just this situation in the discussion of dielectric relaxation in Chapter 3 (see equation (3.32)):

$$N_\mathrm{p}(t) = N_\mathrm{p}(0)\, e^{-t/\tau_\mathrm{d}}, \tag{11.40}$$

where

$$\tau_d = \frac{\varepsilon}{\sigma}. \tag{11.41}$$

From equations (11.40) and (11.41), we see a profound effect resulting from negative conductivity: a space charge fluctuation (such as might occur through random fluctuations in the charge carrier distribution) will *grow* exponentially.

We can now follow the operation of a Gunn-effect device as shown in Figure 11.15. A voltage is placed across the device that exceeds the critical electric field (see Figure 11.15(a)). A dipole space charge distribution forms at a nucleation site, such as a crystal defect, or at the negative electrode itself (see Figure 11.15(b)). This dipole drifts toward the positive electrode as part

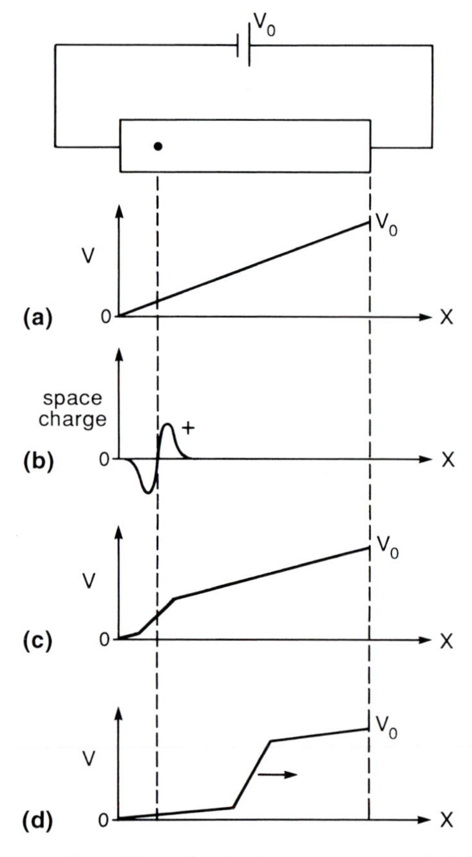

Figure 11.15. The Gunn effect. The physical arrangement is indicated at the top: (a) shows the application of a voltage that exceeds the critical electric field; (b) and (c) illustrate the growth of a space charge dipole at a nucleation site and the resulting voltage profile; and (d) shows the voltage profile as this dipole drifts toward the positive electrode.

of the overall current being carried by the device; as it drifts, however, it grows in strength exponentially – the drifting dipole zone is called a domain. This domain takes most of the voltage drop across the device, reducing the field in the rest of the device below the critical value and suppressing the formation of additional domains (see Figures 11.15(c) and 11.15(d)). This situation holds until the domain has drifted to the positive electrode, where it is collected as a current pulse. The situation is then returned to that in Figure 11.15(a), and the sequence repeats.

The simple Gunn-effect device[2] described above could be used as an oscillator, but it would offer little flexibility in frequency of operation. It is possible to obtain greater control and to extend the performance to higher frequencies by using the negative resistance characteristics provided by this effect in a resonant circuit. In fact, in a common operating mode, *limited space charge accumulation*, the voltage across the Gunn device is varied so rapidly that domains have no opportunity to form, and only the negative resistance is of interest.

To illustrate the application of negative resistance in an oscillator, consider the circuit in Figure 11.16. Starting from point **a** and continuing clockwise around the circuit adding the voltages, we find

$$L\frac{d^2q}{dt^2}+R\frac{dq}{dt}+\frac{q}{C}+V=0. \tag{11.42}$$

Here, L is the inductance of the coil, R the resistance of the resistor, C the capacitance of the capacitor, and V the voltage of the DC voltage supply. Assuming that we start this system with an excess charge of q_i on the capacitor, it can be shown that the charge on the capacitor is

$$q_c=q_i\,e^{-Rt/2L}\cos(\omega t)-VC, \tag{11.43}$$

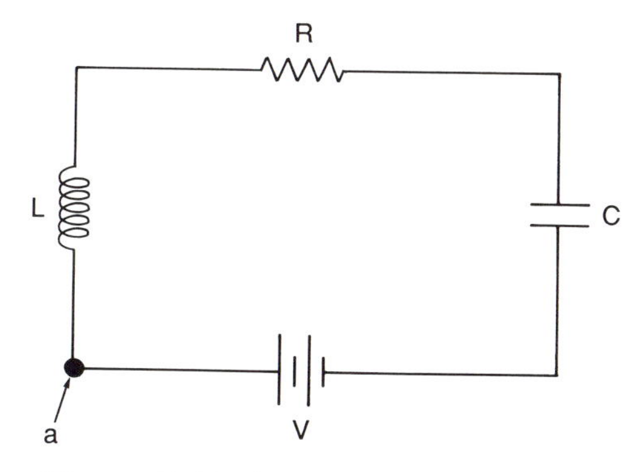

Figure 11.16. A simple oscillator circuit.

where

$$\omega = \left[\frac{1}{LC} - \left(\frac{R}{2L}\right)^2\right]^{1/2}. \tag{11.44}$$

Therefore, if $R<0$, an oscillatory voltage builds up in the circuit until its amplitude is limited by dissipation in elements with positive resistance or in some other way.

Gunn oscillators are convenient LO sources over roughly the 30–150 GHz frequency band, where they can produce powers of ~100 mW. Up to 100 GHz, Gunn devices are fabricated on GaAs; operation at higher frequency can be obtained by making similar devices on InP. This latter material has a band structure like that in GaAs, but it has a shorter time constant for electrons in the central valley of the conduction band to gain or lose energy, resulting in the higher frequency of operation. Gunn oscillators can be readily tuned to the desired frequency and have a long operating life.

11.4.2 Other local oscillators

Impact avalanche transit time (IMPATT) diodes (Streetman 1990) can be used in a similar fashion to Gunn diodes in the construction of high frequency oscillators. IMPATT diode oscillators can operate from roughly 30 to 300 GHz, with power outputs of ~100 mW.

Two functions are combined in the construction of an IMPATT diode: (1) a junction which can be back-biased to provide large localized field strengths; and (2) a drift region with low doping through which charge carriers travel to be collected at an electrode. The diode is operated with a large DC back-bias, which generates a field just under breakdown across the junction. A rapidly oscillating AC voltage is imposed on top of the bias. When the AC voltage increases the back-bias, the junction breaks down due to avalanching of the charge carriers and a large pulse of charge carriers is injected into the drift region. The finite time required for this pulse to cross to the other side and reach the electrode imposes a delay on the output current pulse. The diode is manufactured with a drift region thickness so that this delay is half the period of the AC voltage oscillation. Thus, the current pulse arrives 180° out of phase with the AC voltage; the device conducts a positive-going current when the AC voltage goes negative. This behavior provides a negative resistance at the oscillating frequency, and the resulting instability can drive the AC voltage as we illustrated for the Gunn device.

Traded against their better high frequency limit, IMPATT diode oscillators are much noisier than those using Gunn devices because their operation depends on avalanching. In addition, they are more difficult to tune in frequency because of the specific frequency-dependent adjustments that must be made in their design.

A number of vacuum-tube devices can also provide high frequency local oscillators, such as klystrons and backward wave tubes. However, these tubes are expensive and have relatively short operating lifetimes, so they are increasingly being replaced with solid state devices.

11.4.3 Frequency multiplication

When very high LO frequencies are required, a Gunn or similar oscillator can be used to drive a nonlinear circuit element such as a Schottky diode. The resulting waveform deviates substantially from a sine wave and hence contains significant overtones at higher frequencies than the fundamental of the oscillator. This technique can provide power up to ~ 1000 GHz, but only at levels of 10^{-6} to 10^{-4} W. It is therefore well matched to SIS mixers but generally inadequate to approach quantum-limited performance with Schottky diode mixers.

11.5 Problems

11.1 Use the boundary conditions stated in the text to derive equation (11.13).

11.2 Assuming an insulator thickness of 2×10^{-9} m and an effective series resistance of $R_S = 50\,\Omega$, taking the dielectric constant of PbO to be 26, compute the frequency cutoff of a Pb–PbO–Pb SIS mixer with an area of 1 μm^2. Assuming the series resistance is not associated with the SIS device itself, is there a way to increase the high frequency cutoff (and also increase the saturation voltage) without having to develop smaller devices or making the insulator layer thicker?

11.3 Derive equation (11.43).

11.4 Suppose the output of the receiver analyzed at the end of Chapter 10 is carried by a transmission line with loss of 1 db and at a physical temperature, T, of 290 K. Derive the general expression for the noise temperature of a transmission line:

$$T_N = (L - 1)T,$$

and use it to show that the receiver noise temperature is increased by

$$\Delta T_{\mathrm{N}} = (L-1)(T + T_{\mathrm{N}}^{\mathrm{S}}),$$

where $T_{\mathrm{N}}^{\mathrm{S}}$ is from equation (10.43). Apply this result to correct the noise temperature derived in the example. Hint: Start by using equation (3.42) to show that the maximum power that can be extracted from a resistor at temperature T is

$$P_{\max} = kT\,\mathrm{d}f.$$

11.5 Consider a Schottky diode mixer similar to that in Figure 11.6, with a diameter of 2 μm and a thickness for the epitaxial GaAs layer of 0.5 μm. Assume it is operated at room temperature and at a voltage of $(V_0 - V_b) = 5$ V. Take the mobility of the GaAs to be as in Table 3.1 and the saturation drift velocity of electrons in this material to be $10^7\,\mathrm{cm\,s^{-1}}$. Determine the cutoff frequency and what mechanism determines it.

Notes

1 Unpaired electrons cannot take part in the superconducting behavior and are referred to as normal electrons to distinguish them from the Cooper pairs and quasielectrons, or quasiparticles as they are usually called. The term quasiparticle is applied when the electrons are not completely free of each other and may still exhibit correlated behavior.
2 These devices are commonly called Gunn diodes, although they do not contain a junction. The terminology originates in the fact that they are two-terminal devices and distinguishes them from transistors, which require three terminals.

12

Summary

In Chapter 1, we listed three fundamental types of photon detector, and a variety of characteristics that would help define the applications for which a detector is suited. Since then, we have introduced a vast profusion of detectors with a chaotic variety of performance characteristics. We will now return to the basic detector characteristics to examine them in the light of the potential applications of the detectors we have discussed. In comparing different detector systems, a useful figure of merit is the speed, that is, the inverse of the time required for a system to make a given measurement. In selecting a detector system, the general considerations discussed below can be combined with the measurement requirements for a given situation and the characteristics of competing detector systems to estimate relative speeds of these systems, leading to selection of an optimum approach.

12.1 Quantum efficiency and noise

Two regimes must be distinguished in discussing the effect of quantum efficiency in choosing a detector: (1) photon-noise limited and (2) all other cases. In the first regime, the speed of any incoherent detector is proportional to the detective quantum efficiency.[1] Consequently, detectors with very high DQEs such as bolometers and photodiodes are favored for photon-noise-limited applications.

In the second regime, only if the DQE includes the effect of the total system noise is it a useful parameter. In most cases, the quantum efficiency is quoted for the detector alone. Then, if the limiting noise is the amplifier, the speed goes as the square of the detector responsivity. This situation puts a premium on detector types with gain, such as avalanche diodes, Si:As BIBs, and conventional bulk photoconductors (under operating conditions where their photoconductive gains are large). Photoemissive detectors are also favored because they can be integrated with very low noise amplifiers. If the limiting noise is Johnson noise in the detector, the speed (at a given operating temperature) goes as the detector resistance. No particular detector type is favored, but the detector design and construction must maximize this parameter.

12.2 Linearity and dynamic range

All detectors have some degree of nonlinearity over their useful dynamic range. For most electronic detectors, the effects are modest and highly reproducible, so they can be calibrated and removed from the data with high accuracy. Bolometers by their nature have huge dynamic ranges but can be sufficiently nonlinear that accurate calibration is difficult over the full range. The accuracies in measurement of illumination levels achievable with photographic plates are limited by the nonlinearities of the detector, the variations from sample to sample, and the inability to reset the detector used for the measurement and calibrate it directly. Therefore, photography is the only detection method we have discussed which is limited inherently to modest accuracies. Under some circumstances, however, achieving high accuracies with other detector types can require complex calibration procedures – correcting for source induced backgrounds in image intensifiers is an example. The effects of linearity and dynamic range can be taken into account by estimating the time required for calibration of a given detector to the desired degree of accuracy. Calibration will be much more laborious whenever significant corrections must be made for nonlinearities or multiple observations are required because of limited dynamic range.

12.3 Number of pixels

We assume that spatial positions can be measured either with a single positioning of the image on a large areal array or by multiple positionings on a smaller array with negligible loss of time for moving the image. Then, when many spatial positions need to be observed, the speed for a given total set of measurements scales in proportion to the number of picture elements. Assuming operation in the photon-noise-limited regime, the speed also scales with the detective quantum efficiency. These two relations give a useful guide for selection of an optimum detector type; all other things being equal, the speed of given measurement will be the same for detector systems having equal products of quantum efficiency and number of pixels. This rule accounts for the usefulness of large format detectors of modest quantum efficiency – for example, photographic plates, PtSi arrays.

Different types of electronic detector lead to differing degrees of difficulty in constructing large format arrays. In the ultraviolet, visible, and very near infrared spectral regions, the use of intrinsic silicon detectors, which can be built on the identical structures that are used for high performance readouts, leads to large format arrays with very good low light level performance. A variety of photoemissive detectors can also provide large formats, low noise, and fast response in this spectral region with a modest sacrifice in quantum efficiency compared with silicon-based devices. In the infrared from 1 to 40 μm, the detectors are processed separately from the silicon-based readouts and the two are mated; this construction places an upper bound on the number of pixels that can be provided in an array. From 40 to 200 μm, although high performance germanium photoconductors are available, their properties increase the difficulties of array construction, and as yet there is no well developed array technology. Most bolometric detector types and heterodyne receivers do not lend themselves readily to construction of large format arrays; modest array sizes or single detectors are used in the submillimeter- and millimeter-wave regions and elsewhere when these detector types are employed.

12.4 Time response

In general, we have found that detector optimization for low frequency signal to noise requires sacrifice of rapid time response. For example, the minimization of Johnson noise in both photoconductors and photodiodes requires maximization of the detector resistance and increase of its RC time constant. Bolometer S/N is improved by reducing the thermal conductance to the heat sink, which also slows the detector response.

Surprisingly, over much of the millimeter-wave to ultraviolet spectral range, detectors are available with rapid time response that impose only modest sacrifice in S/N at low light levels. Heterodyne detectors respond at the IF frequency bandwidth, providing for detection in the nanosecond range for the submillimeter- and millimeter-wave regions. This performance is subject to the usual restrictions for these detectors of operation in narrow spectral bands and receiving only a single mode of the signal. The solid state photomultiplier can give response in the nanosecond to microsecond range from the near infrared out to wavelengths of $\sim 27\,\mu$m (albeit with relatively poor quantum efficiency between 1 and $\sim 5\,\mu$m); the dark pulse rate of these detectors corresponds to a dark current only slightly higher than obtained with conventional photoconductors. At wavelengths shorter than 1 μm, photomultipliers can provide nanosecond response times while counting single photons, with a sacrifice of a factor of three to five in quantum efficiency compared with solid state detectors. Only in the far infrared ($\lambda > 27\ \mu m$) and high frequency submillimeter regions do we have to tolerate serious loss of S/N to achieve high frequency response.

The achievable time response degrades with areal detectors. At wavelengths short of 1 μm, various forms of image intensifier can operate with time response in the millisecond range. At longer wavelengths, there is as yet no satisfactory solution to multiplexing solid state photomultipliers, but appropriately designed infrared arrays can be read out in a millisecond or less.

12.5 Spectral response and bandwidth

With the exception of bolometers, nearly all incoherent detectors have similar underlying spectral responses – a responsivity rising in proportion to wavelength until a cutoff wavelength is reached. Bolometers have a very broad response curve and can be favored in some applications for the versatility that results. The bandwidths of incoherent detectors are identical with their response, unless limited by some external optical element such as a filter.

The spectral response of coherent infrared receivers is limited by the response of the photomixer. For coherent submillimeter- and millimeter-wave receivers, the response is limited by the efficiency of the secondary antenna, feed optics, and mixer. However, the spectral bandwidth of these detectors is determined by the intermediate frequency bandwidth at any one time. Furthermore, the IF bandwidth can be divided into sub-bands to provide spectral resolution within the overall spectral bandwidth. This characteristic provides coherent detectors an inherent multiplex advantage for taking spectra, since a single detector element can be used to obtain information at a number of wavelengths simultaneously. As a result, for spectroscopic applications a single heterodyne receiver can achieve a speed advantage over incoherent detectors with nominally similar performance levels for continuum detection.

12.6 Overview

Looking back from the perspective at the end of this text, we have developed a marvelous selection of tools to detect light. From the ultraviolet to well into the infrared, we can make use of arrays of many tens of thousands of detectors that, in virtually all situations, come extremely close in performance to the fundamental limits set by the statistics of the photon signal itself. Future developments can concentrate on the ease of manufacturing and using these devices, as well as on continuing to increase the array formats.

In the far infrared, we can reach similar fundamental performance limits, but as yet only with individual detectors and not with large format arrays. In the submillimeter-wave region, single detectors are becoming available that operate close to the fundamental limits, but modest additional improvements are needed to reach these limits in the most demanding circumstances. In both the far infrared and the submillimeter, further effort is required to develop large format arrays with performance levels per pixel similar to those achieved with single detectors.

12.7 Problems

12.1 Which detector types appear to have the greatest promise for performance improvements? Consider how far the type falls short of performance limits that are fundamental or nearly fundamental for its mode of operation and required construction techniques. You might also give inverse weight to the level of effort that has been applied to advance the performance (where little effort has been applied, one can hope that modest additional effort might produce a significant advance).

12.2 Which detector type would be the best choice to survey one square degree of sky at 0.6 μm with a resolution of 0.25 arcsec (using a suitably large telescope) and looking for point sources down to 10% of the sky surface brightness? Would the choice change if the survey were to reach sources down to 0.1% of the sky surface brightness?

12.3 Consider the CCD of the example in Section 7.4, assume it has a quantum efficiency of 50% and, as in the example, is being read out 5 times a second with a read noise of 963 electrons. Compare its performance with that of an image intensifier system with very high gain and a detective quantum efficiency of 10%, being read out with a similar CCD at the same rate. At what (if any) input photon rate will the two systems have identical signal to noise? Which one will give the better signal to noise at low photon rates?

12.4 Assume you are using a laser operating in the visible to range off layers of the atmosphere, looking at scattered return light as an indication of what is happening in these layers. You can use nonlinear optical elements to offset laser frequencies as necessary. To carry out this experiment in the day, you need very good background suppression and frequency response to distinguish the laser return from the sky background. What detector would you use?

Note

1 The only other relevant parameter is whether the detector has generation–recombination noise (that is, is a conventional photoconductor); the quantum efficiency for such detectors is quoted in a convention where it requires twice as long to make a given measurement as with other types of detector which exhibit only shot noise and have identical quantum efficiencies.

Appendix A
Physical constants

Fundamental constants

k	Boltzmann's constant	$1.381 \times 10^{-23}\,\mathrm{J\,K^{-1}}$
		$8.617 \times 10^{-5}\,\mathrm{eV\,K^{-1}}$
q	electronic charge	$1.602 \times 10^{-19}\,\mathrm{C}$
h	Planck's constant	$6.626 \times 10^{-34}\,\mathrm{J\,s}$
c	speed of light	$2.998 \times 10^{8}\,\mathrm{m\,s^{-1}}$
ε_0	permittivity of free space	$8.854 \times 10^{-12}\,\mathrm{F\,m^{-1}}$
m_e	rest mass of electron	$9.109 \times 10^{-31}\,\mathrm{kg}$
σ	Stefan–Boltzmann constant	$5.669 \times 10^{-8}\,\mathrm{W\,m^{-2}\,K^{-4}}$

Conversions and relations

1 angstrom (1 Å) $= 10^{-10}$ m
1 eV $= 1.602 \times 10^{-19}$ J
mobility $\mu(\mathrm{m^2\,V^{-1}\,s^{-1}}) = 10^{-4}\,\mu(\mathrm{cm^2\,V^{-1}\,s^{-1}})$
density $n(\mathrm{m^{-3}}) = 10^{-6} n(\mathrm{cm^{-3}})$
absorption coefficient $a(\mathrm{m^{-1}}) = 100a(\mathrm{cm^{-1}})$
$\varepsilon = \kappa_0 \varepsilon_0$, where ε is the dielectric permittivity and κ_0 is the dielectric constant
$D = \mu kT/q$, where D is the diffusion coefficient and μ is the mobility
$D(\mathrm{m^2\,s^{-1}}) = 10^{-4} D(\mathrm{cm^2\,s^{-1}})$
$n_0 p_0 = n_i^2 = p_i^2$, where n_0 and p_0 are the free electron and hole concentrations in doped material, and n_i and p_i are the free electron and hole concentrations in the corresponding intrinsic material

Appendix B

Answers to selected problems

1.1 (a) 1 μm: $L = 5.845 \times 10^{-28}$ W m^{-2} Hz^{-1} ster^{-1}
10 μm: $L = 3.307 \times 10^{-12}$ W m^{-2} Hz^{-1} ster^{-1}
(b) 1 μm: $E = 1.836 \times 10^{-35}$ W m^{-2} Hz^{-1}
10 μm: $E = 1.039 \times 10^{-19}$ W m^{-2} Hz^{-1}
(c) 1 μm: $P = 8.647 \times 10^{-27}$ W
10 μm: $P = 4.894 \times 10^{-12}$ W
(d) 1 μm: $N = 4.354 \times 10^{-8}$ s^{-1}
10 μm: $N = 2.464 \times 10^{8}$ s^{-1}

1.2 At 2 μm, 1.10×10^{-10} W
At 20 μm, 1.25×10^{-11} W

1.3 $\lambda_{max} = 30$ μm; Rayleigh–Jeans is within 20% for $\lambda > 40.7$ μm.

1.6 (a) $\lambda_{max} = 0.52$ μm, or 2.39 eV: AlP ($E_g = 2.45$ eV)
AlAs (2.16 eV)
GaP (2.26 eV)
CdS (2.42 eV)
ZnTe (2.25 eV)
AgBr (2.81 eV)
(b) $\lambda_{max} = 5.0$ μm, or 0.25 eV: InSb (0.18 eV)
PbSe (0.27 eV)
PnTe (0.29 eV)
(c) $\lambda_{max} = 21$ μm, or 0.06 eV: none.

2.1 The noise correction is less than 10% for $\lambda < 11.5$ μm. The peak of the blackbody occurs at about 3 μm (from the relation derived in Problem 1.4).

2.2 2.2% will develop.

2.3 $F(u) = (1/2)(\Delta\delta(u) - j/\pi u) + (\pi/10)\,\mathrm{sech}(\pi^2 u/10)$.

2.5 DQE = 0.039.

2.7 IIIa-J has the highest DQE; the DGE for 103a-O is 58% of that for IIIa-J, and that for IIa-O is 28% of that for IIIa-J. Either 103a-O or IIa-O will give a quick detection of the high contrast signal. IIIa-J is the only emulsion capable of detecting the lowest contrast signal (at 3:1 signal to noise).

3.2 $S = 0.040$ A W^{-1}
$\eta G = 0.050$

3.3 DQE = 0.046
$G = 1.06$

3.7 For $T_d = 4$ K, $R_2 = 0.97 \times 10^6\ \Omega$, $S/N = 1$
For $T_d = 77$ K, $R_2 = 0.73 \times 10^6\ \Omega$, $S/N = 0.91$
For $T_d = 300$ K, $R_L = 0.50 \times 10^6\ \Omega$, $S/N = 0.78$

4.1 (a) detector thickness $=450\,\mu$m
(b) $G=0.79$
(c) $S=6.6$ A W^{-1}
(d) assuming $\delta=1$, $R_d=4.3\times10^{11}\,\Omega$
(e) zero background NEP $=5.4\times10^{-18}$ W Hz$^{-1/2}$
(f) detector time constant is 0.1 s

4.4

N_A	M	$M\eta\propto S$	η
1×10^{13}	2.2	1.9	0.88
3×10^{13}	5.1	3.4	0.66
1×10^{14}	8.6	3.2	0.37
3×10^{14}	2.0	0.4	0.18

5.3 $[(\phi\eta q)/(2I_0)]^{1/2}\geqslant 5$.
6.1 (a) signal to noise of 1 with the TIA requires 723 photons s^{-1}
(b) S/N is larger by a factor of 8.45 compared with TIA
(c) S/N is larger by a factor of 17 compared with TIA; note, however, that it would be higher still if we sampled faster
6.2 $V_b=0$ V: 1.88×10^{-9} V electron^{-1}
$V_b=-1$ V: 3.11×10^{-9} V electron^{-1}
$V_b=-2$ V: 3.97×10^{-9} V electron^{-1}
6.4 NEP $=4.0\times10^{-16}$ W Hz$^{-1/2}$
$D^*=2.5\times10^{14}$ cm Hz$^{1/2}$ W^{-1}
DQE $=0.094$
6.5 node capacitance $=0.5$ pF
amplifier read noise $=94$ electrons rms
6.6 $\tau=0.8\,\Delta t$
7.2 (1) Need to use read strategy that avoids kTC noise; (2) must slow readout down from 0.2 to 0.34 s.
7.3

0.099	0.099	0.101	0.099	0.099
0.099	0.099	0.101	0.099	0.099
0.101	0.101	0.101	0.101	0.101
0.099	0.099	0.101	0.099	0.099
0.099	0.099	0.101	0.099	0.099

7.4 0.4%.
7.6 26%.
8.2 The pulse counting will yield S/N of ten times that from current measurement.
8.3 $T<190$ K, or $T<-83$°C.
8.4 Gain $=3$. It makes the light three times brighter but also increases the noise by about a factor of three; for most applications, it is not of much use.

9.6 $C=3\times10^{-9}\,\mathrm{J\,K^{-1}}$.
9.7 Optimum NEP $=1.4\times10^{-14}\,\mathrm{W\,Hz^{-1/2}}$.
10.1 $(S/N)_c/(S/N)_i=0.0033$. The S/N's are equal for $\Delta\nu=1\times10^{9}$ Hz.
11.4 Effective noise temperature becomes 5992 K.
11.5 Frequency response is limited by transit time to $\sim6\times10^{10}$ Hz.

References

American Institute of Physics (1972). *Handbook*. New York: McGraw-Hill.

Archer, J.W. (1986). 'Low-noise receiver technology for near-millimeter wavelengths'. In *Infrared and Millimeter Waves, vol. 15: Millimeter Components and Techniques, Part VI*, ed. K. J. Button. Orlando: Academic Press, pp. 1–86. Discusses the integration of Schottky diode and SIS millimeter-wave mixers into complete high sensitivity receiver systems.

Ashcroft, N. W., and Mermin, N. D. (1976). *Solid State Physics*. New York: Holt, Rinehart, and Winston. A classic and relatively advanced text on general solid state physics.

Bahl, I. J. (1989). 'Transmission lines'. In *Handbook of Microwave and Optical Components, vol. 1*, ed. K. Chang. New York: Wiley, pp. 1–59.

Barbe, D. F., and Campana, S. B. (1977). 'Imaging arrays using the charge-coupled concept'. In *Advances in Image Pickup and Display*, ed. B. Kazan. New York: Academic Press, pp. 171–296.

Bardeen, J., Cooper, L. N., and Schrieffer, J. R. (1957). 'Theory of superconductivity'. *Physical Review*, **108**, 1175–204.

Beynon, J. D. E., and Lamb, D. R. (1980). *Charge-Coupled Devices and their Applications*. London: McGraw-Hill.

Blaney, T. G. (1975). 'Signal-to-noise ratio and other characteristics of heterodyne radiation receivers'. *Space Science Reviews*, **17**, 691–702.

Blouke, M. M., Harp, E. E., Jeffus, C. R., and Williams, R. L. (1972). 'Gain saturation in extrinsic germanium photoconductors operating at low temperatures'. *Journal of Applied Physics*, **43**, 188–94.

Born, M., and Wolf, E. (1975). *Principles of Optics*, 5th edn. Oxford: Pergamon Press.

Boyd, R. W. (1983). *Radiometry and the Detection of Optical Radiation*. New York: Wiley. An extensive treatment of radiometry with general description of optical and infrared detectors and thorough discussion of noise mechanisms.

Bracewell, R. N. (1986). *The Fourier Transform and Its Applications*, 2nd edn. New York: McGraw-Hill. Comprehensive and standard reference on Fourier transforms.

Bratt, P. R. (1977). 'Impurity germanium and silicon infrared detectors'. In *Semiconductors and Semimetals, vol. 12*, eds. R. K. Willardson and A. C. Beer. New York: Academic Press, pp. 39–142. A classic review of the operation of photoconductors, excellent as an introduction and for leading into advanced topics.

Buil, C. (1991). *CCD Astronomy: Construction and Use of an Astronomical CCD Camera*. Richmond, Virginia: Willman-Bell. Description of use of CCD detectors, from basics of operation through hardware and software and image reduction and analysis.

Burt, D. J. (1988). 'Read-out techniques for focal plane arrays'. *Focal Plane Arrays: Technology and Applications, Proc. SPIE*, **865**, 2–16.

Button, K. J., Inguscio, M., and Strumia, F. (eds.) (1984). *Reviews of Infrared and Millimeter Waves, vol. 2: Optically Pumped Far-Infrared Lasers*. New York: Plenum.

Canali, C., Jacoboni, C., Nava, F., Ottaviani, G., and Alberigi-Quaranta, A. (1975).
'Electron drift velocity in silicon' *Physical Review B*, **12**, 2265–84.

Capasso, F. (1985). 'Physics of avalanche diodes'. In *Semiconductors and Semimetals, vol. 22D*, ed. W. T. Tsang. Orlando: Academic Press, pp. 2–172.

Carnes, J. E., and Kosonocky, W. F. (1972). 'Noise sources in charge coupled devices'. *RCA Review*, **33**, 327–43.

Carnes, J. E., Kosonocky, W. F., and Ramberg, E. G. (1972). 'Free charge transfer in charge-coupled devices', *IEEE Transactions on Electronic Devices*, **ED-19**, 798–808.

Christiansen, W. N., and Högbom, J. A. (1985). *Radiotelescopes*, 2nd edn. Cambridge University Press.

Clarke, J., Hoffer, G. I., Richards, P. L., and Yeh, N.-H. (1977). 'Superconductive bolometers for submillimeter wavelengths'. *Journal of Applied Physics*, **48**(12), 4865–79.

Cromwell, R. H. (1986). *Proceedings of SPIE, The International Society for Optical Engineering*, **627**, 610–15.

Csorba, I. P. (1985). *Image Tubes*. Indianapolis, Indiana: Howard Sams. A comprehensive and readable discussion of image intensifiers, photomultipliers, and television tubes; includes both theoretical and practical aspects of these detectors.

Cullum, M. (1988). 'Experience with the MAMA detector'. In *Instrumentation for Groundbased Optical Astronomy*, ed. L. B. Robinson. New York: Springer, pp. 568–81.

Dainty, J. C., and Shaw, R. (1974). *Image Science.* London: Academic Press. An advanced discussion of photography from the standpoint of the information content of a photographic image and the means to extract this information.

Darken, L. S., Sangsingkeow, P., and Jellison, G. E. (1990). 'Hole capture at acceptors in p-type germanium', *Journal of Electronic Materials*, **19**, 105–10.

Dereniak, E. L., and Crowe, D. G. (1984). *Optical Radiation Detectors.* New York: John Wiley and Sons. Basic description of detector operation, with extensive treatment of performance characterization.

DeTemple, T. A. (1987). 'Optically pumped FIR lasers'. In *Handbook of Molecular Lasers*, ed. P. K. Cheo. New York: Dekker, pp. 495–572.

Dolan, G. J., Phillips, T. G., and Woody, D. P. (1979). 'Low-noise 115-GHz mixing in superconducting oxide-barrier tunnel junctions', *Applied Physics Letters*, **34**, 347–9.

Douglas, N. G. (1989). *Millimeter and Submillimeter Wavelength Lasers.* Berlin: Springer.

Eastman Kodak Co. (1987). *Scientific Imaging with KODAK Films and Plates.* Practical guide to the selection and use of photographic materials for scientific applications.

Eccles, M. J., Sim, M. E., and Tritton, K. P. (1983). *Low Light Level Detectors in Astronomy.* Cambridge University Press. Although other subjects are covered, much of this book is devoted to a description of the various techniques and precautions necessary for practical application of photography at low light levels.

Elabd, H., and Kosonocky, W. F. (1982). 'Theory and measurements of photoresponse for thin film Pd_2Si and PtSi infrared Schottky-barrier detectors with optical cavity', *RCA Review,* **43**, 569–89.

Elliott, C. T., Day, D., and Wilson, D. J. (1982). 'An integrating detector for serial scan thermal imaging', *Infrared Physics*, **22**, 31–42.

Escher, J. S. (1981). 'NEA semiconductor photoemitters'. In *Semiconductors and Semimetals, vol. 15*, eds. P. K. Willardson and A. C. Beer. New York: Academic Press, pp. 195–300.

Furenlid, I. (1978). 'Signal-to-noise of photographic emulsions'. In *Modern Techniques in Astronomical Photography*, eds. R. M. West and J. L. Heudier. Munich: European Southern Observatory, pp. 153–64.

Goldsmid, H. J. (1965). *The Thermal Properties of Solids.* New York: Dover Publications.

Goldsmith, P. F., Itoh, T., and Stophan, K. D. (1989). 'Quasi-optical techniques'. In *Handbook of Microwave and Optical Components, vol. 1*, ed. K. Chang. New York: Wiley, pp. 344–63.

Gundlach, K. H. (1989). 'Superconducting tunnel junctions for radioastronomical receivers'. In *Superconducting Quantum Electronics*, ed. V. Kose. Berlin: Springer, pp. 175–204.

Hall, D. N. B., Akens, R. S., Joyce, R., and McCurnin, T. W. (1975). 'Johnson noise limited operation of photovoltaic InSb detectors'. *Applied Optics*, **14**, 450–3.

Haller, E. E., Hueschen, M. R., and Richards, P. L. (1979). 'Ge:Ga photoconductors in low infrared backgrounds'. *Applied Physics Letters*, **34**, 495–7.

Hays, K. M., La Violette, R. A., Stapelbroek, M. G., and Petroff, M. D. (1989). 'The solid state photomultiplier-status of photon counting beyond the near-infrared'. In *Proceedings of the Third Infrared Detector Technology Workshop*, ed. C. R. McCreight. NASA Technical Memorandum 102209, pp. 59–80.

Herter, T., Rowlands, N., Beckwith, S. V. W., Gull, G. E., Reynolds, D. B., Seib, D. H., and Stapelbroek, M. G. (1989). 'Improved Si:As BIBIB hybrid arrays'. In *Proceedings of the Third Infrared Detector Technology Workshop*, ed. C. R. McCreight. NASA Technical Memorandum 102209, pp. 427–38.

Hinken, J. H. (1989). *Superconductor Electronics: Fundamentals and Microwave Applications*. Berlin: Springer. An excellent discussion of superconductivity, SIS junctions, and other devices as used for heterodyne receivers.

Hobson, G. S. (1978). *Charge-Transfer Devices*. New York: Wiley.

Holloway, H. (1986). 'Collection efficiency and crosstalk in closely spaced photodiode arrays'. *Journal of Applied Physics*, **60**, 1091–6.

Horowitz, P., and Hill, W. (1989). *The Art of Electronics*, 2nd edn. Cambridge University Press. An excellent practical discussion of electronic circuitry.

Huggins, H. A., and Gurvitch, M. (1985). 'Preparation and characteristics of Nb/Al-oxide-Nb tunnel junctions'. *Journal of Applied Physics*, **57**, 2103–9.

Jacoby, G. H. (1990). *CCDs in Astronomy*. Astronomical Society of the Pacific Conference Series, vol. 8. San Francisco: Astronomical Society of the Pacific. Conference proceedings that give an up-to-date summary of current research in CCDs.

James, T. H. (1977). *The Theory of the Photographic Process*, 3rd edn. New

York: Macmillan. The standard advanced treatment of the photographic process, consisting of individual articles prepared by members of the staff of Kodak. Not particularly suitable as an introduction.

James, T. H., and Higgins, G. C. (1960). *Fundamentals of Photographic Theory*, 2nd edn. Hastings-on-Hudson, NY: Morgan and Morgan. A thorough description of the fundamentals and practice of photography, useful as a general advanced introduction to the processes involved.

Janesick, J. R., and Elliott, S. T. (1991). 'History and advancement of large area array scientific CCD imagers'. In *Astronomical CCD Observing and Reduction Techniques*, ed. S. B. Howell, Astronomical Society of the Pacific Conference Series, vol. 23. San Francisco: Astronomical Society of the Pacific, pp. 1–67. A detailed review of modern astronomical CCDs, emphasizing state-of-the-art performance and performance limitations.

Janousek, B. K., Daugherty, M. J., Bloss, W. L., Rosenbluth, M. L., O'Loughlin, M. J., Kanter, H., De Luccia, F. J., and Perry, L. E. (1990). 'High-detectivity GaAs quantum well infrared detectors with peak responsivity at 8.2 μm'. *Journal of Applied Physics*, **67**, 7608–11.

Jones, R. C. (1953). 'The general theory of bolometer performance'. *Journal of the Optical Society of America*, **43**, 1–14. A classic development of the theory of the bolometer and relevant figures of merit.

Kaneda, T. (1985). 'Silicon and germanium avalanche photodiodes'. In *Semiconductors and Semimetals, vol. 22D*, ed. W. T. Tsang. Orlando: Academic Press, pp. 247–328.

Katz, J., and Fogel, S. J. (1971). *Photographic Analysis*. Hastings-on-Hudson, NY: Morgan and Morgan. An advanced treatment of the physical and theoretical basis for photography.

Kazanskii, A. G., Richards, P. L., and Haller, E. E. (1977). 'Far-infrared photoconductivity of uniaxially stressed germanium'. *Applied Physics Letters*, **31**, 496–7.

Kazovsky, L. G. (1986). 'Performance analysis and laser linewidth requirements for optical PSK heterodyne communications systems'. *Journal of Lightwave Technology*, **LT-4**, 415–25.

Kelley, R. L., Holt, S. S., Madejsk, G. M., Moseley, S. H., Schoelkopf, R. J., Szymkowiak, H. E., McCammon, D., Edwards, B., Juda, M., Skinner, M., and Zhang, J. (1988). 'High resolution X-ray spectroscopy using microcalorimetry'. In *Conference on X-Ray Instrumentation for Astronomy*, ed. L. Golub. Proc. Soc. Photo-Opt. Inst. Eng., vol. 982, pp. 219–26.

Kelly, W. M., and Wrixon, G. T. (1980). 'Optimization of Schottky-barrier diodes for low-noise, low-conversion loss operation at near-millimeter wavelengths'. In *Infrared and Millimeter Waves, vol. 3: Submillimeter Techniques*, ed. K. J. Button. New York: Academic Press, pp. 77–110.

Keyes, R. J., and Quist, T. M. (1970). 'Low-level coherent/incoherent detection in the infrared'. In *Semiconductors and Semimetals, vol. 5*, eds. R. K. Willardson and A. C. Beer. New York: Academic Press, pp. 321–59.

Kinch, M. A., and Rollin, B. V. (1963). 'Detection of millimetre and submillimetre wave radiation by free charge carrier absorption in a semiconductor'. *British Journal of Applied Physics*, **14**, 672–6.

Kinch, M. A., and Yariv, A. (1989). 'Performance limitations of GaAs/AlGaAs infrared superlattices'. *Applied Physics Letters*, **55**, 2093–5.

Kittel, C. (1986). *Introduction to Solid State Physics*, 6th edn. New York: Wiley. A classic text on general solid state physics.

Kollberg, E. L. (1990). 'Mixers and detectors'. In *Handbook of Microwave and Optical Components,* ed. K. Chang. New York: Wiley Interscience, pp. 57–141.

Kraus, J. D. (1986). *Radio Astronomy*, 2nd edn. Powell, OH: Cygnus-Quasar Books.

Lampton, M. (1981). 'The microchannel image intensifier'. *Scientific American*, **245**, 62–71. A nontechnical explanation of the manufacture and applications of microchannel plates in image intensifiers.

Landolt-Börnstein (1989): *Semiconductors*, eds. O. Madeloug and M. Schulz. New Series, Vol. 22b. Berlin: Springer-Verlag.

La Violette, R. A., and Stapelbroek, M. G. (1989). 'A non-Markovian model of avalanche gain statistics for a solid-state photomultiplier'. *Journal of Applied Physics*, **65**(2), 830–6.

Lehecka, T., Luhmann, N. C., Peebles, W. A., Goldhar, J., and Obenshain, S. P. (1990). 'Gas lasers'. In *Handbook of Microwave and Optical Components, vol. 3*, ed. K. Chang. New York: Wiley Interscience, pp. 451–596.

Levine, B. F. (1990). 'Comment on "Performance limitations of GaAs/AlGaAs infrared superlattices"'. *Applied Physics Letters*, **56**, 2354–6.

Levine, B. F., Bethea, C. G., Hasnain, G., Shen, V. O., Pelve, E., Abbott, R. R., and Hsieh, S. J. (1990). 'High sensitivity low dark current 10 μm GaAs quantum well infrared photodetectors'. *Applied Physics Letters*, **56**, 851–3.

Low, F. J. (1961). 'Low-temperature germanium bolometer'. *Journal of the*

Optical Society of America, **51**, 1300–4. A classic paper that described the first high performance semiconductor bolometer.

Maas, S. A. (1986). *Microwave Mixers*. Dedham, MA: Artech.

McColl, M. (1977). 'Conversion loss limitations on Schottky-barrier mixers'. *IEEE Transactions on Microwave Theory and Techniques*, **MTT-25**, 54–9.

Mather, J. C. (1982). 'Bolometer noise: nonequilibrium theory'. *Applied Optics*, **21**, pp. 1125–9. The complete development of modern theory of bolometer operation – not for the faint of heart.

Mears, C. A., Quing Hu, Richards, P. L., Worsham, A. H., Prober, D. E., and Räisänen, A. V. (1991). 'Quantum limited quasiparticle mixers at 100 GHz'. *IEEE Transactions on Magnetics*, **MAG-27**, 3363–9.

Melen, R., and Buss, D. (eds.) (1977). *Charge Coupled Devices: Technology and Applications*. New York: IEEE Press.

Morgan, J. S., and Timothy, J. G. (1988). 'Status of the MAMA detector development program'. In *Instrumentation for Groundbased Optical Astronomy*, ed. L. B. Robinson. New York: Springer Verlag, pp. 557–67.

Moseley, S. H., Mather, J. C., and McGammon, D. (1984). 'Thermal detectors as X-ray spectrometers'. *Journal of Applied Physics*, **56**, 1257–62.

Nelson, M. D., Johnson, J. F., and Lomheim, T. S. (1991). 'General noise processes in hybrid infrared focal plane arrays'. *Optical Engineering*, **30**, 1682–700.

Newhall, B. (1967). *Latent Image: The Discovery of Photography*. Garden City, NY: Doubleday.

Norton, P. R. (1991). 'Infrared image sensors'. *Optical Engineering*, **30**, 1649–63. A review of current state-of-the-art in infrared arrays.

Norton, P.R., Braggins, T., and Levinstein, H. (1973). 'Impurity and lattice scattering parameters as determined from Hall and mobility analysis in *n*-type silicon'. *Physical Review B*, **8**, 5632–53.

Palaio, N. P., Rodder, M., Haller, E. E., and Kreysa, E. (1983). 'Neutron-transmutation-doped germanium bolometers'. *International Journal of Infrared and Millimeter Waves*, **4**, 933–43.

Pearsall, T. P., and Pollack, M. A. (1985). 'Compound semiconductor photodiodes'. In *Semiconductors and Semimetals, vol. 22D*, ed. W. T. Tsang. Orlando: Academic Press, pp. 173–245.

Petroff, M. D., Stapelbroek, M. G., and Kleinhans, W. A. (1987). 'Detection of individual 0.4–28 μm wavelength photons via impurity-impact ionization in a solid-state photomultiplier'. *Applied Physics Letters*, **51**, 406–8.

Phillips, T. G., and Jefferts, K. B. (1973). 'A low temperature bolometer heterodyne receiver for millimeter wave astronomy'. *Reviews of Scientific Instruments*, **44**, 1009–14.

Phillips, T. G., and Woody, D. P. (1982). 'Millimeter- and submillimeter-wave receivers'. *Annual Review of Astronomy and Astrophysics*, **20**, 285–321.

Pollock, D. D. (1985). *Thermoelectricity: Theory, Thermometry, Tool*, Philadelphia: ASTM Special Publication no. 852.

Press, W. H., Flannery, B. P., Teukolsky, S. A., and Vetterling W. T. (1986). *Numerical Recipes*. Cambridge University Press. A thorough and practical general description of digital methods, including Fourier transformation.

Putley, E. H. (1970). 'The pyroelectric detector'. In *Semiconductors and Semimetals, vol. 5*, eds. R. K. Willardson and A. C. Beer. New York: Academic Press, pp. 259–85.

Putley, E. H. (1977). 'InSb submillimeter photoconductive devices'. In *Semiconductors and Semimetals, vol. 12*, eds. R. K. Willardson and A. C. Beer. New York: Academic Press, 143–68.

Radebaugh, R. (1983). 'Very-low-temperature cooling systems'. In *Cryocoolers, Part 2*, ed. G. Walker. New York: Plenum. A description of helium-3, adiabatic demagnetization, and other low temperature refrigerators.

Ragan, G. L. (ed.) (1948). *Massachusetts Institute of Technology Radiation Laboratory Series, vol. 9: Microwave Transmission Circuits*. New York: McGraw-Hill.

Reine, M. B., Sood, A. K., and Tredwell, T. J. (1981). 'Photovoltaic infrared detectors'. In *Semiconductors and Semimetals, vol. 18*, eds. R. K. Willardson and A. C. Beer. New York: Academic Press, pp. 201–311.

Richards, P. L., Shen, T. M., Harris, R. E., and Lloyd, F. L. (1979). 'Quasiparticle heterodyne mixing in SIS tunnel junctions'. *Applied Physics Letters*, **34**, 345–7.

Rieke, F. F., DeVaux, L. H., and Tuzzolino, A. J. (1959). 'Single-crystal infrared detectors based upon intrinsic absorption'. *Proceedings of the IRE*, **47**, 1475–8.

Rieke, F. M., Lange, A. E., Beeman, J. W., and Haller, E. E. (1989). 'An AC bridge readout for bolometric detectors'. *IEEE Transactions on Nuclear Science*, **NS-36**, 946–9.

Rieke, G. H., Montgomery, E. F., Lebofsky, M. J., and Eisenhardt, P. R. M. (1981). 'High sensitivity operation of discrete solid state detectors at 4K'. *Applied Optics*, **20**, 814–18.

Robinson, F. N. H. (1962). *Noise in Electrical Circuits.* London: Oxford University Press.

Rogalski, A., and Piotrowski, J. (1988). 'Intrinsic infrared detectors'. *Progress in Quantum Electronics*, **12**, 87–289.

Schmit, J. L., and Selzer, E. L. (1969). 'Temperature and alloy compositional dependences of the energy gap of $Hg_{1-x}Cd_xTe$'. *Journal of Applied Physics*, **40**, 4865–9.

Schneider, M. V. (1982). 'Metal–semiconductor junctions as frequency converters'. In *Infrared and Millimeter Waves, vol. 6: Systems and Components*, ed. K. J. Button. New York: Academic Press, pp. 209–75.

Sclar, N. (1984). 'Properties of doped silicon and germanium infrared detectors'. *Progress in Quantum Electronics*, **9**, 149–257. An advanced review addressed specifically to extrinsic photoconductivity with extensive details on behavior with different dopants and on nonlinear effects. Also contains a brief discussion of BIB detectors.

Scott, R. B. (1959). *Cryogenic Engineering.* Princeton: Van Nostrand. A classic description of cryogenic techniques: dated but still very useful.

Scribner, D. A., Kruer, M. R., and Killiany, J. M. (1991). 'Infrared focal plane technology'. *Proceedings of the IEEE*, **79**, 66–85.

Séquin, C. H., and Tompsett, M. F. (1975). 'Charge transfer devices'. *Advances in Electronics and Electron Physics, Supplement 8*. New York: Academic Press. A somewhat dated but very thorough review of the operational principles of CCDs and related devices.

Smith, A. G., and Hoag, A. A. (1979). 'Advances in astronomical photography to low light levels'. *Annual Review of Astronomy and Astrophysics*, **17**, 43–71. This article concentrates on issues involving sensitivity and calibration of plates as used in astronomical applications. It contains an excellent description of hypersensitization techniques.

Solymar, L. (1972). *Superconducting Tunnelling and Applications.* London: Chapman and Hall. A readable yet extensive account of the physical principles behind SIS and Josephson junction mixers.

Solymar, L., and Walsh, D. (1988). *Lectures on the Electrical Properties of Materials.* Oxford University Press. An excellent discussion of a broad range of topics in solid state physics and its applications, introduced from the viewpoint of the underlying physics.

Sommer, A. H. (1968). *Photoemissive Materials.* New York: Wiley.

Spicer, W. E. (1977). 'Negative affinity 3-5 photocathodes: their physics and technology'. *Applied Physics*, **12**, 115–30.

Stark, A. M., Lamport, D. L., and Woodhead, A. W. (1969). 'Calculation of the modulation transfer function of an image tube'. In *Advances in Electronics and Electron Physics, vol. 28B*, eds. J. D. McGee, D. McMullan, E. Kahan, and B. L. Morgan. New York: Academic Press, pp. 567–75.

Stevens, N. B. (1970). 'Radiation thermopiles'. In *Semiconductors and Semimetals, vol. 5*, eds. R. K. Willardson and A. C. Beer. New York: Academic Press, pp. 287–318.

Stillman, G. E., and Wolfe, C. M. (1977). 'Avalanche photodiodes'. In *Semiconductors and Semimetals, vol. 12*, eds. R. K. Willardson and A. C. Beer. New York: Academic Press, pp. 291–393.

Stratton, J. A. (1941). *Electromagnetic Theory*. New York: McGraw-Hill.

Streetman, B. G. (1990). *Solid State Electronic Devices*, 3rd edn. Englewood Cliffs, NJ: Prentice Hall. A standard and excellent text on the operation and construction of solid state electronic devices.

Sze, S. M. (1985). *Semiconductor Devices, Physics and Technology*. New York: Wiley Interscience. A standard and widely used text on solid state physics and the construction and operation of solid state electronic devices.

Szmulowicz, F., and Madarsz, F. L. (1987). 'Blocked impurity band detectors – an analytical model: figures of merit'. *Journal of Applied Physics*, **62**, 2533–40. One of the few descriptions of BIB detectors available in the general literature.

Talley, H. E., and Daugherty, D. G. (1976). *Physical Principles of Semiconductor Devices*. Iowa State University Press. A particularly readable and fairly short discussion of the underlying physical principles governing solid state behavior and electronic devices.

Tani, T. (1989). 'Physics of the photographic latent image'. *Physics Today*, **42**, (9), 36–41.

Teich, M. C. (1970). 'Coherent detection in the infrared'. In *Semiconductors and Semimetals, vol. 5*, eds. R. K. Willardson and A. C. Beer. New York: Academic Press, pp. 361–407.

Timothy, J. G. (1988). 'Photon-counting detector systems'. In *Instrumentation for Groundbased Optical Astronomy*, ed. L. B. Robinson. New York: Springer Verlag, pp. 516–27.

Torrey, H. C., and Whitmer, C. A. (1948). *Massachusetts Institute of Technology Radiation Laboratory Series, vol. 15: Crystal Rectifiers*. New York: McGraw-Hill.

Tucker, J. R. (1979). 'Quantum limited detection in tunnel junction mixers'. *IEEE Journal of Quantum Electronics*, **QE-15**, 1234–58.

Tucker, J. R. (1980). 'Predicted conversion gain in

superconductor–insulator–superconductor quasiparticle mixers'. *Applied Physics Letters*, **36**, 477–9.

Tucker, J. R., and Feldman, M. J. (1985). 'Quantum detection at millimeter wavelengths'. *Review of Modern Physics*, **57**, 1055–113.

van der Ziel, A. (1976). *Noise in Measurements.* New York: Wiley Interscience.

van Vliet, K. M. (1967). 'Noise limitations in solid state photodetectors'. *Applied Optics*, **6**, 1145–69.

Vincent, J. D. (1990). *Fundamentals of Infrared Detector Operation and Testing*, New York: Wiley. A very thorough description of test procedures, applicable with modest change to optical as well as infrared detectors.

Walker, C. K., Kooi, J. W., Chan, M., LeDuc, H. G., Schaffer, P. L., Carlstrom, J. E., and Phillips, T. G. (1992). 'A low noise 492 GHz SIS waveguide receiver'. *Infrared and Millimeter Waves*, **13**, 6.

Wampler, E. J. (1974). 'Phosphor output image tubes'. In *Methods of Experimental Physics, vol. 12: Astrophysics, part A*, ed. N. P. Carleton. New York and London: Academic Press, pp. 237–51. A short discussion of image intensifiers, with examples of considerations in use for astronomical observations.

Wang, J.-Q., Richards, P. L., Beeman, J. W., Haegel, N. M., and Haller, E. E. (1986). 'Optical efficiency of far-infrared photoconductors'. *Applied Optics*, **25**, 4127–34.

Watton, R., Manning, P., and Burgess, D., (1982). 'The pyroelectric/CCD focal plane hybrid'. *Infrared Physics*, **22**, 259–75.

White, G. K. (1987). *Experimental Techniques in Low Temperature Physics*, 3rd edn. Oxford: Clarendon Press.

Wolfe, W. L., and Zissis, G. J. (eds.) (1978). *The Infrared Handbook.* Washington, DC: Office of Naval Research, Department of the Navy. A broad ranging discussion of detectors, optics, atmospheric effects, and many other issues. This handbook is far more general than many assume from its title.

Wyatt, C. L. (1991). *Electro-Optical System Design for Information Processing.* New York: McGraw-Hill. An excellent and thorough description of how optics, detectors, and support electronics are combined into optimized detection systems, particularly recommended for those interested in infrared systems.

Wyatt, C. L., Baker, D. J., and Frodsham, D. G. (1974). 'A direct coupled low-noise amplifier for cryogenically cooled photoconductive IR detectors'. *Infrared Physics*, **14**, 165–76.

Young, A. T. (1974). 'Photomultipliers, their cause and cure'. In *Methods of*

Experimental Physics, vol. 12: Astrophysics part A, ed. N. P. Carleton, New York and London: Academic Press, pp. 1–94. An extensive discussion of the theory, construction, and operation of photomultipliers, with emphasis on their use in accurate photometry.

Zwicker, H. R. (1977). 'Photoemissive detectors'. In *Optical and Infrared Detectors*, 2nd edn., ed. R. J. Keyes. Berlin: Springer-Verlag, pp. 149–96. An extensive discussion of photocathodes, with an emphasis on negative electron affinity materials.

Index

absorption coefficient 22, 57, 77
  in extrinsic semiconductors 83–5
  in intrinsic semiconductors 57,58
  in photodiodes 113
  in photoemissive detectors 205, 231
  in photographic materials 23, 33
absorption length 76, 168, 174, 205, 231
acceptor 15
acoustic-optic spectrometer 266
adiabatic demagnetization refrigerator 253
adjacency effects 40, 41
admittance matrix 310
amplifier noise 144, 145, 149–52, 191–3, 282, 283
antenna temperature 281–3
antenna theorem 280, 290
avalanche diode 123, 124

background-limited infrared photodetector *see* BLIP
backside illumination 93, 94, 168, 173–5
bandgap 12, 14, 109, 129, 130
  and cutoff wavelength 50, 56, 57, 59, 109, 204–7
  and thermal excitation 70–2
bandwidth, electrical 66, 67, 151, 152
bandwidth, intermediate frequency 274–8, 282–4, 286, 293
base fog 32
BCS theory 297, 298
BIB detector 92–9, 102
BLIP 74
blocked impurity band detector *see* BIB detector
blooming 196, 197
Bose–Einstein photon statistics 49
bulk photoconductor 85
bump bonding 165–67
buried channel 186–90

calibration of electronic arrays 193–5
calibration of photography 40–2
capacitance of diode 121, 122
capacitive transimpedance amplifer 153–5
carrier product equation 117
CCD 171–93
channel stops 182, 183
characteristic curve 29–32, 37, 39
charge coupled device *see* CCD
charge injection device *see* CID
charge transfer efficiency 180–3, 185–90, 197, 198
chemical fog 32, 36, 37
CID 175–8
classical mixer 294
coherent detectors 9, 262–318
color photography 28, 29, 40
compensation 89, 90
composite bolometer 251–3
conduction band 12, 13, 15, 16, 70, 82
conduction states 11, 13
conductivity 52, 53
contact potential 105, 117, 118
continuous wave laser 273, 292, 293
contrast in photography 32, 37
conversion gain and loss 269, 286, 287, 294, 302, 308–12
Cooper pair 297, 298
corner cube reflector 290, 291
crosstalk 195, 196
CTE *see* charge transfer efficiency
CTIA *see* capacitive transimpedance amplifier
cutoff wavelength 50, 56, 57, 59, 109, 125, 132, 204–7
Czochralski method 85

D see detectivity
*D** 74, 76
dark current
  calibration of 194, 195
  in CCDs 172, 190
  in extrinsic photoconductor 88, 89, 94
  in intrinsic photoconductor 52, 69–72
  in photodiode 118, 124, 129
  in photoemissive detectors 206–8
Debye temperature 247
density of photographic image 29–33, 35–40, 46
  diffuse 30, 38
  specular 30

depletion region 96, 97, 104–8
in BIB 96, 97
in CCD 171–4, 187–9
in FET 137, 138
in photodiode 104–8, 115, 116, 118–24
detection (in heterodyne receivers) 266, 271–3, 276, 277, 291
detective quantum efficiency *see* DQE
detectivity 74, 76
development (photographic) 21, 22, 26–8, 31, 32, 37, 40
dewar 253, 254
diameter of photographic image 41
Dicke radiometer equation 283
dielectric constants, table of values 54
dielectric relaxation 63–5, 76, 87, 94
diffusion 110–12
diffusion length 112–14, 133, 174, 205, 231
Digicon 227, 228
direct transition 16
distributed floating gate amplifier 192
donor 15
double correlated sampling 148, 149
double sideband 265
DQE 26
of BIB detector 99
measurement of 158
of photoemissive detectors 214, 223
in photography 25–7, 34, 39, 46
dye sensitization 28
dye-sensitized grains 34
dynamic range 9
of integrating amplifier 152, 153
measurement of 155, 157
in photography 22, 40, 41
dynode 211–14

effective wavelength 7, 19
Einstein relation 111
electroluminescence in readouts 167
electron multiplier 211–18
electron optics 209, 211, 218–22
electronographic tube 221, 226
electrothermal feedback 239, 242–4
emissivity 3
emulsion, photographic 21, 28, 29, 31, 34–6, 38–41, 44
epitaxial growth 86
etendue 279, 280
exchangeable power 269
excitation energy 12, 15, 83, 88, 89
table of values 84
exclusion principle 11, 12
extrinsic semiconductors 15, 81–5
table of properties 84

fat zero 186, 187
Fermi level 70, 105, 124, 126, 131, 132
fiber optics 223, 226
filter bank 272
fixed pattern noise 193–5
fixing, photographic 22
flash gate 174, 175
floating gate amplifier 191–3
floating zone growth 85
flux 3, 5, 6
flux density 5, 6
Fourier transform 44–7
table of transforms 46
frame transfer readout 183, 184
frequency multiplication 317
frequency range of heterodyne receiver 306–8
frequency response 61–3
of bolometer 239, 249, 260
of extrinsic photoconductor 87
of heterodyne receiver 278
of intrinsic photoconductor 61–6
of photodiode 121–4
of photoemissive detector 212, 222
of TIA 142–5
fringing fields 182, 183
fundamental mode 290, 292

G–R noise 65–7, 74, 79
gamma of photographic images 31, 32
gelatin, use in photography 20, 21, 25, 27
Gen I, Gen II image intensifiers 220, 221
generation–recombination noise *see* G–R noise
Golay cell 258
granularity of photographic images 27, 38
gross fog 32, 34
Gunn oscillator 312–16
Gurney–Mott hypothesis 25–7

H & D curve 30, 31
heat capacity 235, 247, 248, 260
helium 3 refrigerator 254, 255
HEMT 270, 271, 282
heterojunction 129, 130
high electron mobility transistor *see* HEMT
hook response 90, 91
hot electron bolometer 256, 257, 304
hybrid infrared array 165–70
hypersensitization 34, 35, 36

IBC detector *see* BIB detector
ideality factor 119, 295, 296
image centroiding readout 230
image dissector 212, 227
image intensifier 218–25, 230–2
image signal 265, 273, 309, 312
IMPATT diode 316, 317
impurity band conduction detector *see* BIB detector
ion events 208–10, 215, 225
ion implantation 86
indirect transition 57, 58
integrating amplifiers 145–55
interface trapping noise 185–7
interline transfer readout 183, 184
intermediate frequency 265, 277, 278
irradiance 5, 6, 17

JFET 136, 137
Johnson noise 67, 68, 75, 76
Josephson effect 298, 301, 303

kTC noise 68, 149–51, 154, 155, 191, 192

latent image 26, 37
limited space charge accumulation 315
line address readout 183, 184
line pairs per millimeter 42, 224
linearity 9
  of amplifier 139–41, 147, 154
  measurement of 155, 157
  of photographic plate 32, 40, 41
  of photomultiplier 215, 216
LO *see* local oscillator
load curve 240–2, 256, 259
local oscillator 9, 264, 265, 273, 292, 293, 312–17
low intensity reciprocity failure 31

majority carrier 15, 115–17
MAMA 227
MESFET 270, 271
microchannel 216–18
microchannel plate 220, 226, 227
minimum detectable power 276
minority carrier 15, 115–17
mobility 52–4
  table of values 54
modulation transfer function *see* MTF
MOSFET 136–8, 168, 169
Mott diode 297
MTF 42, 44–7, 195, 196, 224, 225
mu metal 213
multi-anode microchannel array *see* MAMA
multiplexer 169, 170
MUX *see* multiplexer

n-channel 137
NEFD 73, 74
negative electron affinity 206, 209, 214
negative resistance 312–16
NEP 72–6, 79
  of bolometers 242–6, 249, 259, 260
  of photoconductors 72–6
  of photodiodes 109
  of photoemissive detectors 210
neutron transmutation doping 250
noise 9, 65–9, 72–6, 79
  of amplifiers 144, 145, 147–52, 154, 155
  of bolometers 242–6, 255, 256
  of CCDs 185–93
  of extrinsic photoconductors 87, 90, 95, 99, 100
  of heterodyne receivers 273–7, 280–4, 304–6
  of intrinsic photoconductors 65–9
  measurement of 155, 157–9
  of photodiodes 108, 109, 118, 122–4
  of photoemissive detectors 210, 213–15, 223, 225
  in photography 24–7, 32, 37–9, 46, 47, 49
noise equivalent flux density *see* NEFD
noise equivalent power *see* NEP
noise temperature 280–7, 304–6
Norton equivalent circuit 142

opacity of photographic image 30, 31
operational amplifier 138, 139
overexposure of photographic image 32

p-channel 137
periodic table 10, 11, 14
phosphor 218–20, 223, 231
photocathode 202–9, 231
photoconductive gain 56, 60, 65, 75, 76, 90
  of BIB detector 98, 99
photoexcitation 11, 14
photographic film 21
photographic plate 20, 21
photoionization cross section 84
  table of values 84
photomixer 263, 266–8, 285–7
photomultiplier 211–16
photon detectors 8
photon noise 24–6, 49, 65–7, 74
PIN diode 122, 123
platinum silicide 124–6
Plumbicon 229
point spread function 44, 45, 47
pulse counting 124, 214, 215
pyroelectric detector 258

quantum efficiency 9, 23–5
  of bolometer 251, 252, 257
  of CCD 173–5, 193
  of extrinsic photoconductor 84, 85, 94, 96, 97
  of intrinsic photoconductor 50, 55–8, 60, 74–7
  of photodiode 112–14, 126–9, 133, 134
  of photoemissive detector 208, 210, 212, 213, 231
quantum limit 273–7, 280–2, 284, 286, 287, 304, 306
quantum mixer 269, 294
quantum well detector 129–32
quantum yield 175

radiance 1–8, 16
radiant exitance 2, 3, 6
radiation effects on photoconductors 90, 91, 94
radiometry 1–8
  effective wavelength 7, 19
  flux 3, 5, 6
  flux density 5, 6
  irradiance 5, 6, 17
  radiance 1–8, 16
  radiant exitance 2, 3, 6
  spectral radiance 1–6, 16
read noise 149–52
reciprocity failure 31–4, 36, 37, 40
reciprocity law 31
recombination center 15
recombination time 55, 56, 64, 65, 76, 87
  table of values 54
reflection 23
reset noise 68
responsivity 55–7, 60, 73–6
  of bolometer 239–42, 246
  measurement of 156–9
Reticon 227

saturation current 114–16, 119
Schottky diode 124–9, 294–6
Schottky diode mixers 294–7, 304–12
SEC vidicon 230
secondary antenna 290–2, 304
secondary electron conduction vidicon *see* SEC vidicon
Seebeck coefficient 257
shot noise
  of BIB detector 99
  of photodiode 108
  of photoemissive detector 210
Si:As BIB 93, 97, 98
signal processing in the element *see* SPRITE
silicon intensified target vidicon 229, 230
silicon target vidicon 229
silver halide grains 20–3
silver speck 22
single mode detector 290
single sideband 265
SIS mixer 297–303, 306–8
SIT vidicon *see* silicon intensified target vidicon
skipper readout 192, 193
solid state photomultiplier *see* SSPM
source induced background 220, 225
spatial frequency 42–7
spectral radiance 1–6, 16
speed in photography 32–4, 36–8
spot sensitometer 40
SPRITE 59, 60
square law detector 271, 272
square law mixer 264, 266
SSPM 99–101
ST vidicon *see* silicon target vidicon
stressed detector 92

TDI *see* time-delay integration
T-grains 34, 39
thermal conductance 235, 248, 249, 258, 259
thermal detector 8, 9, 234–61
thermal excitation 12–15, 52, 69–72
  in extrinsic material 88, 89
  in intrinsic material 69–72
  in photodiode 118
  in photoemissive detector 206–8
  in photography 26, 34
thermal limit 273–7, 280, 281
thermopile detector 257
Thévenin equivalent circuit 142
thinning of detectors 168, 174
TIA *see* transimpedance amplifier
time-delay integration 60, 184
transimpedance amplifier 140–5
trap 15
  in CCDs 182, 185–7, 190
  role in photography 25, 26
triple correlated sampling 149, 150
tunneling 91, 95, 106, 118, 130, 131, 298–301
turbidity 41

ultraviolet flooding of CCDs 174
underexposure in photography 41

valence band 12, 15, 16, 82
valence states 11
vidicon 228, 229

waveform factor 157
waveguide 292, 293
well depth 152, 154, 175, 176

Y-factor 285, 287

zone refining 85
z-plane construction 200, 201

1/f noise 68
2-phase CCD 179, 180
3-phase CCD 179, 180
4-phase CCD 179, 180